Carbohydrate-Based Vaccines

ACS SYMPOSIUM SERIES **989**

Carbohydrate-Based Vaccines

René Roy, Editor
Université du Québec à Montréal

**Sponsored by the
ACS Division of Carbohydrate Chemistry**

American Chemical Society, Washington, DC

ISBN: 978-0-8412-3983-8

The paper used in this publication meets the minimum requirements of American National Standard for Information Sciences—Permanence of Paper for Printed Library Materials, ANSI Z39.48-1984.

Copyright © 2008 American Chemical Society

Distributed by Oxford University Press

All Rights Reserved. Reprographic copying beyond that permitted by Sections 107 or 108 of the U.S. Copyright Act is allowed for internal use only, provided that a per-chapter fee of $36.50 plus $0.75 per page is paid to the Copyright Clearance Center, Inc., 222 Rosewood Drive, Danvers, MA 01923, USA. Republication or reproduction for sale of pages in this book is permitted only under license from ACS. Direct these and other permission requests to ACS Copyright Office, Publications Division, 1155 16th Street, N.W., Washington, DC 20036.

The citation of trade names and/or names of manufacturers in this publication is not to be construed as an endorsement or as approval by ACS of the commercial products or services referenced herein; nor should the mere reference herein to any drawing, specification, chemical process, or other data be regarded as a license or as a conveyance of any right or permission to the holder, reader, or any other person or corporation, to manufacture, reproduce, use, or sell any patented invention or copyrighted work that may in any way be related thereto. Registered names, trademarks, etc., used in this publication, even without specific indication thereof, are not to be considered unprotected by law.

PRINTED IN THE UNITED STATES OF AMERICA

Foreword

The ACS Symposium Series was first published in 1974 to provide a mechanism for publishing symposia quickly in book form. The purpose of the series is to publish timely, comprehensive books developed from ACS sponsored symposia based on current scientific research. Occasionally, books are developed from symposia sponsored by other organizations when the topic is of keen interest to the chemistry audience.

Before agreeing to publish a book, the proposed table of contents is reviewed for appropriate and comprehensive coverage and for interest to the audience. Some papers may be excluded to better focus the book; others may be added to provide comprehensiveness. When appropriate, overview or introductory chapters are added. Drafts of chapters are peer-reviewed prior to final acceptance or rejection, and manuscripts are prepared in camera-ready format.

As a rule, only original research papers and original review papers are included in the volumes. Verbatim reproductions of previously published papers are not accepted.

ACS Books Department

Contents

Preface ix

Acknowledgments.. xiii

1. T-Cell Immunity of Carbohydrates 1
   Luis Pena Icart, Violeta Fernandez-Santana,
   Roberto C. Veloso, Tania Carmenate, Suzanne Sirois,
   René Roy, and Vicente Verez Bencomo

2. The Regulatory Framework for Glycoconjugate Vaccines 21
   Christopher Jones

3. Conjugation Methods toward Synthetic Vaccines 36
   V. Pozsgay and J. Kubler-Kielb

4. *Haemophilus influenzae* Type b Conjugate Vaccine
   with a Synthetic Capsular Polysaccharide Antigen:
   Chemical View 71
   Vicente Verez Bencomo, René Roy, Maria C. Rodriguez,
   Annete Villar, Violeta Fernandez-Santana, Ernesto Garcia,
   Yury Valdes, Lazaro Heynngnezz, Ivan Sosa, and Ernesto Medina

5. Immunology of Experimental Synthetic Carbohydrate–Protein
   Conjugate Vaccines against *Streptococcus pneumoniae* Serotypes 85
   Harm Snippe, Wouter T. M. Jansen,
   and Johannis P. Kamerling

6. From Epitope Characterization to the Design of Semisynthetic
   Glycoconjugate Vaccines against *Shigella flexneri* 2a Infection 105
   Laurence A. Mulard and Armelle Phalipon

7. Automated Oligosaccharide Synthesis to Create Vaccines
   for Malaria and Other Parasites 137
   Bridget L. Stocker, Alexandra Hölemann,
   and Peter H. Seeberger

8. A Uniquely Small, Protective Carbohydrate Epitope
   May Yield a Conjugate Vaccine for *Candida albicans* 163
   David R. Bundle, M. Nitz, Xiangyang Wu,
   and Joanna M. Sadowska

9. Studies toward the Development of Anti-Tuberculosis
   Vaccines Based on Mycobacterial Lipoarabinomannan 184
   Hyo-Sun Kim, Ella S. M. Ng, Ruixiang Blake Zheng,
   Randy M. Whittal, David C. Schriemer, and Todd L. Lowary

10. The Lipoarabinomannan Glycolipid of *Mycobacterium
    tuberculosis*: Progress in Total Synthesis via n-Pentenyl
    Orthoesters 199
    Bert Fraser-Reid, Jun Lu, K. N. Jayaprakash,
    and Siddhartha Ray Chaudhuri

11. Lipopolysaccharide Antigens of *Chlamydia* 239
    P. Kosma, H. Brade, and S. V. Evans

12. Synthetic Carbohydrate-Based Antitumor Vaccines 258
    Rebecca M. Wilson, J. David Warren, Ouathek Ouerfelli,
    and Samuel J. Danishefsky

13. Synthetic Glycopeptides for the Construction of Anticancer
    Vaccines 293
    Horst Kunz, Sebastian Dziadek, Sven Wittrock,
    and Torsten Becker

14. Glycopeptide-Based Cancer Vaccines: The Role of Synthesis
    and Structural Definition 311
    R. Rao Koganty, Damayanthi Yalamati, and Zi-Hua Jiang

15. Peptide Mimics of Bacterial Polysaccharides: Potential
    for Discriminating Vaccines 335
    Silvia Borrelli, Margaret A. Johnson, Rehana B. Hossany,
    and B. Mario Pinto

Indexes

Author Index 359

Subject Index 361

Preface

Given the increasing incidence of microbial resistance to antibiotics, the failure to adequately treat certain diseases such as parasite infections, the resurgences of some bacteria that were thought to be eradicated in developed countries, and the magnificent progress recently made in glycobiology, the time has come to better adequately (re)address the issue of prophylactic treatments through vaccination programs. There was a time when killed or attenuated bacteria or viruses constituted the first vaccine generation. A partial understanding of how the immune system handles carbohydrate antigens has been the causative event that triggered the use of bacterial components such as capsular polysaccharide as one of the structural elements against which our bodies could mount potent humoral immune responses. These purified antigens represented vaccines of the second generation. Although extremely successful in the adult population, they were inefficient for young infants under the age of two. It was later appreciated that capsular polysaccharides alone were T-independent antigens; that is, the immature immune systems of infants could not stimulate antibody formation beyond the level of B-cells. Consequently, T-cell stimulations for the required memory effect and antibody maturation and affinity were not achieved. It took some time to realize that capsular polysaccharides necessitated the conjugation to protein carriers in order for the new vaccine generation (3rd) to trigger proficient T-cell activation.

It is our hope that this book, the first to be entirely dedicated to carbohydrate-based vaccines, will shed some light on the state-of-the-art advances that have been made in this fascinating field. For the first time, this topic has been explored at length, both in terms of the handling of carbohydrate antigens by the immune system, the regulatory issues related to vaccine accreditation, as well as by a compilation of chemical processes that have been developed in glycoconjugation chemistry. This book is intended to serve new as well as experienced practitioners in this rapidly evolving discipline. Several practical examples, both academic and industrial, are provided including the one that recently culminated in the first commercial vaccine against *Haemophilus influenza* type b produced by the stepwise chemical synthesis of the repeating capsular polysaccharide. This semi-synthetic vaccine represents the first case of a fourth generation vaccine.

This book is the result of three important symposia held consecutively as part of the program of the American Chemical Society's Division of Carbohydrate Chemistry within the 229th American Chemical Society (ACS) National Meetings (March 13–17, 2005, San Diego, California). A section of the first Symposium on *Frontiers in Modern Carbohydrate Chemistry* (ACS Symposium Series 960), organized by Professor A. V. Demchenko, dealt with recent methods towards the syntheses of complex oligosaccharides and may therefore represent a preparatory phase into our Symposium on *Carbohydrate-Based Vaccines,* thus forming the core of the present volume. Few speakers in the above section were also keynote speakers in the vaccine Symposium. Additionally, the *Hudson Award Symposium,* initiated by the award recipient lecture given by Dr. D. R. Bundle, dealt with the chemical synthesis of a vaccine candidate against *Candida albicans*. Altogether, the Symposium on *Carbohydrate-Based Vaccines* encompassed several plenary lectures which were completed by experts in the field who could not attend the meeting.

This book contains 15 chapters which could be divided into 5 sections. We have tried to cover major aspects dealing with glycoconjugate vaccines forming the basis of an in depth understanding of the field. The first section (Chapters 1–3) describes T-cell immunity of carbohydrates (R. Roy, V. Verez-Bencomo et al.), regulatory issues (C. Jones), and modern synthetic methods to covalently attach carbohydrate residues to protein carriers (V. Pozsgay and J. Kubler-Kielb). The second section (Chapters 4–6) presents progress towards the preparation of bacterial polysaccharide vaccines (J. P. Kamerling et al.; L. A. Mulard and A. Phalipon), including the recently disclosed first semi-synthetic *Haemophilus influenza* type b vaccine that has become a commercial product, presently sold in Cuba (V. Verez-Bencomo, R. Roy et al.). The third section (Chapters 7–11) illustrates examples extracted from malaria, parasites, and fungal infections (P. H. Seeberger et al.; D. R. Bundle et al.; T. L. Lowary et al.; B. Fraser-Reid et al.; P. Kosma, H. Brade, and S. V. Evans). The fourth section (Chapters 12–14) exposes major problems encountered en route to antitumor vaccines (S. J. Danishefsky et al.; H. Kunz et al.; R. Rao Koganty et al.). This class of carbohydrate-based vaccine surely represents as yet incompletely resolved targets that may rely on cellular immunity rather than simply humoral immunity easily reached by the former bacterial polysaccharide vaccines. The last section concludes with an excellent report by Dr. Pinto et al. which illustrates the possibility of replacing carbohydrate antigens by suitably chosen peptide mimics.

Readers of this text may originate from various disciplines, since fundamental aspects of immunology and immunochemistry are covered throughout. Organic chemists will also find this book of interest since several conjugation methods—starting from the traditional reductive amination through Diels–Alder cycloaddition and "click chemistry," chemical ligation, and even olefin metathesis reactions—have all been used for the covalent attachment of complex carbohydrate moieties to protein carriers or peptides used as T-cell epitopes. Moreover, modern aspects of oligosaccharide syntheses are nicely covered throughout. Peptide synthesis has not been left aside in this book since several protein carriers have been replaced by synthetic peptides towards vaccine preparations.

In conclusion, it is the opinion of the editor that the next vaccine generation will be constitutionally derived from all synthetic materials. This book has intended to unify researchers who were either active in synthetic carbohydrate antigens or synthetic "protein/peptide" epitopes, but rarely both. Issues not raised in the present text but which are the subject of intensive research activities concern the use of "lipopeptides," liposomes, and dendrimers bearing mannosides residues to target immune cells, such as macrophages and dendritic cells.

René Roy
Department of Chemistry
Université du Québec à Montréal
P.O. Box 8888, Succ. Centre-ville
Montréal, Québec H3C 3P8, Canada
Email: roy.rene@uqam.ca

Acknowledgments

It is implicit that such a successful symposium could not have taken place without the participation of all dedicated scientists and their collaborators to whom I am thankful for their assistance and patience in editing this book. The financial contributions from their funding agencies and respective institutions are also greatly appreciated, given the recent difficulties in raising moneys for the organization of such meetings. I am similarly indebted to participants of the meetings who could not find the time or resources to participate in the writing of this book. Finally, I am thankful to the generous sponsors who graciously participated in making this event possible: Dr. F. Michon from BioVeris Corporation (MD); Dr. V. Pavliak from the Bacterial Vaccine Discovery section of Wyeth Research (NY); Dr. J. Kihlberg from AstraZeneca (Sweden); R. Rao Koganty from Biomira Inc. (Alberta, Canada), and the ACS Division of Carbohydrate Chemistry. The contributions of anonymous peer reviewers are also particularly appreciated.

Carbohydrate-Based Vaccines

Chapter 1

T-Cell Immunity of Carbohydrates

Luis Pena Icart[1], Violeta Fernandez-Santana[1], Roberto C. Veloso[1], Tania Carmenate[1], Suzanne Sirois[2], René Roy[2,*], and Vicente Verez Bencomo[1,*]

[1]Center for the Study of Synthetic Antigens, Faculty of Chemistry, University of Havana, Ciudad Habana, Cuba 10400
[2]Department of Chemistry, Université du Québec à Montréal, P.O. Box 8888, Succ. Centre-Ville, Montréal, Québec H3C 3P8, Canada

> Glycoconjugate vaccines were developed as efficient therapies against infectious diseases. The immune responses towards glycoconjugates rely on diverse T-cell triggering mechanisms depending on the nature of the carbohydrate antigens. Unfortunately, carbohydrates alone are usually T-cell independent but natural glycoconjugates, such as glycoproteins and glycolipids, are increasingly accepted as antigens directly recognized in T-cell restricted mode. In this review, different types of T-cell activation processes and possible involvements of carbohydrate antigens will be discussed.

Introduction

Bacterial polysaccharides, lipopolysaccharides (LPS), and other glycolconjugates from bacteria, parasites, virus, and even cancer cells are important recognition markers for the immune system. Unspecific recognitions of these antigens are mediated by a family of receptors through innate immunity referred to as Pattern Recognition Receptors (PRR). The best-established mechanism for the recognition of these antigens depends on immunoglobulin (Ig) molecules that are either freely circulating antibodies or present as receptors on B-cells. The recognition of these carbohydrate antigens by Ig receptors on B-cells generates the first signals for their transformation into specific antibody-producing cells. However, this process typically depends on second separate signals habitually provided by T-lymphocytes.

Bacterial polysaccharides, due to their polymeric nature, can cross-link several IgM receptors at the surface of B-cells. This activation provides the required signals and is sufficient for the production of anti-polysaccharide antibodies without the participation of T-cells, hence with no memory effect.

The capacity of bacterial polysaccharides to stimulate the immune response directly without participation of T-cells was employed in some of the subunit vaccines widely established since the seventies. The lack of booster responses and the inability of infants to respond to non-conjugated polysaccharide antigens are two very important limitations for the use of polysaccharides as vaccines in infants. Despite this drawback, they were widely and successfully employed in vaccines against bacterial disease such as meningitis and pneumonia.

Carbohydrate antigens are common cell surface molecules playing important roles in the recognition of foreign and self-antigens by the adaptative immune system. They constitute complex structures that are presented and processed by T-lymphocytes. The presentation/binding mechanisms could be very diverse and, in principle, will define the type and intensity of the resulting immune responses. The presentation and processing of different carbohydrate antigens cover several different overwhelming mechanisms that are still poorly understood. It is generally accepted that the following steps are representative: a) recognition and binding of the carbohydrate antigens to antigen presenting cells (APCs); b) internalization; c) processing by proteases, oxygen and nitrogen species, and possibly glycosidases; d) binding to major histocompatibility complex molecules (MHC); e) cell surface presentation; f) recognition by T-cell receptors (TCR) in restricted modes, and g) activation of effector mechanisms by T-cells. In the present chapter, these steps will be briefly discussed, with an emphasis on carbohydrate-antigen involvements and the possible implications for anti-carbohydrate vaccine developments.

There are two accepted and interrelated classical mechanisms for the presentation and processing of carbohydrate antigens: the pathway by which endogenous antigens are degraded for presentation within class I MHC molecules (cytosolic pathway) and the pathway by witch exogenous antigens are internalized and degraded for presentation within class II MHC molecules (endocytic pathway). In the last case, intact antigens are recognized by circulating antibody molecules or by the mannose-binding protein (MBP), both of which activate the complement system. The antigens can also be recognized by cell surface immunoglobulins on B-cells. Antibody-tagged antigens are themselves recognized by specific receptors on dendritic cells or macrophages through antibody's Fc fragments.

In the endogenous (cytosolic) pathway, the antigens are usually the results of degradation products inside the cells by proteasomes. Proteins destined for destruction are conjugated to a molecule of ubiquitin that binds to the terminal amino group of a lysine residue. Sets of peptides, averaging about 8 amino acids long are produced as a result of the process that follows (*1*). The resulting antigenic peptides are transported into the ER where they bind to MHC-I. While MHC-I are recognized by a subset of CD8+ T-lymphocytes, MHC-II bind to CD4+ T-lymphocytes.

CD1 antigen presentation provides a non-classic pathway of T cell stimulation independent of MHC class I and II. Antigen presentation by CD1 molecules differs from that by MHC molecules by having distinct requirements for intracellular trafficking, processing, and loading of lipid antigens. The recognition and internalization of pathogens are mediated by multiple mechanisms including phagocytosis, endocytosis, or via receptors that are referred to as pattern-recognition receptors (PRRs).

The MHC-II pathway

Processing and presentation

Molecules from exogenous pathways are endocytized by one of the cells described below and internalized into endocytic vesicles. Vesicles with increasing acidity then fuse to form endosomes possessing a wide variety of enzymes including proteases and phosphatases (*2*). The paradigm for the exogenous antigen presentation states that only proteins from the endocytic pathways are processed by this mechanism. In spite of the absence of identified glycosidase activity capable of depolymerizing capsular polysaccharides, other more indirect mechanisms are possible for the cleavage of polysaccharide as for example chemical oxidation by nitric oxide NO (*3*).

MHC-II and Li (invariant chain) are assembled into the endoplasmic reticulum (ER) and transported to the exocytic vesicles through the Golgi apparatus. The fundamental role of Li in MHC class II molecules is to provide a stopper for the peptide-binding cleft that prevents access of short MHC class I destined peptides, longer polypeptides, or even unfolded proteins to the open-ended peptide-binding-cleft. Then, antigen-rich endo/lysosomes fuse with exocytic vesicles with an increasing acidification that activates proteases responsible for processing protein antigens and for the cleavage of Li. Processed antigens are loaded onto MHC-II and then shuttled to the cell surface for recognition by $\alpha\beta$ TCR (*4*) (see Fig. 1).

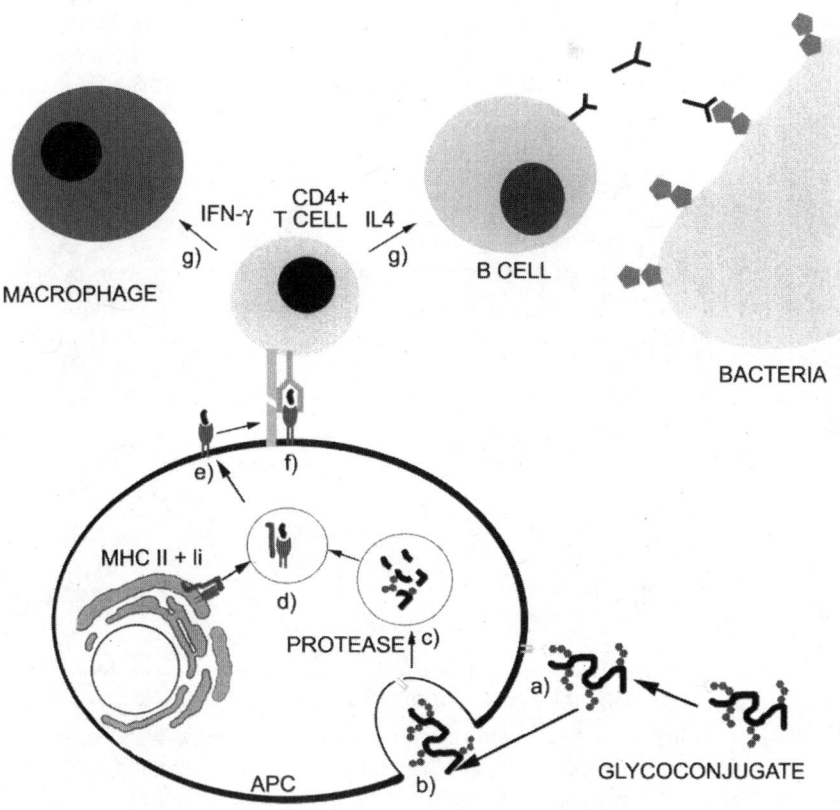

Figure 1. Processing and presentation of antigens by MHC-II pathway.
a) glycoconjugates tagged by antibodies are recognized by receptors on the cell membrane of APCs; b) internalization; c) processing in proteasomes; d) binding to the MHC-II molecules and formation of a complex; e) presentation; f) recognition by T-cells in a restricted way; g) activation of two different effector mechanisms by T-cells.

MHC class II molecules are heterodimers consisting of two α and β chains of ~32 kDa (230 amino acids (AA)). Both chains are encoded within the MHC complex and are transmembrane glycoproteins consisting of two extracellular domains (α1, α2 and β1, β2), a transmembrane and a cytoplasmic domain (~30 AA). Thus, complete MHC class II molecules possess four extracellular domains, each being approximately ninety amino acids long and encoded by a distinct exon. The α2 and β2 domains, have an Ig constant region domain-like structure. The α1 and β1 domains contain regions that are α helices and other regions that are β strands. The α1 and β1 domains essentially form mirror images to each other. The complex results in a region enclosed by helical structures on two sides, β strands on another and an open hydrophobic binding pocket at the top. These two domains form the walls of a cleft into which a 12-18 amino acids peptide antigens can bind in an elongated conformation. The appearance of the peptide-binding site sandwiched between TCR and MHC-II complex is illustrated in Fig. 2 (*5*).

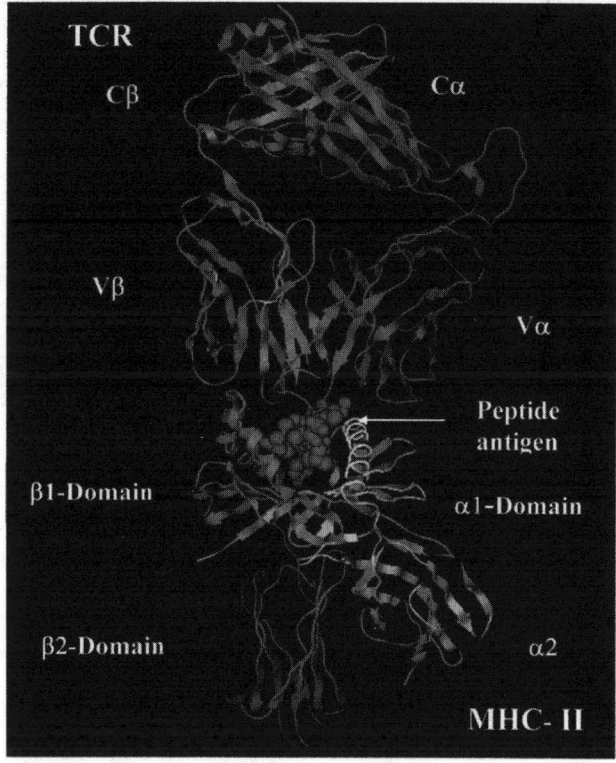

Figure 2. T-cell receptor (TCR) and MHC Class II complex with an antigenic peptide. Atom coordinates were obtained from PDB file 1FYT.

For many years, peptides were believed to be the only type of molecules presented by MHC-II. Evidences now showed that this is possible also for some glycopeptides and even fragments of some specific bacterial polysaccharides that could interact through electrostatic interactions with the binding sites.

Glycopeptides, similarly to peptides, are bound to MHC-II in elongated conformations. The ends of MHC class II molecules are relatively open as shown in Fig. 3. As a result, size-restrictions for peptides that can bind to MHC class II molecules are less stringent when compared to MHC-I. Peptide contacts with MHC class II molecules can also go beyond the α-helical region (6-8).

Most eukaryotic proteins and several viral proteins carry oligosaccharide chains; a fact that has generated interest in the role of the saccharides during antigen processing, presentation, and recognition by T cells (9-12).

Model studies performed with synthetic neoglycopeptides have revealed that mono- or small oligosaccharides can be attached to T-cell immunogenic peptides which retain MHC binding if the position for the glycan is chosen carefully (13-17). These studies also demonstrated that immunization of mice with neoglycopeptides can elicit T cells which specifically recognize the carbohydrate moiety if it is located in the center of the peptides. Figure 3 shows an immunodominant glycopeptide in which a hydroxylysine residue carries a β-D-galactopyranosyl moiety in the center of the glycan (8). Recently, some reports also described that glycoproteins found in nature can be degraded to glycopeptides in antigen presenting cells and presented to T-cells as complexes with MHC molecules. In most, but not all, of the cases that involved natural glycopeptides, the carbohydrate moieties attached to the peptides have been found to be small, *i.e.* mono- or disaccharides. Therefore the processing and effective presentation in these cases could be function of the carbohydrates length (Fig. 3). Absence or over expression of glycosyltransferases associated with diseases such as cancer could affect the length of carbohydrate chains attached to glycoproteins. This can result in glycopeptides that are presented in different ways in the disease states as opposed to the case in healthy situations (18).

The flexibility of MHC-II presenting structures is such that it can even accommodate fragments of polysaccharides. The common characteristics and probably the minimal requirements for these antigens are the presence of both positive (e.g. free amines) and negative (e.g. phosphates or carboxylates) charges within a structural repeating unit (19). Structural studies have shown several bacterial polysaccharides to contain zwitterionic motifs (e.g. types 5 and 8 polysaccharides of *Staphylococcus aureus* (19)). The two most studied among this family of molecules are the capsular polysaccharides from *Bacteroides fragilis* and from *Streptococcus pneumoniae* type 1. (20)

Zwiterionic polysaccharides (ZPS) are endocytized and delivered to the vesicles like other antigens via Toll-like receptors or by phagocytosis or endocytosis. Following this event, ZPS are processed by reactive nitrogen

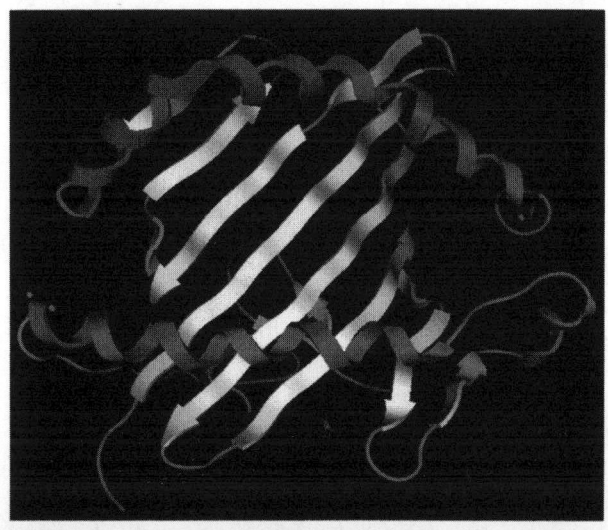

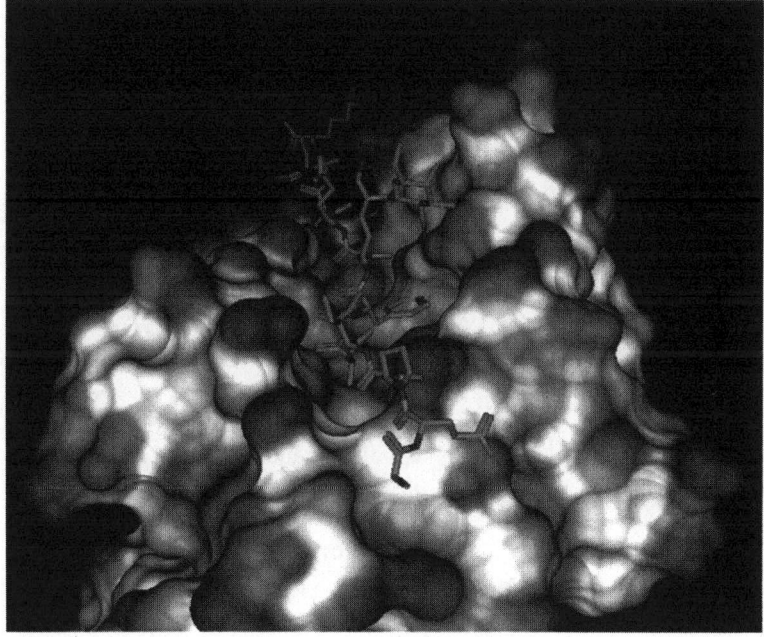

Figure 3. **Top**: *Top view of a truncated MHC-II antigen binding site (PDB:1DLH) (7).* **Bottom:** *Model of a complex between MHC-II H-2A^q and an immunodominant glycopeptide from type II collagen (8). Carbon atoms of the carbohydrate moiety are colored light grey and the oxygen atoms are dark grey.*

species (RNS) such as free NO radicals in order to provide suitable sizes to be presented by MHC II molecules. Presumably, the oxidative process for ZPS occurs before the formation of endo/lysosomes. The role of pH in the binding of ZPS is not completely clarified. However, strong evidences support that ZPS cleavage cannot occur under acidic pH, demonstrating that processing must occur before acidification and protease activation. (*3*) Therefore, ZPS antigens must be internalized and processed at neutral pH through oxidation, and localized together with MHC-II proteins in an acidic environment to facilitate MHC-II binding and presentation (*3*).

T-cell activation and consequences

The structures for the complex between MHC-loaded peptides and TCR reveal that there are extensive contacts between the two molecules confined to particular regions (*21*) (see Fig. 2). TCRs are actually aligned at a diagonal relative to the MHC-peptide complexes (*22*). Most of the contacts between TCRs and MHC molecules are indeed between the complementarity determining regions (CDR) of the TCR in Vα and Vβ domains and the MHC residues surrounding the peptides and the peptides themselves (*23*). The side chains of several of the residues of the peptides project up into the TCR (Fig. 2). It is not known whether most of the binding energies of association come from the contacts with MHC or with the peptides. There are no contact with MHC and either the TCR constant regions or the β strands of the variable regions. There is no contact between the TCR to the α3 domain, or the β strands on the floor of the MHC molecules. Changes of a single residue of either the peptides or the MHC molecules can completely abolish the productive interactions between TCR and MHC-peptides (*24*). All these contacts confirm the dynamic processes in which the organization of the molecules creates the scenery for the interactions between peptide-MHCs and TCRs (*25*).

Naive T-helper (Th) lymphocytes recognize peptides into the cleft of MHC-II on the surface of APCs. The recognition induces the expression of CD40L (Ligand). Dendritic cells expose B7 molecules that interact with T-lymphocytes CD27, thus amplifying the signals. Upon activation, the lymphocytes could develop into Th1 or Th2, depending on the cytokine patterns secreted by APCs. Th1 lymphocytes can further involve cell-mediated immunity through macrophage activation by IL-2 (Interleukin), γ-INF (INterFeron) and β-TNF (Tumor Necrosis Factor). Th2 lymphocytes are involved in humoral and allergic reactions through IL-4, 5, 6, 10, and 13 secretions (*26*).

In the case of fragments of capsular polysaccharide antigens that could be processed and presented in MHC-II, the recognition occurs by $\alpha\beta$ TCRs on CD4+ T cells (*19*). In some cases, T-cell clones were described with specificity against, for example, *Streptococcus pneumoniae* capsular polysaccharide serotype 1 (*27*).

The MHC-I pathway

Processing and presentation

Ubiquitin-tagged (glyco)proteins are degraded into the cytosol by proteasomes. The 8-18 amino acids containing peptides produced during this process are translocated to the ER through transporter associated with antigen processing (TAP) (*28*). In the ER, peptides are assembled to MHCs with the help of chaperones such as calnexin, calreticulin, and tapasin. The binding of peptides to MHC class I is followed by dissociation of the complex with the chaperones and liberation of TAP (*29*). The complex peptide-MHC-I is then transported to the membranes (Fig. 4).

Class I MHC molecules are heterodimers consisting of a 12 kDa β-2 microglobulin and a 30 kDa heavy chain composed of three extracellular, one transmembrane, and one cytosolic domains. These molecules are expressed on T- and B-lymphocytes, macrophages, and dendritic cells (Fig. 5). The binding site of MHC-I includes two α-helices flanking a β-sheet at the bottom. It contains six pockets and is closed on both sides allowing more define specificity for short peptides. As for the MHC-II, the peptides bind in extended conformations within MHC-I.

Because of the relatively strong interactions between MHC-I and peptides, it has been possible to identify the sequence of several peptides that can bind to particular MHC class I molecules. Comparisons of the bound peptide sequences reveal that for each MHC class I molecule, there are consensus residues that a peptide must possess in order to bind to a particular MHC class I molecule. Crystallographic analyses have demonstrated that these residues are important for binding into a pocket formed within the class I molecules. These pockets are different in each class I molecule, however there are usually two or three pockets required for peptide binding.

Glycosylations are common posttranslational modifications of proteins to be secreted or transported onto cell surfaces (*10*). Nevertheless, evidences for the processing of glycoproteins to glycopeptides and subsequent glycopeptide presentation by MHC class I receptors during normal immune responses are scarce (*30, 31*). In addition to the N- and O-linked glycosylations that occur in the ER and Golgi apparatus, O-linked β-N-acetylglucosamine (O-GlcNAc) substitutions on serine and threonine residues are found abundantly on proteins in the cytosol and nucleus of mammalian cells (*14, 32*).

Since cytosolic proteins are the preferred sources of peptides for presentation by MHC class I molecules, peptides carrying O-GlcNAc residues could, in principle, enter into the class I presentation pathway (33). O-GlcNAc substituted peptides can be efficiently translocated across the ER membrane by TAP. Synthetic glycopeptides with GlcNAc substitutions on serine residues are

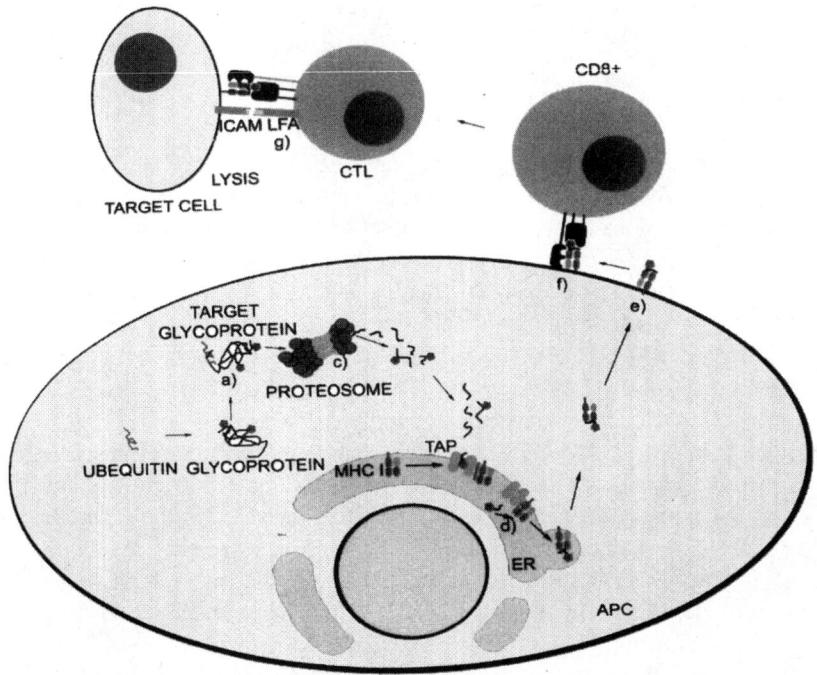

Figure 4. Processing and presentation of antigens by MHC-I pathway. a) glycoprotein is tagged by ubiquitin; c) processing in proteasomes; d) transport by TAP inside the ER, and binding to MHC-I molecules with formation of a complex; e) presentation; f) recognition by T-cells in a restricted way; g) activation of the effector mechanism by T-cells.

efficiently transported by TAP into the ER *in vivo*. Small glycopeptides mainly those substituted with O-GlcNAc could be accommodated into the peptide binding groove (34). The presence of mono- or disaccharides attached to short peptides and placed in the center could satisfy α/β TCR antigen binding preferences. The same type of carbohydrate attached to slightly longer peptides may also satisfy the requirements for γ/δ TCR antigen binding, leading to glycopeptide and carbohydrate-specific T cell responses (31) (see Fig. 6).

T-cell activation and consequences

MHC I-peptide complexes are externalized to the surface of APCs where they are recognized by the TCR of CD8+ T lymphocytes. The TCR is formed by

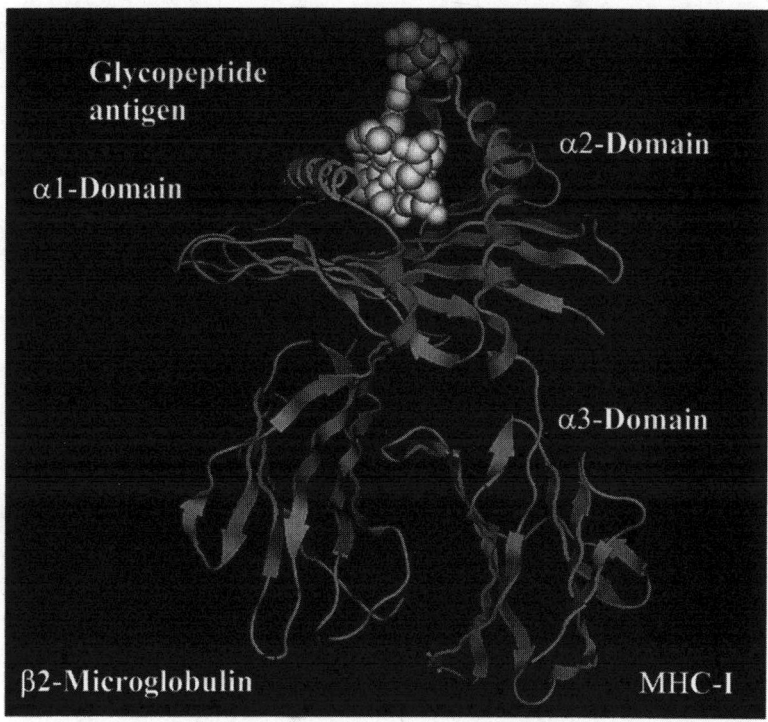

Figure 5. MHC class I molecule bound to a glycopeptide antigen (PDB 1KGB). The bound peptide antigen is light grey and the sugar dark grey-medium grey.

two α and one β chains. The CDR is composed of three α and three β hypervariable loops on the N-terminal domains.

Evidences reveal that there are extensive contacts between MHC-I peptide complex and TCR, however the contacts are confined to particular regions of the molecules. The TCR is actually aligned at a diagonal relative to the MHC-I peptide molecule. The α and β chains CDR2 loops contact only the MHC-I molecules, while the CDR1 and CDR3 loops contact both peptides and MHC-I molecules. Furthermore, the Vα domain is always closer to the N-terminal end of the bound glycopeptides and the Vβ domain to the C-terminal end. The surface of the TCR that contact peptide/MHC-I is relatively flat, sometimes with a central cavity. By contrast, the MHC-I surface contains two higher curvatures. To achieve a large interface, the TCR apparently binds between these two high points on the MHC-I/glycopeptide surface (29) (Fig. 6).

As for MHC-II, there are no contact between MHC-I and either the TCR constant regions or the β strands of the variable regions. Neither between TCR

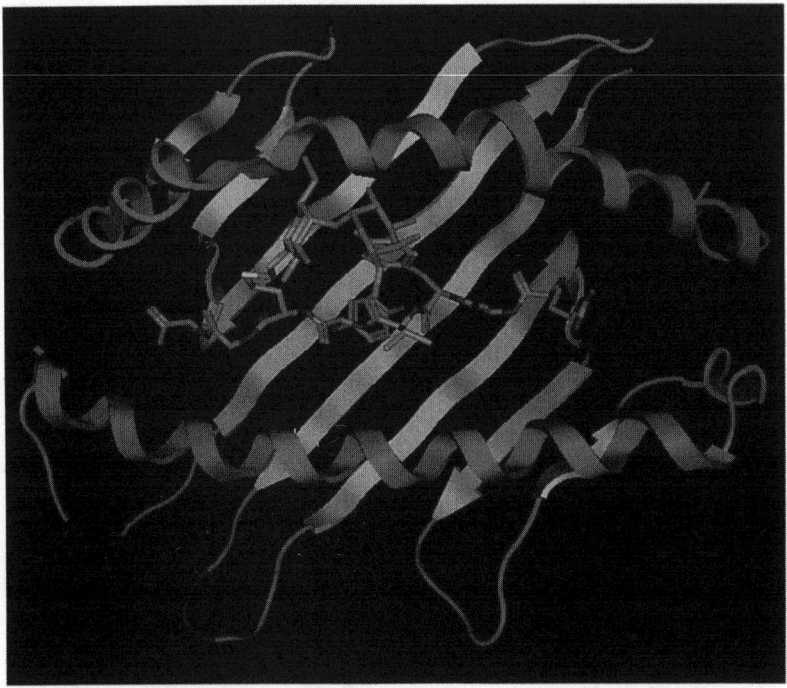

Figure 6. Top view of a portion of MHC Class I H-2KB presenting glycopeptide RGY8-6H-Gal2 PDB 1KBG (35).

and the α3 domains, or the β strands at the bottom of the MHC-I molecules. A change in a single residue of either the peptides or the MHC-I molecules can completely abolish the productive interactions between TCRs and MHC-peptide complexes (36).

There are several models that explain the activation of TCR but the most accepted is the serial model of TCR activation. According to this model, activation is the consequence of a high number and frequency of sequential interactions between the T cells and several dendritic cells that present specific MHC/peptide complexes. High signal frequency would reflect high antigen load in the peripheral tissues represented by a high number of antigen presenting dendritic cells in the draining lymphoid compartments, supporting, sequential productive short-term encounters for T-cell commitment to activation and proliferation (37).

Activated CD8+ T-lymphocytes become T-cytotoxic. The effectors function through the following steps: a) adhesion and formation of the conjugates between the cytotoxic T lymphocytes (CTL) and the target cells. As a result of interactions between the LFA-1 (Lymphocyte Function Associated antigen: CTL) and ICAM-1 (Intercellular-Adhesion Molecule: target cells), the lymphocytes are programmed for lysis; b) membrane attack, the granulosoma contents are exocytized through contact points to the intercellular space; c) the complexes dissociate just before lysis of the target cells; d) destruction of the target cells. Destruction may occur by several mechanisms such as secretion of perforins that lyses the cells and granzymes that affect DNA inducing cells to apoptosis.

CD1-mediated glycolipids presentation

CD1 includes a family of presenting molecules that evolved for presentation of lipids and glycolipids to T-cells (38). Human CD1 is related to five different genes. The first group is composed of three different isoforms CD1a, CD1b, and CD1c that are recognized by α–β classical T-lymphocytes. Group II is composed of CD1d that is recognized only by γ–δ T-lymphocytes known also as NKT (Natural Killer T cells) (38).

The recognition and internalization of pathogens are mediated by one of the receptors described below. These receptors recognize conserved molecular patterns, which are shared by a large group of microorganisms. Toll-like receptors (TLRs), as parts of PRRs, play an important role in the recognition of microbial components. TLR 2 mediates the recognition of whole mycobacterium, mycobacterium lipoarabinomannan, glycolipids, and glycophosphatidylinositols. Captured pathogens are then processed into the phagosomes and the endosomes vesicles where interactions with the specific CD1 molecules occur (Fig. 7).

In the ER, the α-chain (folded into three domains: $\alpha 1$, $\alpha 2$, and $\alpha 3$) is associated non-covalently to $\beta 2$-microglobulins through ER chaperones. A putative antigen-binding super domain is composed of $\alpha 1$ and $\alpha 2$ helices that sit atop and cross an eight-stranded antiparallel β-pleated sheet platform. Compared to MHC class I, the $\alpha 2$ helix is kinked upward and the $\alpha 1$ helix is raised 4–6 Å above the β-sheets, resulting in a deeper groove (See Fig. 8). In addition, the β-stranded platform displays substantially greater curvature and is more bowl-shaped compared to MHC class I or class II platforms. The relative positions of the $\alpha 1$ and $\alpha 2$ helices are closer together along with their longitudinal axes, resulting in a narrower groove (14 Å) compared to MHC- I or II (Fig. 5 and 6).

Assembled CD1 molecules are sorted into different cellular compartments. CD1a is then sorted into recycling endosomes and trafficked to the plasma membrane in an AFR6-(a GTPase)-dependent manner. It is largely excluded from lysosomes. In contrast, CD1b is sorted to late endosomes and directed to lysosomes by binding the adaptor protein AP3. CD1a and CD1b then acquire distinct self- or foreign lipid antigens in the recycling endosomes or lysosomes, respectively. However, CD1c broadly traffics in both early recycling endosomes as well as in late endosomes and lysosomes. CD1d traffics mainly in early and late endosomes but not in recycling endosomes and only partially localizes in lysosomes. Neither CD1c nor CD1d associates with the adaptor protein AP3. CD1c and CD1d then acquire distinct self- or foreign lipid recognition in the intracellular compartments described previously (Fig. 7).

The aliphatic chains of glycolipids usually interact with CD1 molecules exposing the carbohydrates. This has led to the suggestion that each hydrophobic tail fits into a pocket of the CD1 groove. Consistent with this, binding studies demonstrated that the lipid portion of antigens is required for CD1 binding. Nevertheless, several unanswered questions remain opened such as for example: what happens to the lipid fragments that cannot fit inside the groove, the selectivity of different CD1 molecules in binding particular lipids, and the stability of CD1-lipids interactions (39).

CD1-glycolipid complexes are expressed mainly on dendritic cells and other APC and some subsets of CD1 cells that express only CD1c as a unique group. The TCR of CD1-restricted T-cells is highly discriminating, so that even monosaccharides that differ in the orientation of a single hydroxyl group can be distinguished. TCR presumably recognizes molecular surfaces composed of α-helical amino acids of the CD1 molecules along with exposed portions of the carbohydrates. Consistent with this, the CDR3 regions of TCRs that are reactive with lipoglycans and mycolic acids show a high frequency of charged amino acids (40).

CD1-restricted T cells that are specific for foreign microbial lipids are stimulated to carry out effector functions, including the secretion of cytolytic granules containing perforin and granulysin, which promote the lysis of infected cells or have direct antimicrobial effects, and the production of IFN-γ and TNF-β, which activate the microbicidal functions of macrophages.

Strong evidences also suggest that antigen-specific CD1-restricted T cells play a role in resistance to microbial infections *in vivo*. It comes from the recent finding that CD1-restricted T-cells specific for mycobacterial antigens derived from leprosy patient lesions and effective cell-mediated immunity is correlated with expression of group I CD1 molecules by dendritic cells in the leprosy lesions. Furthermore CD1b-restricted T-cell clones originally derived from a leprosy patient were specific for a form of polysaccharide produced by pathogenic *Mycobacterium leprosy* (41).

Other results came from patients infected with *Mycobacterium tuberculosis* in which mycobacterial-antigen-specific CD1-restricted T-cell clones were also biased towards secretion of Th1 type cytokines (42).

However, the features that might most clearly distinguish the group I CD1 restricted T-cells from MHC-restricted T-cells could be the timing of their activation in an immunological response. CD1d-restricted T-cells interact with

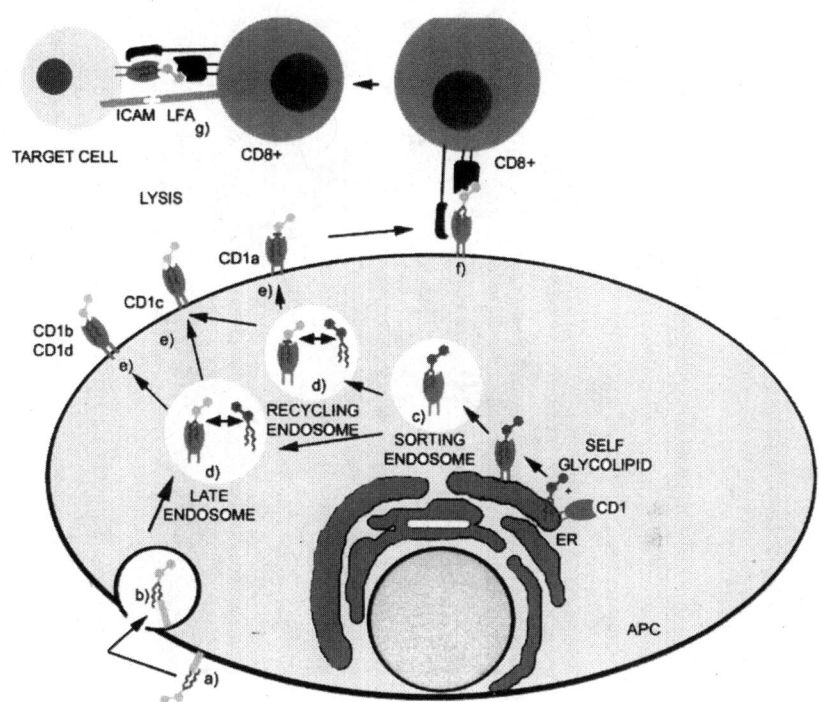

*Figure 7. Processing and presentation of antigens by CD1 pathway.
a) glycolipids are recognized by Toll-like receptors at the cell surface of APCs;
b) internalization; c) processing either on sorting or recycling endosomes;
d) binding to the CD1 molecules by displacing self glycolipids and formation
of a complex; e) presentation; f) recognition by T-cells in a restricted way;
g) activation of effector mechanism by T-cells.*

monocytes, leading to the production of cytokines like INF-γ and IL-2 that enhance NK-cell activations immediately after infection. CD1d and CD1c restricted T-cells can also interact with B-cells and secrete cytokines that led to B-cell activations and antibody secretions. As the immune response continues, monocytes differentiate into immature dendritic cells, which express CD1a, CD1b, and CD1c that can present mycrobial lipid antigens to CD1-restricted T-cells, activating their effector functions (43).

Most of the CD1-restricted microbial antigen-specific T-cells described so far recognize lipid antigens from mycobacteria. A prototypic antigen presented in the context of CD1 is lipoarabinomannan (LAM), a mannose polymer

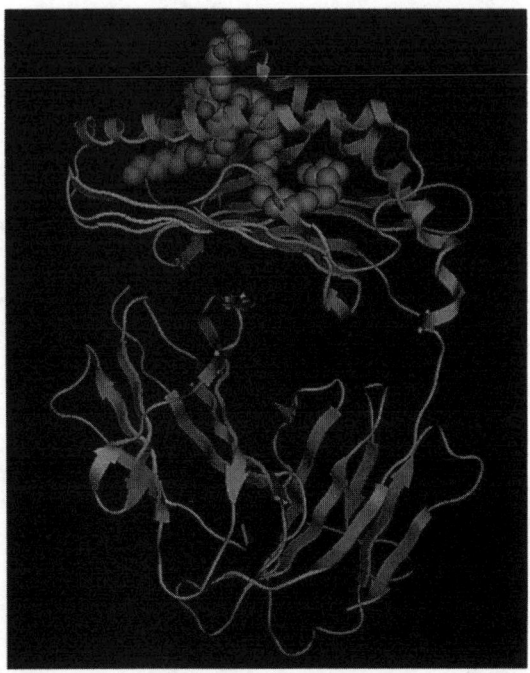

Figure 8. Crystal structure of human CD1d bound to a-Galactosylceramide. The sphingosine chain is bound in the C' pocket while the longer acyl chain is anchored into the A' pocket (46) (PDB 1ZT4).

substituted at one end with arabinose and at the other with a phosphatidic acid containing tuberculostearic and palmitic acid (44).

Unlike group I CD1-restricted T-cells, CD1d-restricted T-cells account for most of the small T-cell subsets that expresses NK1.1 antigens, a marker that is found on all NKs. These unusual T-cells were denominated NKT cells. Murine CD1d restricted NKT cells recognize α-galactosylceramide (α-GalCer) (Fig. 8) an unusual α-glycosylated sphingolipid derived from a marine sponge. The presence of this pattern recognition receptor (PPR) of the innate immune system suggest that NKT cells may be involved to facilitate rapid T-cells recognition of a class of characteristic microbial antigens; CD1d-restricted T-cells with this TCRs may function as a kind of hybrid innate-adaptive lymphocyte subsets (45).

In contrast to the group I CD1-restricted T-cells, for which there are strong evidences of specificity for bacterial glycolipid antigens, it is not yet clear if CD1d-restricted T-cells respond to microbial antigens *in vivo*. Although NKT cells recognition of microbial antigens has not yet been clearly demonstrated *in vivo*, specific activation of these T-cell subsets by administration of α-GalCer lipid antigen appears to have a dramatic effect on a variety of microbial infections. Some recent studies suggest that NKT cells can indeed recognize certain microbial glycolipids, in particular phosphatydylinositol mannosides that are components of bacterial glycoconjugates. Further studies are needed in order to fully understand the role of this T-restricted recognition in the general immune responses.

Conclusions

The above remarks constitute useful information for the rational design of modern glycoconjugate vaccines (47-55). Ongoing activities in this area are growing and as a result several semi-synthetic vaccines have emerged (see other chapters in this book), one of which has resulted in the first chemical synthesis of the complex repeating unit of the capsular polysaccharide of *Haemophilus influenza* type b that has become a commercial product (56). It is clear that the next challenges will have to address the issues of carbohydrate-based vaccines against various cancers that, most likely, will be derived from MHC-I pathways.

References

1. Reist, E. A.; Vos, J. C.; Gromme, M.; Neefjes, J. *Nature* **2000**, *404*, 774-778.
2. Hiltbold, E. M.; Roche, P. A. *Curr. Opin. Immunol.* **2002**, *14*, 30-35.
3. Cobb, B. A.; Wang, Q.; Tzianabos, A. O.; Kasper, D. L. *Cell* **2004**, *117*, 677-687.
4. Schubert, U.; Antón, L. C.; Gibbs, J. R.; Norbury, C. C.; Yewdell, J. W.; Bennink, J. R. *Nature* **2000**, *404*, 770-774.
5. Reits, E. A.; Griekspoor, A.; Neijssen, J.; Groothuis, T.; Jalink, K.; van Veelen, P.; Janssen, H.; Calafat, J.; Drijfhout, J. W.; Neefjes, J. *Immunity* **2003**, *18*, 97-108.
6. Garcia, K. C.; Taiton, L.; Wilson, I. A. *Annu. Rev. Immunol.* **1999**, *17*, 369-397.
7. Stern, L.J.; Brown, J.H.; Jardetzky, T.S. ; Gorga, J.C. ; Urban, R.G. ; Strominger, J.L.; Wiley, D.C. *Nature.* **1994**, *368*, 215-221.
8. Kjellen, P.; Brunsberg, U.; Broddefalk, J.; Hansen, B.; Vestberg, M.; Ivarsson, I.; Engstrom, A.; Svejgaard, A.; Kihlberg, J.; Fugger, L.; Holmdahl, R. *Eur. J. Immunol.* **1998**, *28*, 755-767.
9. Kihlberg, J.; Elofsson, M.; Salvador, L. A. *Methods Enzymol.* **1997**, *289*, 221-45
10. Carbone, F. R.; Gleeson, P. A. *Glycobiology* **1997**, *7*, 725-730.

11. Rudd, P. M.; Elliot, T.; Cresswell, P.; Wilson, I. A.; Dwek, R. A. *Science* **2001**, *291*, 2370-2376.
12. Werdelin, O.; Meldal, M.; Jensen, T. *Proc. Natl. Acad. Sci.* **2002**, *99*, 9611-9613.
13. Ishioka, G. Y.; Lamont, A. G.; Thomson, D.; Bulbow, N.; Gaeta, F. C. A.; Sette, A.; Grey, H. M. *J. Immunol.* **1992**, *148*, 2446-2451.
14. Haurum, J. S.; Arsequell, G.; Lellouch, A. C.; Wong, S. Y. C.; Dwek, R. A.; Mc Michael, A. J.; Elliot, T. *J. Exp. Med.* **1994**, *180*, 739-744.
15. Deck, B.; Elofsson, M.; Kihlberg, J.; Unanue, E. R. *J. Immunol.* **1995**, *155*, 1074-1078.
16. Abdel-Motal, U. M.; Berg, L.; Rosén, A.; Bengtsson, M.; Thorpe, C. J.; Kihlberg, J.; Dahmén, J.; Magnusson, G.; Karlsson, K. A.; Jondal, M. *Eur. J. Immunol.* **1996**, *26*, 544-551.
17. Jensen, T.; Hansen, P.; Galli-Stampino, L.; Mouritsen, S.; Frische, K.; Meinjohanns, E.; Meldal, M.; Werdelin, O. *J. Immunol.* **1997**, *158*, 3769-3778.
18. Tzianabos, A. O.; Onderdonk, A. B.; Rosner, B.; Cisneros, R. L.; Kasper, D. L. *Science* **1993**, *262*, 416-424.
19. Tzianabos, A. O.; Wang, J. Y.; Lee, J. C. *Proc. Natl. Acad. Sci.* **2001**, *98*, 9365-9370.
20. Kalka-Mol, W. M.; Tzianabos, A. O.; Bryant, P. W.; Niemeye, M.; Ploegh, H. L.; and Kasper, D. L. *J. Immunol.* **2002**, *169*, 6149-6153.
21. Reinherz, E. L.; Tan., K.; Tang, L.; Kern, P.; Liu, J.; Xiong, Y.; Hussey, R. E.; Smolyar, A.; Hare, B.; Zhang, R.; Joachimiak, A.; Chang, H, C.; Wagner, G.; Wang, J. *Science* **1999**, *286*, 1913-1921.
22. Garboczi, D. N.; Ghosh, P.; Utz, U.; Fan, Q. R.; Biddison, W. E.; Wiley, D. C. *Nature* **1996**, *384*, 134-141
23. Garcia, K. C.; Degano, M.; Stanfield, R. L.; Brunmark, A.; Jackson, M. R.; Peterson, P. A.; Teyton, L.; Wilson, I. A. *Science* **1996**, *274*, 209-219.
24. Rudolph, G. M.; Wilson, I. A. *Curr. Opin. Immunol.* **2002**, *14*, 52-65.
25. Bromley, S. H. K.; Burack, R. W.; Johnson, K. G.; Somersalo, K.; Sims, T. N.; Suemen, C.; Davis, M. M.; Shaw, A, S.; Allen, P. M.; Dustin, M. L. *Annu. Rev. Immunol.* **2001**, *19*, 375-396.
26. Wuorimaa, T.; Kayhty, H. *Scand. J. Immunol.* **2002**, *56*, 111-129.
27. Stingele, F.; Corthesy, B.; Kusy, N.; Porcelli, S. A.; Kasper, D. L.; Tzianabos, A. O. *J. Immunol.* **2004**, *172*, 1483-1490.
28. Ubel, S.; Tampe, R. *Curr. Opin. Immunol.* **1999**, *11*, 203–208.
29. Endert, P. M. *Curr. Opin. Immunol.* **1999**, *11*, 82-88.
30. Varki, A. *Glycobiology* **1993**, *3*, 97-130.
31. Kihlberg, J.; Elofsson, M. *Curr. Med. Chem.* **1997**, *4*, 85-116.
32. Haurum, J. S. **1996**. D. Phil. Thesis, University of Oxford, Oxford, UK.
33. Haurum, J. S.; Tan, L.; Arsequell, G.; Frodsham, P.; Lellouch, A. C.; Moss, P. A.; Dwek, R. A.; Mc Michael, A. J.; Elliott, T. *Eur. J. Immunol.* **1995**, *25*, 3270-3276.

34. Glithero, A.; Tormo, J.; Haurum, J. S.; Arsequell, G.; Valencia, G.; Edwards, J.; Springer, S.; Townsend, A.; Pao, Y. L.; Wormald, M.; Dwek, R. A.; Jones, E. Y.; Elliott, T. *Immunity* **1999**, *10*, 63-74.
35. Speir, J.A.; Abdel-Motal, U. M.; Jondal, M.; Wilson., I. A. *Immunity* **1999**, *10*, 51-61.
36. Hennecke, J.; Wiley, D. C. *Cell* **2001**, *104*, 1-4.
37. Friedl, P.; Gunzer, M. *TRENDS in Immunol.* **2001**, *22*, 187-191.
38. Martin, L. H.; Calabi, F.; Lefebvre, F. A.; Bilsland, C. A.; Milstein, C. *Proc. Natl. Acad. Sci. USA.* **1987**, *84*, 9189-9193.
39. Matsuda, J. L.; Kronenberg, M. *Curr. Opin. Immunol.* **2001**, *13*, 19-25.
40. Grant, E. P.; Degano, M.; Rosat, J. P.; Stenger, S.; Modlin, R. L.; Wilson, I. A.; Porcelli, S. A.; Brenner, M. B. *Exp. Med.* **1999**, *189*, 195-205.
41. Gumperz, J. E.; Brenner, M. B. *Curr.Opin. Immunol.* **2001**, *13*, 471-478.
42. Sieling, P. A.; Ochoa, M. T.; Jullien, D.; Leslie, D. S.; Sabet, S.; Rosat, J. P.; Burdick, A. E.; Rea, T. H.; Brenner, M. B.; Porcelli, S. A.; Modlin, R. L. *J. Immunol* **2000**, *164*, 4790-4796.
43. Porcelli, S.; Morita, C. T.; Brenner, M. B. *Nature* **1992**, *360*, 593-597.
44. Brigl, M.; Brenne, M. B. *Annu. Rev. Immunol.* **2004**, *22*, 817-890.
45. Kawano, T.; Cui, J.; Koezuka, Y.; Toura, I.; Kaneko, Y.; Motoki, K.; Ueno, H.; Nakagawa, R.; Sato, H.; Kondo, E.; Koseki, H.; Taniguchi, M. *Science* **1997**, *278*, 1626-1629.
46. Koch, M.; Stronge, V. S.; Shepherd, D.; Gadola, S. D.; Mathew, B.; Ritter, G.; Fersht, A. R.; Besra, G. S.; Schmidt, R. R.; Jones, E. Y.; Cerundolo, V. *Nat. Immunol.* **2005**, *6*, 819-826.
47. Roy, R. *Drug Discovery Today: Technologies* **2004**, *1*, 327-336.
48. Kuberan, B.; Linhardt, R. J. *Curr. Org. Chem.* **2000**, *4*, 653-677.
49. Pozsgay, V. *Adv. Carbohydr. Chem. Biochem.* **2000**, *56*, 153-199.
50. Jennings, H. J.; Sood, R. K. Synthetic glycoconjugates as human vaccines. In *Neoglycoconjugates. Preparation and Applications*, Lee, Y. C., Lee, R. T., Eds.; Academic Press: New York, 1994; pp 325-371.
51. Dick, W. E., Jr.; Beurret, M. *Contrib. Microbiol. Immunol.* **1989**, *10*, 48-114.
52. Danishefsky, S. J.; Allen, J. R. *Angew. Chem. Int. Ed.* **2000**, *39*, 836-863.
53. Keding, S. J.; Danishefsky, S. J. in *Carbohydrate-Based Drug Discovery*; Ed. Wong, C.-H.; Wiley-VCH: Weinheim, Germany, 2003, Vol. 1, 381-406.
54. Ouerfelli, O; Warren, J. D.; Wilson, R. M.; Danishefsky, S. J. *Expert Rev. Vaccines* **2005**, *4*, 677-685.
55. Jennings, H. J. *Adv. Carbohydr. Chem. Biochem.* **1983**, *41*, 155-208.
56. Verez-Bencomo, V.; Fernández-Santana, V.; Hardy, E.; Toledo, M. E.; Rodríguez, M. C.; Heynngnezz, L.; Rodriguez, A.; Baly, A.; Herrera, L.; Izquierdo, M.; Villar, A.; Valdés, Y.; Cosme, K.; Deler, M. L.; Montane, M.; Garcia, E.; Ramos, A.; Aguilar, A.; Medina, E.; Toraño, G.; Sosa, I.; Hernandez, I.; Martínez, R.; Muzachio, A.; Carmenates, A.; Costa, L.; Cardoso, F.; Campa, C.; Diaz, M.; Roy, R. *Science* **2004**, *305*, 522-524.

Chapter 2

The Regulatory Framework for Glycoconjugate Vaccines

Christopher Jones

Current address: Laboratory for Molecular Structure, National Institute for Biological Standards and Control, Blanche Lane, South Mimms, Herts EN6 3QG, United Kingdom

Vaccines, including glycoconjugate vaccines, are frequently administered to healthy infants in government-sponsored immunisation schemes to prevent diseases which are now rare, due largely to the effectiveness of these vaccines. High levels of confidence in the efficacy and safety of vaccines are demanded by the public, so that vaccine development, production, licensing and sale are strictly regulated processes. Glycoconjugate vaccines against three pathogens have been licensed. These have shaped detailed requirements within the existing regulatory framework, and similar principles are likely to be applied when assessing novel glycoconjugate vaccines. As glycoconjugate vaccines modify bacterial carriage and alter natural exposure to the organisms, surveillance after vaccine introduction may modify immunisation schedules. This article discusses the framework within which glycoconjugate vaccines are regulated and those factors which have been shown to be important in ensuring their efficacy.

Introduction[1]

Glycoconjugate vaccines, in which an appropriate glycan is covalently attached to a carrier protein, have proven the most effective means to elicit protective immunity against a wide range of bacterial pathogens which express a glycocalyx and where antibodies directed against cell surface saccharides are protective. Vaccines against three such pathogens have been licensed, and a large number of similar vaccines are in development *(1)*. Glycoconjugate technology appears to be generic and development of novel vaccines of this type is relatively predictable. Vaccines are used in government-sponsored mass immunisation campaigns directed at healthy infants and the public demands high regulatory standards to ensure that the safety and efficacy of these products are assured. Because vaccines have proved successful in reducing the incidence or eliminating many devastating diseases *(2)*, memories of the severity of these diseases has faded and tolerance of even very rare vaccine-associated side effects has dissipated.

Whilst the framework for the regulation of vaccine development, manufacture, clinical trial, licensing and supply is well established, specific types of vaccine such as the glycoconjugates require different detailed considerations, a fresh analysis of the factors influencing vaccine safety and efficacy, and appropriate reference materials and reference methods. These differ somewhat between markets and regulatory environments. Glycoconjugate vaccines, unlike poly-saccharide vaccines, reduce carriage of the pathogen by healthy individuals, interrupting normal transmission of disease and helping to protect unvaccinated individuals *(3-4)*. This reduces natural exposure of infants to these organisms, which can influence the optimal immunization strategy in a manner that can only be determined by population surveillance after vaccine introduction.

This chapter is intended to provide an introduction to the process whereby a vaccine can be brought to market, sources of information available to the developer of a potential vaccine and an overview of those factors which are most likely to have to be controlled during manufacture.

Guidelines, Recomendations, reagents and advice

The International Conference on Harmonisation of Technical Requirements for Registration of Pharmaceuticals for Human Use (ICH) is a mechanism to

---

[1] This review is a personal opinion, and should not be considered the formal position of NIBSC.

harmonise regulatory processes in the three major pharmaceutical markets (the US, Europe and Japan) and so simplify licensing of a single product in multiple regions. The ICH develops and publishes concensus Guidelines which define, at a high level, the appropriate methodology for the production, clinical trial, testing and manufacture of pharmaceuticals. For example, the Q series of documents relate to product quality, the Q5 document series to biotechnology products, the Q2 series concerns analytical method validation, and Q6B specification setting *(5)*. World Health Organisation (WHO) Recommendations exist for the manufacture and quality control (QC) of *Haemophilus influenzae* type b (Hib), meningococcal Group C and pneumococcal glycoconjugate vaccines *(6-9)*. Recommendations for future products are likely to be based on similar principles. These documents are often adopted as formal quality standards by regulators in developing countries. The regulatory framework for the licensing of pharmaceuticals, including vaccines, in the USA is formalised in the Code of Federal Regulations Title 21 (21 CFR), whilst 45 CFR Part 46 covers protection of human subjects involved in clinical trials *(10)*. The FDA and the Center for Biological Evaluation and Research (CBER) produce various guidance documents *(11)*, although currently none refer explicitly to glycoconjugate vaccines. The regulatory framework for medicinal products in the EU is available on-line at EudraLex *(12)*. Clinical trials in Europe are regulated by the Clinical Trials Directive (2001/20/EC) *(13)*. Some agencies, such as the MHRA, advise on the implementation of the directive *(14)*. The European Medicines Evaluation Agency (the EMEA) also produces advice documents, such as the Notes for Guidance *(15)*. The European Pharmacopoeia (EP) publishes monographs defining legally binding minimum standards for products, including glycoconjugate vaccines *(16-18)*, used within its jurisdiction. Both the EP and the Pharmacopeia of the United States of America (USP) define general standards expected for vaccines and validated protocols for relevant tests. Where glycoconjugate immunogens are components of combination vaccines, specifications for bulk components are often the same, but testing of the final fill may prove more complex and immunogenicity may be reduced *(19)* or enhanced. Almost all regulatory agencies will provide advice on the licensing of potential new products, at varying degrees of formality. Some charge a fee for this.

Due to the great structural variation between different products, no reference vaccines are available, although reagents to assist assay standardization *(20,21)* are available from the National Institute for Biological Standards and Control (NIBSC) in the UK *(22)*. Some materials are also available through the US Centers for Disease Control and Prevention (CDC).

Vaccine development and licensing

Vaccine and manufacturing process design:

Early attention to regulatory requirements for characterization and sourcing of strains, raw materials and appropriate quality systems can reduce the need to repeat work at a later stage. The choice of an appropriate glycan and carrier protein will be made early in product development, and should include consideration of manufacturing and testing requirements of these components. For example, batches of chemically or genetically toxoided toxins should be assessed for reversion to toxicity *(6-9)*, or may be subject to existing pharmacopoeial monographs. The disease target will define the appropriate glycan chain and hence influence the preferred manufacturing method and conjugation chemistry to be used *(1)*. When glycans are prepared from bacterial sources, the strain should be from a reliable source or well characterized in-house, and animal-derived culture components should be avoided *(23)*. Experience indicates that a wide variety of structural types can lead to effective glycoconjugate vaccines, although the detailed responses differ *(1)*. These can be described as neo-glycoconjugate, crosslinked network or vesicle vaccines, which are illustrated in Figure 1.

Development, characterisation and preclinical testing:

Preclinical development of a prototype vaccine is likely to concentrate on validating the vaccine model (i.e. that the proposed molecule will elicit protective immunity), developing and validating assays, optimizing immunogenicity as determined by functional antibody assays, and assessing the quality of the antibody response (antibody isotype, T cell involvement and specificity) *(25)*. If an animal challenge model that mimics disease in humans is available, it should be used in an appropriate manner. The stability of the vaccine can be assessed at this stage, to highlight degradation pathways and allow optimization of the manufacturing process and formulation conditions.

The USP 28th Revision (USP 28) contains guidance for recombinant protein manufacture and characterisation *(26)*. Mass spectrometry is valuable to confirm protein sequence, define post-translational modifications and identify process, product-related and host cell-derived impurities.

Glycan characterization methodology centres around NMR spectroscopy and molecular sizing *(27,28)*. It should not be assumed that reported structures of bacterial polysaccharides are correct *(29,30)*. NMR is also useful to characterise activated polysaccharide, including mean molecular weight for oligo-saccharides and quantitation of the degree of activation *(31)*. Simple tests may prove more

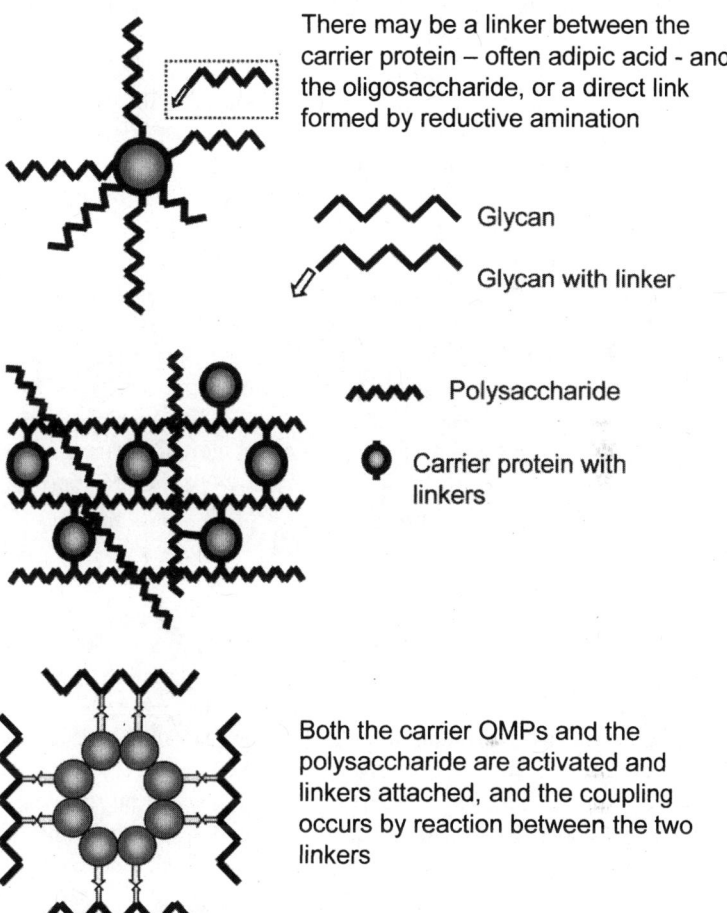

Figure 1. Cartoon representing three basic structural types of glycoconjugate vaccine. (a) a neoglycoconjugate vaccine in which oligosaccharide chains are attached to a carrier. Crosslinking may occur if the glycan chain is bifunctional. (b) A cross-linked matrix structure with multiple activation of a high mass polysaccharide and attachment to multiple carrier proteins. Each carrier protein is attached to several polysaccharide chains. (c) A vaccine with size-reduced glycan chains attached to an LPS-depleted vesicle of outer membrane proteins (Adapted from reference 24.)

useful in manufacturing, but instrumental approaches provide validation and calibration of these assays. Whilst methods for protein analysis are reasonably standard, glycan methodology must be matched to the polysaccharide structure. Whilst meningococcal CPSs are acid labile and can be quantified after hydrolysis, *Salmonella enterica* serotype Typhi Vi and *Staphylococcus aureus* Types 5 and 8 CPSs resist hydrolysis. "Characterisation" methodology used for QC purposes, such as HPSEC-MALLS, NMR or mass spectrometry *(32-33)*, should be validated in line with ICH Guidelines.

The extent of characterization which can be carried out on the final bulk conjugate depends on the type of vaccine being developed – neoglycoconjugate vaccines which use recombinant carrier proteins not chemically toxoided are simplest *(34)*. The integrity of the glycan chains in the conjugate (eg. loss of O-acetyl groups during conjugation) can be confirmed by NMR spectroscopy, and possibly to determine polysaccharide-protein ratios *(27)*. Neoglycoconjugate vaccines are characterized principally by size, composition and stability. Some methods produce a result that is an average over a diverse molecular population, which may not always be appropriate *(35)*. As vaccine QC focuses on product consistency, validation of the proposed manufacturing process is important.

Phase 1 and Phase 2 studies [2]

Vaccines undergo initial assessments of their safety and immunogenicity in humans in small scale trials (20-80 volunteers) in healthy adults. This is also an opportunity to optimise dosages or formulation conditions. Detailed serological analysis provides data on the immunological responses *(36-42)*. In subsequent stages, the dosage and formulation are usually fixed, and the recipients are progressively younger (if a paediatric vaccine is envisaged), the study is carried out with larger numbers of recipients, or in subpopulations where the immune response may be different (eg. HIV infected individuals, recipients who have had bone marrow transplants, the elderly etc.) *(43-49)*. It should be noted that the immune response of adults to conjugate vaccines often different from that of infants, in that an antibody response can be generated by one immunization because an adult will probably have prior exposure to the organism through carriage. Related studies may also be carried out at about this stage to answer specific queries about a product, or to assess responses in subpopulations. For example, restricted immunoglobulin V_H domain usage in anti-CPS antibodies means that some people are genetically incapable of responding to some of the pneumococcal serotypes in CPS vaccines. It was not clear whether these people would respond to pneumococcal conjugate vaccines *(50)*. The end-point of these trials, however, usually remains safety and immuno-genicity, rather than protection from disease.

---

[2] Clinical Trial Phases are defined in 21 CFR 312.21

Phase 3 Clinical trial.

In general, the Phase 3 trial is a double-blind experiment used to demonstrate the efficacy of the vaccine in the final target population *(51-52)*. It is the "Gold Standard" bioassay, carried out in the target population and directly assessing the key endpoint – protection from disease. For most glycoconjugate vaccines, the expected target population is infants, and the principal endpoint is a reduction in disease in this population, although secondary endpoints such as functional antibody levels or a reduction in disease severity may be helpful. The Phase 3 clinical trial is an extremely valuable stage at which to collect additional information relevant to the vaccine, such as the proportion of patients who seroconvert, relationships between total and functionally active antibody, uncommon adverse events etc. The accumulation of additional safety data is another key outcome of the Phase 3 trial. Material used in clinical trials will usually have been characterized more extensively than appropriate for subsequent manufacturing batches, to allow correlations between the analytical data and clinical efficacy to be investigated. Detailed information exceeding that required for routine release, including clinical efficacy data, is generally required on a number of production batches ("consistency lots"), to ensure that the manufacturing process is sufficiently robust. In the case of the meningococcal Group C conjugate vaccines, the relatively low incidence of disease in the UK and the well understood relationship between bacteriocidal antibody levels and protection from disease allowed measurements of functional antibody levels to be used as a surrogate correlate of protection *(53)*. Correlations between total antibody levels and bacteriocidal antibody levels were also established and the role of the *O*-acetylation of the meningococcal Group C polysaccharide investigated. This was the first time a surrogate had been used to allow licensing of a vaccine in the UK. Post-licensing surveillance confirmed that the vaccine protected against disease and was used to estimate clinical efficacy. For ethical reasons, designing Phase 3 clinical trials is more complex when a product already exists, and surrogate correlates of protection become more important. The National Institutes of Health maintain a web-accessible database of clinical trials *(54)*.

Controls applied during vaccine manufacture, specification setting, release tests and independent testing of batches ("batch release" or "lot release")

A key part of the strategy to ensure the quality of all pharmaceutical products is adherence to Good Manufacturing Practice (GMP), with a fully validated manufacturing process, raw material testing, in-process controls and final release testing. Release tests are carried out by manufacturers to ensure that the final material meets the specifications agreed with the licensing authority within whose jurisdiction the material will be sold.

ICH Guideline Q6B *(55)*, agreed in 1999, defines the procedure for setting and justifying acceptance specifications for biotechnological and biological products. To cite Q6B, "Specifications are chosen to confirm the quality of the drug substance or drug product rather than establish full characterization and should focus on those molecular and biological characteristics found useful in ensuring the safety and efficacy of the product".

Tests apply to both finished product and intermediates, with the aim of ensuring consistency between production batches and, above all, with the material used in clinical trials. Because animal models for glycoconjugate vaccines are poor, there is a greater reliance on physicochemical approaches than for other vaccines. A direct correlation between the result of the test and either clinical efficacy or safety is rarely available. EP and WHO Recommendations no longer cite the need for *in vivo* testing of Hib conjugate vaccines *(6,16)*, although an animal-based General Safety Test is still required in the US. A consequence of this is that dosages are specified in SI mass units (μg of saccharide per dose) rather than arbitrary units of biological activity. The complexity of polyvalent glycoconjugate vaccines requires that much of the testing is performed on monovalent conjugates rather than a blended bulk or the final fill. When used, *in vivo* bioassays demonstrate that the immunogen is generating T cell-dependent immune responses (i.e that it is a conjugate immunogen), typified by lack of antibodies after the first dose and a booster response after the second dose.

European, US and Japan Regulatory Authorities require independent testing of batches of vaccine before they are marketed ("batch release"). Testing is carried out by CBER in the USA and by Official Medicines Control Laboratories (OMCLs) in the EU. Compliance to a formal quality standard is an increasingly significant expectation for OMCLs. National control laboratories (NCLs) and manufacturers need to collaborate at this stage, to allow access to production material prior to licensing so that in-house tests can be established and validated, to allow product characterisation and to develop sufficient expertise to allow NCLs to make informed decisions on the quality of the vaccine.

Licensing

ICH Guideline M4 describes a Common Technical Document for submission to Regulatory Authorities in the three ICH regions, supplemented with region-specific administrative particulars *(56)*. License applications (called Marketing Authorization Applications (MAAs) in Europe) containing details of the manu-facturing process, QC procedures and clinicals trials data are submitted to the regulatory agency appropriate for the country in which the product will be sold. In the EU, two procedures are available: licensing by one of the national authorities with subsequent acceptance by other EU countries through a mutual recognition process, or through a "centralized procedure" involving the EMEA. Choice of the appropriate mechanism is complex: conjugates using recombinant carrier proteins are obliged to use the centralized procedure whilst meningococcal Group C conjugates were licensed by the UK national authority. In the centralized procedure, the EMEA secretariat appoint national authorities of Rapporteur and co-Rapporteur countries to lead a consultative process. Almost invariably, the license application (or MAA) will provoke a number of questions requiring clarification or additional data. Based on that process and advice from its Working Parties, the Committee on Human Medicinal Products (CHMP), which includes experts from member states, will issue an "opinion", which, if positive, will allow the European Commission to issue a marketing authorization. This will contain a provision for review.

In cases where a license is sought for a new vaccine where a licensed product already exists, the FDA have a concept of "non-inferiority" *(57,58)*, i.e. the new product must not be less effective than the existing product. For glycoconjugate vaccines, this is likely to be based on the total or functional antibody levels elicited, and require rigorous standardisation of the assays measuring antibody levels. Conceptual difficulties exist when additional components are added to a multivalent vaccine, as a reduced antibody response to individual serotypes could be less important than protection against additional serotypes.

Phase 4 and post licensing surveillance – the importance of carriage

Phase 3 clinical trials are rarely large enough to show up low frequency adverse events, and continued monitoring by companies and public health authorities is required after vaccine introduction. Formal reporting systems exist for this. Mass paediatric vaccination is concerned with protection both of the individual and the general population. Widespread use of a vaccine leads to changes in patterns of disease in a way that may not have been foreseen, especially when, as for the glycoconjugate vaccines, it interferes with normal carriage of the organism and, hence, transmission between individuals. Continued monitoring of both vaccinated and unvaccinated populations is common, including serology and carriage rates. For some vaccines, special studies may be triggered by changes in disease rates or estimates of vaccine efficacy *(59)*. In the glycoconjugate vaccine field, there is ongoing concern that elimination of carriage of "vaccine" serotypes will lead to their replacement by "non-vaccine" serotypes, resulting in disease rates dropping initially but increasing again due to serotype replacement. Immunisation of infants and young children with pneumococcal glycoconjugates led to significant reductions in the incidence of disease due to vaccine serotypes in unvaccinated populations

– parents and grandparents. This probably reflects reduced carriage by children diminishing transmission to those with whom they are in daily contact *(4)*. The observed reduction in the efficacy of the Hib conjugate vaccine in the UK since its introduction reflected low antibody levels and carriage rates, suggesting that the children's immune system is not being re-stimulated by natural contact with the organism *(3)*, and that the first estimate of efficacy reflected conditions different to those which now exist. The vaccination schedule in the UK has been revised to include a booster dose.

License variations

Changes in the sources of materials, in the manufacturing process, QC procedures, product formulation or intended vaccination schedule require a variation in the license. Traditionally, for a biological such as a vaccine, the product and the production process are considered inseparable, and a change in the manufacturing process would require that the product be re-examined in a clinical setting (a "bridging trial") to ensure equivalent biological responses in the target population, probably monitored by a surrogate correlate of protection (e.g. antibody levels). The rationale for this was that, as a biological cannot be completely defined by physicochemical methods, only use in the target population could confirm biological identity. Process improvements therefore carry financial overheads which discourage change and tend to lock-in old processes. Attitudes to glycoconjugate vaccines are somewhat different, and it was accepted for Hib, at least, that physicochemical characterisation of the polysaccharide is sufficient to ensure identity, whilst changes in the carrier protein or conjugate production probably would require some sort of bridging clinical trial *(60)*.

Factors affecting the efficacy of glycoconjugate vaccines

Identity of the individual glycan and carrier protein components (34) Glycan identity is best confirmed by NMR spectroscopy as the method is able to distinguish subtle structural changes and provide quantitative information on *O*-acetylation. Immunological approaches, combined with compositional data, can be used but immunological cross-reactivity can exist between serotypes. Immunochemical methods demonstrate identity of the carrier protein, but peptide mapping, mass spectrometry and size exclusion chromatography are valuable. The optimal size of an oligosaccharide hapten is determined in preclinical development and suitable manufacturing procedures developed. Confirmation that the material in each batch meets the specification will be required *(27)*. For conjugates prepared from high molecular weight polysaccharides, specifications for the glycan in the polysaccharide vaccine are often used. Assessment of the degree of substitution of high molecular weight polysaccharides randomly activated is required *(33)*.

Molecular size of the final conjugate, as a marker of the integrity of the complex. Whilst the molecular size of a glycoconjugate vaccine has not been shown to modulate immunogenicity, molecular sizing by HPSEC is a simple method to ensure consistency in the manufacturing process and that degradation has not occurred *(61)*.

Saccharide content of the vial. Glycoconjugate vaccines are dosed by mass, not potency-related units. Whilst monovalent Hib and Men C conjugates typically contain 10 µg of saccharide per dose, multi-valent vaccines are formulated with less of each serotype. The sugar content of the vial can be measured chemically, but immunochemical methods may have advantages for multivalent products. Chemical approaches may be problematic for some vaccines.

Identity of saccharide and protein components in the conjugate. Confirmation is required that the correct material is being delivered to the patient and, in the case of multivalent vaccines, that all of the components have indeed been included, is required. Immunological assays have been preferred, although NMR can be applied to bulk conjugate and can confirm the level of *O*-acetylation. The carrier protein in the final conjugate is usually determined immunologically.

Polysaccharide-protein ratio. An optimal polysaccharide-protein ratio will have been determined in preclinical development and a specification set. The value for each production batch can be determined at the bulk conjugate stage. Separate measurement of the saccharide and protein contents is the most common approach, but methods which can give a direct measurement would seem preferable. The value obtained is an average over all the molecules, but a mixture of components with high and low polysaccharide-protein ratio can have lower immunogenicity *(35)*.

Free, unconjugated saccharide. Unconjugated saccharide reduces the immuno-genicity of glycoconjugate vaccines, perhaps by competing for saccharide-specific sIg on the surface of B cells *(62)*. Limit specifications exist for the proportion of unconjugated vs. total saccharide in final fills, and values are determined by separation of the conjugate from the saccharide - by size or hydrophobicity or with carrier-specific antibodies - and separate quantitation of the fractions. Both Hib PRP and the meningococcal CPSs are labile and increases in free saccharide can determine shelf-life *(63)*.

Absence of potentially toxic byproducts from the conjugation reactions. Many conjugation procedures create low molecular weight, potentially toxic by-products that are removed later in the manufacturing process. Complete removal of these materials should be verified *(64)* unless the process has been validated.

Discussion

Glycoconjugate vaccines are regulated within the same framework as other vaccines. They differ from other vaccines in that the conjugate technology seems to be robust to major structural variation, animal models of infection tend to be poor, and functional antibody levels frequently correlate with protection. This has influenced development and the choice of endpoints in many studies and resulted in an unusually strong dependence on physicochemical methods of QC. Experience with existing glycoconjugate vaccines is likely to influence the detailed approaches for future glycoconjugate vaccines.

Acknowledgements

I thank Drs Ian Feavers, Jim Robertson and Stephen Inglis for critical reading of this manuscript and for helpful suggestions, and Dr Neil Ravenscroft and other colleagues in regulatory agencies and vaccine manufacturers for discussions over a period of many years.

References

1. Jones, C. *An. Bras. Acad. Cienc.* **2005**, *77*, 293-324.
2. Wenger, J. D. *Pediatr. Infect. Dis. J.* **1998**, *17*, S132-S136.
3. Ramsay, M. E.; Andrews, N. J.; Trotter, C. L.; Kaczmarski, E. B.; Miller, E. *BMJ* **2003**, *326*, 365-366.
4. Whitney, C. G.; Farley, M. M.; Hadler, J.; Harrison, L. H.; Bennett, N. M.; Lynfield, R.; Reingold, A.; Cieslak, P. R.; Pilivili, T.; Jackson D.; Facklam R. R.; Jorgensen J. H.; Schuchat A. *N. Eng. J. Med.* **2003**, *348*, 1737-1746.
5. *ICH Guidelines,* http://www.ich.org/cache/compo/276-254-1.html
6. *World Health Organ. Tech. Rep. Ser.* **2000**, *897*, 27-60.
7. *World Health Organ. Tech. Rep. Ser.* **2004**, *924*, 102-128.
8. *World Health Organ. Tech. Rep. Ser.* **2004**, *926*, 90-94.
9. *World Health Organ. Tech. Rep. Ser.* in press.
10. *Code of Federal Regulation (CFR)* http://www.accessdata.fda.gov/scripts/cdrh/cfdocs/cfcfr/CFRSearch.cfm.
11. *FDA guidance documents* http://www.fda.gov/cber/guidelines.htm
12. *EudraLex:* http://pharmacos.eudra.org/F2/eudralex/index.htm.
13. *EU Clinical trials Directive* http://europa.eu.int/eur-lex/pri/en/oj/dat/2001/l_121/l_12120010501en00340044.pdf
14. *MHRA information on the implementation of EU Clinical Trials Directive* http://medicines.mhra.gov.uk/ourwork/licensingmeds/types/clintrialdir.htm.
15. *EMEA Notes for Guidances* http://www.emea.eu.int/index/indexh1.htm.

16. *European Pharmacopoeia 5th Edition*, Strasbourg, 2005, pp 662-664.
17. *European Pharmacopoeia 5th Edition*, Strasbourg, 2005, pp 680-682.
18. *European Pharmacopoeia 5th Edition*, Strasbourg, 2005, pp 2851-2852.
19. Ball, L. K.; Falk, L. A.; Horne, A. D.; Finn, T. M. *J. Infect. Dis.* **2001**, *33*, S299-S305.
20. Sikkema, D. J.; Ziembiec, N. A.; Jones, T. R.; Hildreth, S. W.; Madore, D. V.; Quataert, S. A. *Clin. Diagn. Lab. Immunol.* **2005**, *12*, 218-223.
21. Quataert, S. A.; Rittenhouse-Olson, K.; Kirch, C. S.; Hu, B.; Secor, S.; Strong, N.; Madore, D. V. *Clin. Diagn. Lab. Immunol.* **2004**, *11*, 1064-1069.
22. *Reference materials:* http://www.nibsc.ac.uk/products/catalogue.html. Some pneumococcal reference materials, whilst held and distributed by NIBSC were prepared by Dr David Goldblatt (Institute of Child Health, UK: see http://www.ich.ucl.ac.uk/ich/academicunits/Immunobiology/StaffList).
23. *EMEA Guidance on use of bovine-derived materials* http://www.emea.eu.int/pdfs/human/bwp/TSE%20NFG%20410-rev2.pdf
24. Ward, J.; Lieberman J. M.; Cochi, S. L. in "Vaccines"; Plotkin S. A.; Mortimer, E. A. Eds.; W.B. Saunders and Co., Philadelphia, PA, 1994, pp 337-386.
25. Fattom, A. I.; Horwith, G.; Fuller, S.; Propst, M.; Naso, R. *Vaccine,* **2004**, *22*, 880-887.
26. *Pharmacopeia of the United States of America, 28th Revision*, Rockville, MD, 2005, pp 2186, 2540-2550, 2587-2607.
27. Ravenscroft, N. *Pharmeuropa Spec. Issue Biol.* **2000**, 131-144.
28. Macnair, J. E.; Desai, T.; Teyral, J.; Abeygunawardana, C.; Hennessey, J. P. Jr. *Biologicals* **2005**, *33*, 49-58.
29. Jones, C. *Carbohydr. Res.* **2005**, *340*, 1097-1106.
30. Jones C.; Lemercinier, X. *Carbohydr. Res.* **2005**, *340*, 403-409.
31. Xu, Q.; Klees, J.; Teyral, J.; Capen, R.; Huang, M.; Sturgess, A.W.; Hennessey, J. P. Jr; Washabaugh, M.; Sitrin R.; Abeygunawardana C. *Anal. Biochem.* **2005**, *337*, 235-245.
32. Jones C.; Lemercinier, X. *J. Pharm. Biomed. Anal.* **2002**, *30*, 1233-1247.
33. Xu, Q.; Abeygunawardana, C.; Ng, A. S.; Sturgess, A. W.; Harmon, B. J.; Hennessey, J. P. Jr. *Anal. Biochem.* **2005**, *336*, 262-272.
34. Jones, C.; Lemercinier, X.; Crane, D. T.; Gee C. K.; Austin, S. *Dev. Biol. (Basel)* **2000**, *103*, 121-136.
35. Egan, W.; Frasch, C. E.; Anthony, B. F. *JAMA* **1995**, *273*, 888-889.
36. Berkowitz, C. D.; Ward, J. I.; Meier, K.; Hendley, J. O.; Brunell, P. A.; Barkin, R. A.; Zahradnik, J. M.; Samuelson, J.; Gordon, L., *J. Pediatr.* **1987**, *110*, 509-514.
37. Lepow, M. L.; Samuelson, J. S.; Gordon, L. K. *J. Infect. Dis.* **1984**, *150*, 402-406.
38. Richmond, P.; Goldblatt, D.; Fusco, P. C.; Fusco, J. D.; Heron, I.; Clark, S.; Borrow, R.; Michon, F. *Vaccine* **1999**, *18*, 641-646.

39. Anderson, E. L.; Bowers, T.; Mink, C. M.; Kennedy, D. J.; Belshe, R. B.; Harakeh, H.; Pais, L.; Holder, P.; Carlone, G. M. *Infect. Immun.* **1994**, *62*, 3391-3395.
40. Bruge, J.; Bouveret-Le Cam, N.; Danve, B.; Rougon, G.; Schulz, D. *Vaccine* **2004**, *22*, 1087-1096.
41. Campbell, J. D.; Edelman, R.; King, J.C. Jr; Papa, T.; Ryall, R.; Rennels, M.B. *J. Infect. Dis.* **2002**, *186*, 1848-1851.
42. Wuorimaa, T.; Käyhty, H.; Leroy, O.; Eskola, J. *Vaccine* **2001**, *19*, 1863-1869.
43. Lepow, M. L.; Samuelson, J. S.; Gordon, L. K. *J. Pediatr.* **1985**, *106*, 185-189.
44. Wuorimaa, T.; Dagan, R.; Eskola, J.; Janco, J.; Ahman, H.; Leroy, O.; Käyhty, H. *Pediatr. Infect. Dis. J.* **2001**, *20*, 272-277.
45. Madore, D. V.; Johnson, C. L.; Phipps, D. C.; Myers, M. G.; Eby, R.; Smith, D. H. *Pediatrics* **1990**, *86*, 527-534.
46. Vadheim, C. M.; Greenberg, D. P.; Marcy, S. M.; Froeschle, J.; Ward, J. I. *Pediatr. Infect. Dis. J.* **1990**, *9*, 555-561.
47. Käyhty, H.; Peltola, H.; Eskola, J.; Rönnberg, P. R.; Kela, E.; Karanko, V.; Mäkelä, P. H. *Pediatrics* **1989**, *84*, 995-999.
48. Granoff, D. M.; Weinberg, G. A.; Shackelford, P. G. *Pediatr. Res.* **1988**, *24*, 180-185.
49. Shinefield, H. R.; Black, S.; Ray, P.; Chang, I.; Lewis, N.; Fireman, B.; Hackell, J.; Paradiso, P. R.; Siber, G.; Kohberger R.; Madore, D. V.; Malinowski, F. J.; Kimura, A.; Le, C.; Landaw, I.; Aguilar, J.; Hansen, J. *Pediatr. Infect. Dis. J.* **1999**, *18*, 757-763.
50. Musher D. M.; Groover J. E.; Watson D. A.; Rodriguez-Barradas M. C.; Baughn, R. E. *Clin. Infect. Dis.* **1998**, *27*, 1487-90.
51. Shinefield, H. R.; Black, S. *Pediatr. Infect. Dis. J.* **2000**, *19*, 394-397.
52. Trotter, C. L.; Andrews, N. J.; Kaczmarski, E. B.; Miller, E.; Ramsay, M. E. *Lancet* **2004**, *364*, 365-367.
53. Miller, E.; Salisbury, D.; Ramsay, M. *Vaccine* **2001**, *20*, S58-S67.
54. *NIH database of clinical trials*: http://www.clinicaltrials.gov.
55. *ICH Guidelines on specification setting for biotechnology products* http://www.ich.org/MediaServer.jser?@_ID=432&@_MODE=GLB
56. *The ICH Common Technical Document* http://www.ich.org/cache/compo/276-254-1.html
57. Jodar, L.; Butler, J.; Carlone G.; Dagan R.; Goldblatt, D.; Kayhty H.; Klugman K.; Plikyatis B.; Siber G.; Kohberger R.; Chang I.; Cherain T. *Vaccine* **2003**, *21*, 3265-3272.
58. Lee L. H.; Frasch C. E.; Falk, L. A.; Klein D. L.; Deal, C. D. *Vaccine* **2003**, *21*, 2190-2196.
59. McVernon, J.; Howard, A.J.; Slack, M. P. E.; Ramsay, M. E. *Epidemiol. Infect.* **2004**, *132*, 765-767.
60. Holliday, M. R.; Jones, C. *Biologicals* **1999**, *27*, 51-53.

61. von Hunolstein, C.; Parisi, L.; Recchia, S. *Vaccine* **1999**, *17*, *118-125*.
62. Peeters, C. C. A. M.; Tenbergen-Meekes, A.-M. J.; Poolman, J. T.; Zegers, B. J. M.; Rijkers, G. T. *Vaccine,* **1992**, *10*, 833-840.
63. Sturgess, A. W.; Rush, K.; Charbonneau, R. J.; Lee, J. I.; West, D. J.; Sitrin, R. D.; Hennessy, J. P. Jr *Vaccine*, **1999**, *17*, 1169-1178.
64. Lei, Q. P; Lamb, D. H.; Shannon, A. G.; Cai, X.; Heller, R. K.; Huang, M.; Zablackis, E.; Ryall, R.; Cash, P. *J. Chromatogr. B Analyt. Technol. Biomed. Life Sci.* **2004**, *813*, 103-112.

Chapter 3

Conjugation Methods toward Synthetic Vaccines

V. Pozsgay and J. Kubler-Kielb

National Institute of Child Health and Human Development, National Institutes of Health, 6 Center Drive, MSC 2423, Bethesda, MD 20892 2423

Introduction

The observation at the beginning of the twentieth century that purified capsular polysaccharides (CPSs) of pneumococci are immunogenic in adults was followed by the discovery that the immune serum so formed can offer specific protection against infection by homologous bacteria.[1] The protective effect of pneumococcal CPSs was firmly established in clinical trials reported in 1945.[1] These findings were the basis for the development of polysaccharide (PS) based vaccines that include vaccines against *Haemophilus influenzae* type b, a 23-valent vaccine against pneumococci, a vaccine against *Neisseria meningitides* type A, C, Y, and W135, and the Vi vaccine against *Salmonella typhi*. The human immune response to PSs is developmentally regulated and, in contrast to proteins that can produce antibodies in infants, maturation of antibody production against PSs does not usually occur in young children before two years of age.[2,3] The reason for this evolutionary difference remains unclear.[4] Known exceptions to the above findings are the pneumococcal type 3 and the group A meningococcal polysaccharides.[5]

The development of glycoconjugate vaccines that can produce saccharide-specific antibodies in infants has its roots in the discovery that low molecular-weight, non-immunogenic compounds termed haptens can produce hapten-specific antibodies in rabbits if covalently attached to horse gamma globulin.[6] Application of this principle to carbohydrates led to the development of glycoconjugate vaccines licensed for immunization against *H. influenzae* type b, *N. meningitidis* types A, C, Y, and W135, and seven serotypes of pneumococci.[7,8]

While the exact mechanism by which PSs and their protein conjugates elicit humoral immune response is not fully understood, the prevailing concept is that high molecular weight PSs that are T-cell independent antigens, simultaneously interact with immunoglobulin receptors on the surface of B-cells. This cross-linking initiates a cascade of events leading to the secretion of PS-specific IgG and IgM antibodies. An experimental support for this view comes from the fact that "small" polysaccharides, e. g. the O-polysaccharide region of LPSs, are not immunogenic. Reexposure to PSs does not produce a booster (memory) response. In contrast to pure PSs, PS-protein conjugates are recognized and internalized by the B-cells, where the protein is fragmented into peptides of 15-20 amino acids. The peptide fragments, bearing the saccharide haptens are reexposed on the antigen presenting cells together with major histocompatibility complex (MHC) class II molecules to $CD4^+$ helper T-lymphocytes. Interaction of the antigen-MHC complex with its specific receptors on the T-cells leads to differentiation of B-cells into antibody-secreting cells and antigen-specific memory cells. An indirect support to this concept is provided by the observation that small peptides corresponding to B-cell epitopes, can convert non-immunogenic saccharides to immunogenic ones (see the chapter on the PADRE antigen). Repeated exposure to the PS-protein or peptide conjugates results in a booster response of mainly IgG isotype. The saccharide-specific antibodies bind to extracellular antigens such as the PSs on bacterial surface that can result in killing the bacteria.

Numerous protocols have been proposed for the covalent attachment of carbohydrates to proteins.[9-15] Accumulated evidence indicates that the conjugation strategy including the types of chemical linkages between the saccharide and the protein as well as the geometry of the linker moiety, the density and the length of saccharide chains influence the immune response to the carbohydrate moiety. For example, in a project aimed at developing a conjugate vaccine against *Shigella dysenteriae* type 1, the highest antibody response to the O-specific polysaccharide (O-SP) of this bacterium was achieved when an average of ten copies of a hexadecasaccharide fragment of the O-SP was linked to human serum albumin.[16] Conjugates of smaller fragments as well as both lower and higher saccharide loadings resulted in diminished responses. The importance of the linker has also been demonstrated. It was reported that the 4-(maleimidomethyl)cyclohexane-1-carboxylate linker connecting a synthetic Lewis^y tetrasaccharide to keyhole limpet hemocyanin (KLH) proved to be the immunodominant portion whereas the immune response against the carbohydrate portion was low. On the contrary, the use of the shorter 3-(bromo-acetamido)propionate linker resulted in a diminished anti-linker response and an increased one against the carbohydrate epitope.[17] That a shorter linker does not necessarily confer better anti-saccharide immunogenicity in oligosaccharide-protein conjugates is reported in Ref. 18. The importance of the linker is further demonstrated by the finding that in mice conjugates of synthetic oligomannose clusters and KLH elicited most of the IgG antibodies against the linker and only a small fraction of the antibodies was directed to the carbohydrate part.[19] The choice of the methods for conjugation of carbohydrates to proteins is restricted because of their pH and temperature sensitivity and their limited solubility in most organic solvents. The procedure should be performed under mild conditions to prevent denaturation of protein and structural changes of saccharide. Therefore, buffer at or near neutral pH is the only solvent in almost all cases. The coupling reagent should be chemoselective to allow functionalization at the desired site only without affecting the hydroxyl groups of the saccharide and the intermediates should be sufficiently stable to survive purification steps. An additional requirement is that the chemical linkages connecting the hapten to the protein should be appropriate for human use. Ideally, the conjugation procedure should be high-yielding and should allow the recovery of the uncoupled material in its original chemical form. Few, if any of the published protocols fulfill all the requirements as evidenced by the frequency with which new procedures are being proposed.

This review illustrates the most often used saccharide-protein conjugation methods together with some recent developments. No attempts were made to exhaustive literature coverage and the authors apologize to those colleagues whose work was inadvertently omitted. Since most protocols employ multiple chemical reactions, classification of the procedures is sometimes arbitrary and individual sections may be overlapping.

Reductive Amination

Reductive amination continues to be the one of the most popular methods for binding free oligo- and poly-saccharides[20-22] through their reducing-end, a latent carbonyl group, to the ε-amino group of the lysine residues in proteins, as first reported by Gray in 1974.[23] For example, the labile aldimine (Schiff-base) **2** that is initially formed in an equilibrium reaction from glucose **1**, is reduced with $NaCNBH_3$ to afford a stable secondary amine **3**. Other reducing agents, such as BH_3[24] and $NaBH(OAc)_3$[25] have also been proposed. This spacerless conjugation procedure was originally shown to be effective for simple sugars containing a masked aldehydo function like glucose or lactose but failed to occur with sugars containing a keto function like fructose, 2-deoxy-oxulosonic acid (Kdo), or sialic acid.[26] A recent report demonstrated successful application of reductive amidation for conjugation of meningococcal lipopolysaccharide (LPS) derived oligosaccharides through their reducing Kdo residue to protein carriers.[27]

A similar mechanism, including formation of Schiff-base, followed by Amadori rearrangement was proposed for dry thermal glycation of bovine serum

albumin (BSA).[28] Using this procedure, an average of 19 lactose molecules were bound to BSA.

Reductive amination has found numerous applications for linking synthetic oligosaccharides equipped with an aldehydo group to proteins. The most often used methods to create an aldehydo group at the terminus of the spacer include (1) acid-catalyzed hydrolysis of an acetal-protected, latent aldehyde in the aglycon moiety,[29] (2) chemical oxidation of the spacer's terminal hydroxymethyl group,[30] (3) ozonolysis of an aglycon's terminally positioned carbon-carbon double bond,[31] and (4) introduction of a secondary spacer.[24,32] For example, the oligosaccharide hydrazide **4**, representing a fragment of the O-specific PS of *Shigella dysenteriae* type 1 was acylated with the heterobifunctional linker **5** to afford acetal **6**, from which the aldehyde **7** was obtained by mild acid treatment.[32]

In another example, glycosyl azide **8** (GM3 ganglioside epitope) was converted to the amine **9** which was acylated with 4-pentenoic anhydride. Ozonolysis of the resulting olefin **10** afforded the aldehyde **11**.[33] The use of a dec-9-enyl spacer as the aldehyde precursor for linking Leishmania-related phospholipooligosaccharides has been reported.[34] The scope of this technique is illustrated by the fact that the sensitive phosphodiester linkages were completely preserved. The application of the crotyl group has also been successful.[35] Reductive amination has been utilized for the introduction of heterobifunctional spacers. For example, the aldehyde **12** was reacted with the maleimido-hydrazide **13** in the presence of $NaCNBH_3$ to afford the intermediate **14** which was coupled to thiolated keyhole limpet hemocyanin (KLH).[35] Although KLH is still one of the most frequently used proteins for the preparation of neoglycoconjugates, its continued use has recently been questioned[36] because of its cross-reactivity with the sera of *Schistosoma*-infected patients.[37] In an other example, a trisaccharide-aldehyde construct (**15**) was coupled with the hydrazide **13** under reductive amination conditions to afford the maleimide derivative **16** that can be coupled to thiolated proteins through a stable thioether linkage.[38]

Reductive amination was used for the synthesis of neoglycoconjugates with Kdo-containing tetra- (**17**) and pentasaccharide fragments of deep rough LPSs of *E. coli*.[39] For example, treatment of the tetrasaccharide **17** with allylamine **18** under reductive conditions afforded the allyl derivative **19**.

Addition of cysteamine **21** introduced a primary amino group (→ **22**) for subsequent conversion to an isothiocyanate derivative. An unexpected side reaction was the reduction of the allyl double bound during aminoallylation to produce the propylamino derivative **20** that is no longer reactive.[39]

Reductive amination has also been used to conjugate chemically modified bacterial oligo- and poly-saccharides to proteins. For example, *O*-deacylated meningococcal LPSs were efficiently coupled to protein carriers (→ **24**) through the reducing-end glucosamine residue (**23**) of their Lipid A region.[27,40]

Other examples include formation of aldehydo group on the saccharide by controlled periodate oxidation as described for conjugation of meningococcal group A, B and C PSs[41], pneumococcal glycoconjugates[42] and *E. coli* and *Salmonella typhimurium* LPSs' core oligosaccharides.[43] Alternatively, an aldehyde group can be formed by nitrous acid deamination if a non-acetylated aminosugar is present in the carbohydrate chain as applied for the preparation of *Bordetella pertussis* lipooligosaccharide-tetanus toxoid (TT) conjugates.[44]

Reductive amination is a mild and experimentally simple procedure, and is suitable to couple large oligosaccharides to proteins at various degrees of incorporation[30,32] while retaining the antigenicity of the carrier.[16] However, direct coupling of aldehydo-spacered saccharides to proteins is sometimes only marginally successful[38] while in other cases large excess of the hapten required to achieve "respectable degrees of conjugation".[45] A disadvantage of reductive amination is that for monosaccharides and for the reducing-end saccharide unit of oligomers the cyclic structure is converted to a polyhydroxy-alkyl chain. These drawbacks coupled with long reaction times and with the disadvantage that unreacted hapten can not be fully recovered in its original form detract from the efficiency of this procedure.

Conjugation Through Thioether Linkages

A frequently used mild method for bioconjugation involves the formation of a stable thioether linkage between the carbohydrate and the protein counterparts. In one application, a glucosylamine **25** was treated with iodoacetic acid to yield the iodoacetamide **26** from which the unprotected derivative **27** was obtained by saponification.[46] This electrophilic derivative was coupled with the free thiol group of the single cysteine residue in BSA yielding a glycosylated protein **28**. The progress of the derivatization was monitored by using Ellman's reagent that reacts with free sulfhydryl groups with the release of 5-thio-2-nitrobenzoic acid, showing absorption at 412 nm. It was confirmed that the iodoacetamide **27** reacts selectively with sulfhydryl groups and leaves both α and γ amino groups in peptides unchanged.

In a recent variation, an amino group was introduced to the reducing end monosaccharide moiety of dextrans **29** (MWs 20 to 150 kDa) by their treatment with a 200-fold excess of ammonium chloride in the presence of $NaCNBH_3$.[47] Next, the aminated saccharide **30** was reacted with 2-iminothiolane (Traut's reagent) **31** to furnish the thiol **32**. Alternatively, treatment of PS **29** with cystamine **33** under reducing conditions and simultaneous cleavage of the disulfide bond with mercaptoethanol **34** afforded the thiol **35**. The thiolated saccharides, e.g. **35** were covalently bound to bromoacetylated protein **36** under slightly basic conditions to yield the thioether-linked construct **37**.

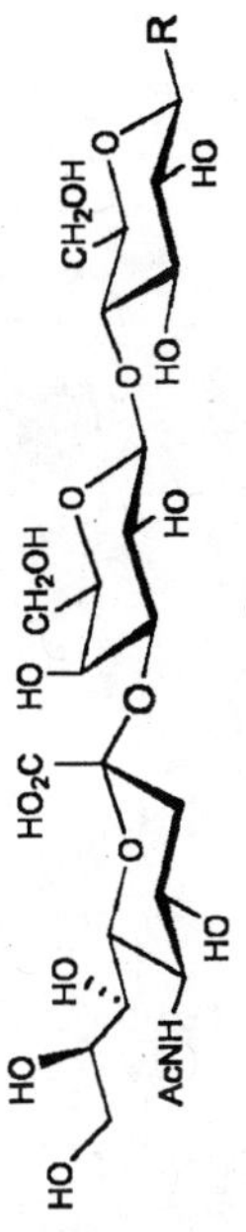

8 R = N₃
9 R = NH₂

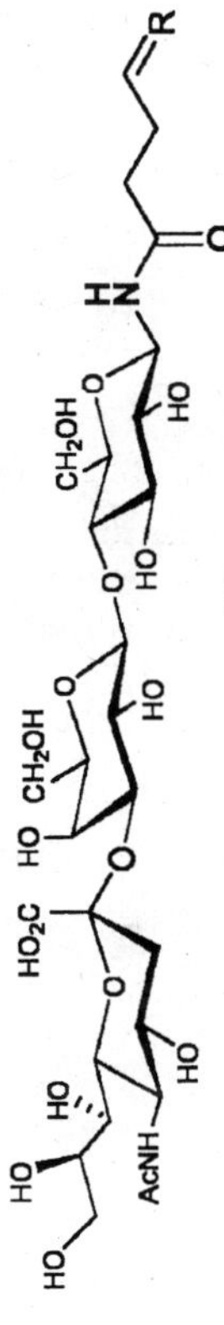

10 R = CH₂
11 R = O

44

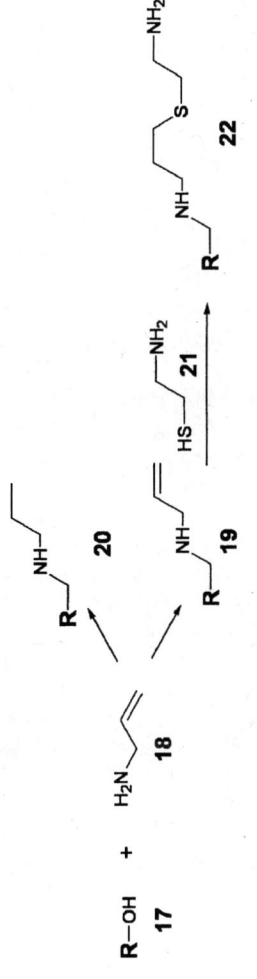

*FA - 3-hydroxy-fatty acid

This approach was successfully used to prepare pneumococcal CPS conjugates after their partial depolymerization by sonication or irradiation in an electron beam accelerator.[48]

In another application, bromoacetylated BSA **38** was coupled with the heterobifunctional aminooxy-thiol reagent[49] **39** to furnish the thioether-bridged aminooxy-albumin **40** for attachment to saccharides.[50]

The irreversible reaction of maleimide with thiols to afford stable linkages has been employed frequently for the preparation of neoglycoconjugates. For example, starting from the aminopropyl lactoside **41** the maleimide **43** was prepared using the commercially available reagent **42**.[38] Next, the saccharide-maleimide construct was allowed to react with thiolated human serum albumin (HSA) resulting in a stable glycoconjugate (**44**).

Alternatively, the thiol group may be introduced to the saccharide component. Following this approach, chitosan, (**45**) a β-1,4-linked polymer of 2-amino-2-deoxy-D-glucose, was thiolated with 2-iminothiolane **31** to yield the modified derivative **46** that was coupled to maleimide-modified BSA. The reaction can be followed spectrophotometrically by measuring the decrease of the maleimide absorbance.[51] A disadvantage of the maleimide-based conjugations is the instability of the maleimide moiety in basic aqueous media giving rise to an *N*-acyl derivative that no longer reacts with thiols.[52]

Another study employed oxidative coupling of thioaldoses (**47**) corresponding to mammalian glycoproteins to cysteine containing peptides and proteins to give disulfide-linked neoglycoconjugates (**48**).[53]

Bioconjugation Through Oxime Linkages

Condensation of oxo compounds with aminooxy group-containing reagents leads, selectively, to oximes in high yields, usually at or slightly below pH 7. In

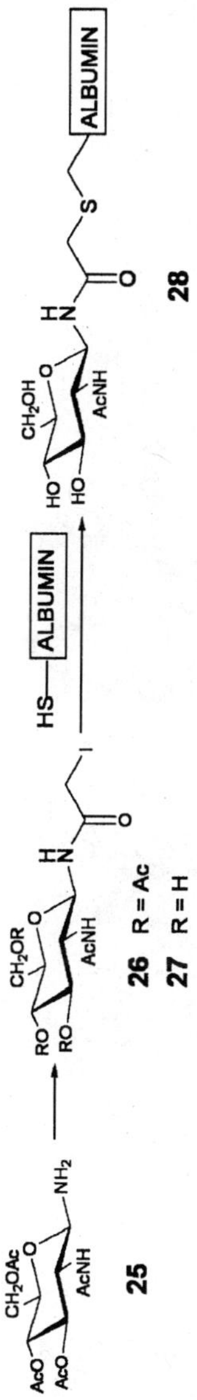

26 R = Ac
27 R = H

25

28

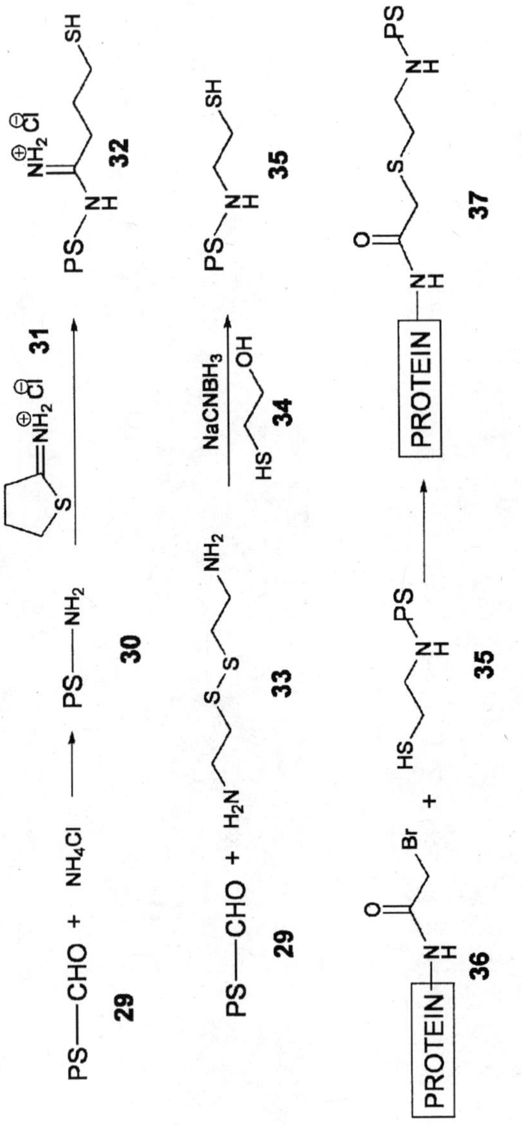

PS—CHO + NH₄Cl ⟶ PS—NH₂
29   **30**

S=C(cyclic)—NH₂⁺ Cl⁻ **31**

PS—NH—C(=NH₂⁺ Cl⁻)—(CH₂)₃—SH **32**

PS—CHO + H₂N—CH₂CH₂—S—S—CH₂CH₂—NH₂ **29** + **33**

HS—CH₂CH₂—OH **34**, NaCNBH₃

PS—NH—CH₂CH₂—SH **35**

PROTEIN—NH—C(=O)—CH₂—Br **36** + HS—CH₂CH₂—NH—PS **35**

⟶ PROTEIN—NH—C(=O)—CH₂—S—CH₂CH₂—NH—PS **37**

38 + 39 → 40

50

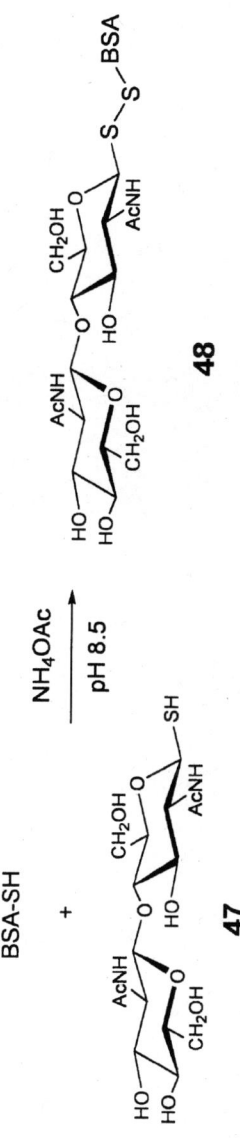

52

a recently reported protocol[50] aminooxylated BSA 40 was condensed with saccharides equipped with either aldehydo or keto groups. For example, tetrasaccharide 49 was acylated with 5-oxo-hexanoic anhydride 50 to give the ketone 51 which was condensed with the aminooxylated albumin 52 to yield neoglycoproteins containing up to 22 saccharide chains. An advantage of the procedure is that unreacted hapten can be recovered unchanged and may be reused. In control experiments, no conjugates were formed between the native protein and the carbonylated saccharides, indicating complete chemoselectivity. The procedure is also suitable for the attachment of phosphodiester-containing saccharides to proteins without disruption of the sensitive phosphodiester linkages[50] as well as LPS-derived oligosaccharides containing aldehyde moieties. The reaction proceeds under very mild conditions and in much shorter times compared to the reductive amination approach. Preparation of protein conjugates of oxidized bacterial PSs by the oxime technique has been reported.[54]

Homobifunctional Linkers

The squarate method

In the first step of this attractive approach,[55-57] the electrophilic homobifunctional reagent squaric acid dimethyl 54 (or diethyl) ester is condensed with a saccharide (53) containing a nucleophilic amino group in its aglycon. At pH 7 the condensation stops at the monoester-monoamide stage (55). Subsequent exposure of a solution of the monoester 55 and a protein to pH 9 activates the second electrophilic group in the intermediate and anchors the saccharide-squaric acid construct to the ε-amino groups of the protein (→ 56). A one-pot version has also been reported.[58] After the initial enthusiastic reception by a number of laboratories[59-63] it was recently found that square-bridged conjugates elicited decreased immune response relative to those that contained an adipic acid linker.[64] Although the reason for this is not understood, one possible explanation may be insufficient distance between the saccharide and the protein in the squarate-anchored constructs.

Adipic acid-derived linkers

Adipic acid dihydrazide (ADH), (57) was first introduced in 1980 in glycoconjugate synthesis to prepare *Haemophilus influenzae* type b PS-protein conjugates[65] and continues to be a popular reagent for building lattice-like constructs of PSs with proteins.[66,67] Following activation of the PS component 58 with cyanogen bromide 59 to form a cyanate ester 60 an excess of ADH is added

54

that introduces the strongly nucleophilic hydrazide moiety in the PSs (→ **61**). Subsequently, the hydrazide-derivatized saccharide is linked to the carboxyl groups of the protein using a water-soluble carbodiimide resulting in cross-linked conjugates.

Recently another cyanylating reagent, 1-cyano-4-dimethylaminopyridinium tetraflouroborate was introduced. It is easier to use, can be employed in lower pH, and has fewer side reactions.[68]

N,N'-Bis-hydroxysuccinimide ester of adipic acid (**62**) was used to prepare conjugates of synthetic saccharide fragments of *Streptococcus pneumoniae* type 14 with a mutant (CRM$_{197}$, **65**) of diphtheria toxin.[64] The reagent **62** was attached to oligosaccharide **63** featuring a primary amino group, in a dimethyl sulfoxide solution. The activated oligosaccharide **64** was isolated by precipitation with acetone and was condensed with CRM$_{197}$ (**65**) in a phosphate-buffered aqueous solution to afford neoglycoproteins (**66**) containing an average of 4 and 8.5 chains of oligosaccharides per protein molecule. The use of the related disuccinimidyl substrate (**67**) and disuccinimidyl glutarate (**68**) has also been reported.[69] Critical in these procedures is the exclusion of water during the first step to avoid hydrolysis of the half-ester intermediate. The highest average level of incorporation was 25 hexasaccharide chains per HSA molecule. A disadvantage of this method is the sensitivity of the succinimide moiety to hydrolysis that precludes chromatographic purification of the intermediate.[45]

The 4-nitrophenyl diester of adipic acid (**69**) was reported to be an efficient linker that is easy to prepare.[45] After reaction with the terminal amine of β-linked oligomannoside **70** the half-ester intermediate **71** was purified by silica gel chromatography and reverse-phase isolation methods under slightly acidic conditions.[70] Subsequent incubation of the half ester with BSA in a pH 7.5 buffered solution afforded neoglycoconjugates **72** containing up to 13 saccharide chains per protein molecule, with 30-45 % incorporation efficiency. The method does not appear to permit the recovery of the unreacted half ester in its reactive form **71**.

Glutardialdehyde

This popular reagent has been used for many years for the coupling carbohydrate and on protein counterparts through their free amino groups. Major applications include generation of monoclonal antibodies against Lipid A by glutardialdehyde mediated reaction between de-*O* and de-*N*-acylated glucosamine disaccharide backbone of *E. coli* J-5 Lipid A bound to BSA[71] and against phosphorylated Kdo moiety in deep rough mutant of *H. influenzae* LPS also bound to BSA.[72]

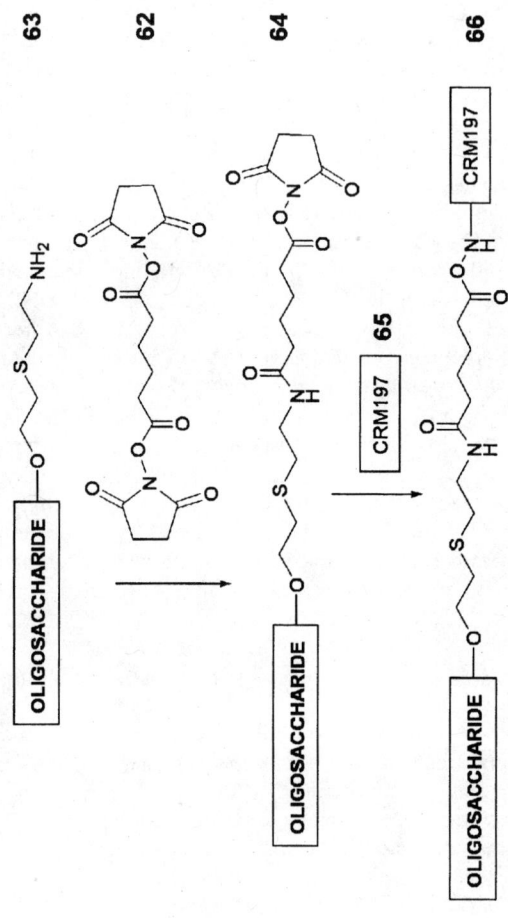

57

Conjugation Through Cycloaddition

The Diels-Alder reaction

Diels-Alder cycloaddition reactions between an electronically matched pair of a diene and a dienophile have been used for glycoconjugation in water at ambient temperature.[73] In a conjugation protocol reported by us the carbohydrate part was equipped with a diene moiety and the protein counterpart was derivatized with a dienophile unit.[73] In one approach, the protected L-rhamnosyl donor **73** was condensed with *trans,trans*-2,4-hexadien-1-ol followed by deprotection to afford the diene-modified rhamnose **74**. Alternatively, saccharide hydrazide **75** was acylated with diene-acid **76** to yield the dienylated derivative **77**. Maleimide was selected as the dienophile counterpart for its well-known reactivity in Diels-Alder reactions and was installed in albumin by using the commercially available reagent **42**. Cycloaddition of the diene **77** and the dienophile **78** partners in an aqueous solution afforded glycoconjugate **79** containing an average of 13 hexasaccharide chains. Higher incorporation levels were achieved when the monosaccharide **74** was used as the diene component. The level of incorporation was markedly affected by the pH: at pH 5.7 the average incorporation was twofold relative to the one achieved at pH 8.5 under otherwise identical conditions.[73] A remarkable feature of the method is that uncoupled saccharide can be recovered by diafiltration in an unchanged, pure form. Because of the known propensity of maleimides to undergo hydrolysis giving rise to an open chain derivative that does not function as a dienophile, this approach needs to be fine-tuned by using a more stable dienophile moiety. The Diels-Alder cycloaddition was also used for the immobilization of peptides[74] and proteins[75] to solid surfaces using a quinine as the dienophile and a cyclopentadiene derivative as the diene moiety, respectively.

Azide-alkyne [3 + 2] cycloaddition

The application of the irreversible Huisgen-cycloaddition to conjugation involving an acetylene moiety and an azide appears to be well-suited for the preparation of a wide variety of biomolecules.[76-78] In a recent application, a virus-PS construct was prepared using two Huisgen-cycloaddition protocols in tandem. First, an azide-functionalized glycopolymer (**81**) was prepared from methacryloxyethyl glucoside **80** by atom transfer radical polymerization. The polymer was coupled to a di-acetylene-functionalized fluorescein **82** used in excess, to afford the fluorescein-saccharide construct **83**. Cowpea mosaic virus

59

was derivatized with azido-succinimide **84** that installed azide groups on the ε-amino groups of its lysine moieties (→ **85**). The azido-modified virus was then coupled with the alkyne-polymer **83** in the presence of copper(I)triflate and sulfonated bathophenanthroline as the catalysts to afford a virus-polymer construct **86** having approximately 125 polymer chains per particle. This catalyst combination appears to be superior to an earlier reported[77] method using Cu(I) catalyst only. The azide-alkyne cycloaddition appears to hold great promise for the preparation of other types of bioconjugates, too.

Staudinger Ligation

Recently, the traceless Staudinger coupling[79,80] was adapted for the synthesis of neoglycoproteins.[81] In this protocol reaction of an azide and a phosphine yields an aza-ylide which is stabilized by intramolecular *N*-acylation and the simultaneous release of the phosphine moiety. Using the activated carboxylic acid **87** as the linker, the phosphino-thioester derivative **88** of tetanus toxoid (TT) was prepared. In compound **88** the borane moiety increases stability to oxidation and enhances reactivity. The degree of derivatization was in the range of 10-20. The derivatized protein was treated with azido-saccharide **89** in 5:1 DMF – 0.05 M aq sodium chloride solution in the presence of diazabicyclooctane to yield conjugate **90** that underwent to a three-step sequence, involving decomplexation (release of the borane), iminophosphorane formation (→ **91**) and acyl transfer, to afford glycoconjugate **92**. The yield of the conjugation was in the 45-80 % range, resulting in constructs containing 3 to 7 saccharide chains per protein molecule. In a similar fashion, tetanus toxoid conjugates of an undecasaccharide portion of the lipopolysaccharide of *Vibrio cholerae* O1 serotype were also prepared.[81] The method appears to be of general utility if low levels of saccharide incorporations are needed. In this protocol, uncoupled saccharide is unlikely to be recoverable as the starting azido derivative because of its loss due to hydrolysis.

Transglycosylation

The principle for synthesis of neoglycoconjugates using transglycosylation is based on the hydrolytic and transglycosylation activities of the enzyme endo-β-*N*-acetylglucosaminidase from *Arthrobacter protophormiae* (Endo-A). In the first step the enzyme releases *N*-linked oligosaccharides from glycoproteins by cleaving the di-*N*-acetylchitobiose unit. Next, it transfers the oligosaccharide either to water, leading to hydrolysis, or to an acceptor, leading to transglycosylation.[82]

61

In one example, a high mannose-type oligosaccharide released from Asn-linked N-glycan **93** was transferred to the p-isothiocyanatophenyl-β-D-glucopyranoside acceptor **94**, yielding compound **95**, which in the next step was coupled to amino groups on the protein providing **96**.[83] The efficiency of coupling the mannose-type oligosaccharide **93** to ribonuclease A, lysozyme, or lactalbumin was in the 10-40 % range.

Olefin Metathesis

The application of the olefin metathesis in glycobiology has been recently reviewed and examples of complex neoglycoconjugates, including glycopolymers, glycoclusters, glycopeptide and glycolipids were presented.[84] Cross-metathesis of alkene-containing carbohydrates and amino acids has gained much attention recently. For instance, metabolically stable glycopeptide analogues can be generated this way. As an example, alkenyl C-glycoside (**97**) was efficiently crosslinked with alkene-containing oligopeptide (**98**) to give a C-glycosyl oligopeptide (**99**).[85] Olefin metathesis was also applied in the construction of a carbohydrate-based polyvalent antitumor vaccine.[86]

Acknowledgment

The authors express their gratitude to Drs. John B. Robbins and Rachel Schneerson for their valuable comments. This work is supported by the National Institutes of Health.

References

1. MacLead, C. M.; Hodges, R. G.; Heidelberger, M.; Bernhard, W. G. *J. Exp. Med.* **1945**, *82*, 445-465.
2. Egan, W. *Ann. Rep. Med. Chem.* **1993**, *28*, 257-265.
3. Robbins, J. B.; Schneerson, R.; Gotschlich, E. C. *Lancet* **1997**, *350*, 1709.
4. Vinuesa, C. G.; MacLennan, I. C. M. Antibody responses to polysaccharides. In *Immunobiology of Carbohydrates*, Wong, S. Y. C., Arsequell, G., Eds.; Landes Bioscience/Eurekah.com: Georgetown, TX; New York, NY, 2003; pp 128-147.
5. Robbins, J. B.; Schneerson, R.; Gotschlich, E. C. *Pediatr. Infect. Dis. J.* **2000**, *19*, 945-953.
6. Landsteiner, K.; Lampl, H. *Biochem. Zeitschr.* **1918**, *86*, 343-394.
7. http://www.fda.gov/cber/vaccine/licvacc.htm **2005**.

97 + **98** →(Grubb's catalyst)→ **99**

8. Moingeon, P.; Moreau, M.; Lindberg, A. A. Challenges and opportunities in the development of new conjugate vaccines against infectious diseases. In *Immunobiology of Carbohydrates*, Wong, S. Y. C., Arsequell, G., Eds.; Landes Bioscience/Eurekah.com: Georgetown, TX; New York, NY, 2003.
9. Pozsgay, V. *Adv. Carbohydr. Chem. Biochem.* **2000**, *56*, 153-199.
10. Pozsgay, V. Chemical synthesis of bacterial carbohydrates. In *Immunobiology of Carbohydrates*, Wong, S. Y. C., Arsequell, G., Eds.; Landes Bioscience/Eurekah.com: Georgetown, TX; New York, NY, 2003; pp 192-273.
11. Jennings, H. J.; Sood, R. K. Synthetic glycoconjugates as human vaccines. In *Neoglycoconjugates. Preparation and Applications*, Lee, Y. C., Lee, R. T., Eds.; Academic Press: New York, 1994; pp 325-371.
12. Dick, W. E., Jr.; Beurret, M. *Contrib. Microbiol. Immunol.* **1989**, *10*, 48-114.
13. Stowell, C. P.; Lee, Y. C. *Adv. Carbohydr. Chem. Biochem.* **1981**, *37*, 225-281.
14. Langenhan, J. M.; Thorson, J. S. *Curr. Org. Synth.* **2005**, *2*, 59-81.
15. Davis, B. G. *Chem. Rev.* **2002**, *102*, 579-601.
16. Pozsgay, V.; Chu, C.; Pannell, L.; Wolfe, J.; Robbins, J. B.; Schneerson, R. *Proc. Natl. Acad. Sci. U. S. A.* **1999**, *96*, 5194-5197.
17. Buskas, T.; Liao, J.; Boons, G.-J. *Chem. Eur. J.* **2004**, *10*, 3517-3524s.
18. Mawas, F.; Niggemann, J.; Jones, C.; Corbel, M. J.; Kamerling, J. P.; Vliegenthart, J. F. G. *Infect. Immun.* **2002**, *70*, 5107-5114.
19. Ni, J.; Song, H.; Wang, Y.; Stamatos, N. M.; Wang, L.-X. *Bioconj. Chem.* **2006**, *17*, 493-500s.
20. Fernandez-Santana, V.; Cardoso, F.; Rodriguez, A.; Carmenate, T.; Pena, L.; Valdes, Y.; Hardy, E.; Mawas, F.; Heynngnezz, L.; Rodriguez, M. C.; Figueroa, I.; Chang, J. N.; Toledo, M. E.; Musacchio, A.; Hernandez, I.; Izquierdo, M.; Cosme, K.; Roy, R.; Verez-Bencomo, V. *Infect. Immun.* **2004**, *72*, 7115-7123.
21. Laferriere, C. A.; Sood, R. K.; de Muys, J. M.; Michon, F.; Jennings, H. J. *Infect. Immun.* **1998**, *66*, 2441-2446.
22. Zhang, J.; Yergey, A.; Kowalak, J.; Kovac, P. *Tetrahedron* **1998**, *54*, 11783-11792.
23. Gray, G. R. *Arch. Biochem. Biophys.* **1974**, *163*, 426-428.
24. Lee, R. T.; Wong, T.-C.; Lee, R.; Yue, L.; Lee, Y. C. *Biochemistry* **1989**, *28*, 1856-1861.
25. Dalpathado, D. S.; Jiang, H.; Kater, M. A.; Desaire, H. *Anal. Bioanal. Chem.* **2005**, *381*, 1130-1137.
26. Roy, R.; Katzenellenbogen, E.; Jennings, H. J. *Can. J. Biochem. Cell. B.* **1984**, *62*, 270-275.
27. Mieszala, M.; Kogan, G.; Jennings, H. J. *Carbohydr. Res.* **2003**, *338*, 167-175.

28. Boratynski, J.; Roy, R. *Glycoconj. J.* **1998**, *15*, 131-138.
29. Zhang, J.; Kovac, P. *Tetrahedron Lett.* **1998**, *39*, 1091-1094.
30. Pozsgay, V. *Glycoconj. J.* **1993**, *10*, 133-141.
31. Bernstein, M. A.; Hall, L. D. *Carbohydr. Res.* **1980**, *78*, C1-C3.
32. Pozsgay, V. *J. Org. Chem.* **1998**, *63*, 5983-5999.
33. Xue, J.; Pan, Y. B.; Guo, Z. W. *Tetrahedron Lett.* **2002**, *43*, 1599-1602.
34. Routier, F. H.; Nikolaev, A. V.; Ferguson, M. A. J. *Glycoconj. J.* **1999**, *16*, 773-780.
35. Ragupathi, G.; Koganty, R. R.; Qiu, D. X.; Lloyd, K. O.; Livingston, P. O. *Glycoconj. J.* **1998**, *15*, 217-221.
36. Roy, R. *Drug Discovery Today:Technologies* **2004**, *1*, 327-336.
37. Nyame, A. K.; Kawar, Z. S.; Cummings, R. D. *Arch. Biochem. Biophys.* **2004**, *426*, 182-200.
38. Zou, W.; Abraham, M.; Gilbert, M.; Wakarchuk, W. W.; Jennings, H. J. *Glycoconj. J.* **1999**, *16*, 507-515.
39. Muller-Loennies, S.; Grimmecke, D.; Brade, L.; Lindner, B.; Kosma, P.; Brade, H. *J. Endotoxin. Res.* **2002**, *8*, 295-305.
40. Cox, A. D.; Zou, W.; Gidney, M. A. J.; Lacelle, S.; Plested, J. S.; Makepeace, K.; Wright, J. C.; Coull, P. A.; Moxon, E. R.; Richards, J. C. *Vaccine* **2005**, *23*, 5045-5054.
41. Jennings, H. J.; Lugowski, C. *J. Immunol.* **1981**, *127*, 1011-1018.
42. Lee, C. J.; Wang, T. R.; Frasch, C. E. *Vaccine* **2001**, *19*, 3216-3225.
43. Lukasiewicz, J.; Jachymek, W.; Niedziela, T.; Dzieciatkowska, M.; Lakomska, J.; Miedzybrodzki, R.; Fortuna, W.; Szymaniec, S.; Misiuk-Hojlo, M.; Lugowski, C. *FEMS Immunol. Med. Mic.* **2003**, *37*, 59-67.
44. Niedziela, T.; Letowska, W.; Lukasiewicz, J.; Kaszowska, M.; Czarnecka, A.; Kenne, L.; Lugowski, C. *Infect. Immun.* **2005**, *73*, 7381-7389.
45. Wu, X. Y.; Ling, C. C.; Bundle, D. R. *Org. Lett.* **2004**, *6*, 4407-4410.
46. Davis, N. J.; Flitsch, S. L. *Tetrahedron Lett.* **1991**, *32*, 6793-6796.
47. Pawlowski, A.; Kallenius, G.; Svenson, S. B. *Vaccine* **1999**, *17*, 1474-1483.
48. Pawlowski, A.; Kallenius, G.; Svenson, S. B. *Vaccine* **2000**, *18*, 1873-1885.
49. Bauer, L.; Suresh, K. S.; Ghosh, B. K. *J. Org. Chem.* **1965**, *30*, 949-951.
50. Kubler-Kielb, J.; Pozsgay, V. *J. Org. Chem.* **2005**, *70*, 6987-6990, **2006**, *71*, 5422.
51. Masuko, T.; Minami, A.; Iwasaki, N.; Majima, T.; Nishimura, S. I.; Lee, Y. C. *Biomacromolecules* **2005**, *6*, 880-884.
52. Thibaudeau, K.; Leger, R.; Huang, X. C.; Robitaille, M.; Quraishi, O.; Soucy, C.; Bousquet-Gagnon, N.; van Wyk, P.; Paradis, V.; Castaigne, J. P.; Bridon, D. *Bioconj. Chem.* **2005**, *16*, 1000-1008.
53. Watt, G. M.; Boons, G. J. *Carbohydr. Res.* **2004**, *339*, 181-193.
54. Lees, A.; Sen, G.; LopezAcosta, A. *Vaccine* **2006**, *24*, 716-729.
55. Tietze, L. F.; Schröter, C.; Gabius, S.; Brinck, U.; Goerlach-Graw, A.; Gabius, H.-J. *Bioconj. Chem.* **1991**, *2*, 148-153.

56. Tietze, L. F.; Arlt, M.; Beller, M.; Glüsenkamp, K.-H.; Jähde, E.; Rajewsky, M. F. *Chem. Ber.* **1991**, *124*, 1215-1221.
57. Kamath, V. P.; Diedrich, P.; Hindsgaul, O. *Glycoconj. J.* **1996**, *13*, 315-319.
58. Pozsgay, V.; Dubois, E. P.; Pannell, L. *J. Org. Chem.* **1997**, *62*, 2832-2846.
59. Lefeber, D. J.; Kamerling, J. P.; Vliegenthart, J. F. G. *Chem. Eur. J.* **2001**, *7*, 4411-4421.
60. Nitz, M.; Bundle, D. R. *J. Org. Chem.* **2001**, *66*, 8411-8423.
61. Saksena, R.; Zhang, H.; Kovac, P. *Tetrahedron-Asymmetry* **2005**, *16*, 187-197.
62. Hossany, R. B.; Johnson, M. A.; Eniade, A. A.; Pinto, B. M. *Bioorg. Med. Chem.* **2004**, *12*, 3743-3754.
63. Wang, J. Y.; Chang, A. H. C.; Guttormsen, H. K.; Rosas, A. L.; Kasper, D. L. *Vaccine* **2003**, *21*, 1112-1117.
64. Mawas, F.; Niggemann, J.; Jones, C.; Corbel, M. J.; Kamerling, J. P.; Vliegenthart, J. F. G. *Infect. Immun.* **2002**, *70*, 5107-5114.
65. Schneerson, R.; Barrera, O.; Sutton, A.; Robbins, J. B. *J. Exp. Med.* **1980**, *152*, 361-376.
66. Kubler-Kielb, J.; Coxon, B.; Schneerson, R. *J. Bacteriol.* **2004**, *186*, 6891-6901.
67. Carmenate, T.; Canaan, L.; Alvarez, A.; Delgado, M.; Gonzalez, S.; Menendez, T.; Rodes, L.; Guillen, G. *FEMS Immunol. Med. Mic.* **2004**, *40*, 193-199.
68. Lees, A.; Nelson, B. L.; Mond, J. J. *Vaccine* **1996**, *14*, 190-198.
69. Lahmann, M.; Bulow, L.; Teodorovic, P.; Gyback, H.; Oscarson, S. *Glycoconj. J.* **2004**, *21*, 251-256.
70. Wu, X. Y.; Bundle, D. R. *J. Org. Chem.* **2005**, *70*, 7381-7388.
71. Brade, L.; Holst, O.; Brade, H. *Infect. Immun.* **1993**, *61*, 4514-4517.
72. Muller-Loennies, S.; Brade, L.; Brade, H. *Eur. J. Biochem.* **2002**, *269*, 1237-1242.
73. Pozsgay, V.; Vieira, N. E.; Yergey, A. *Org. Lett.* **2002**, *4*, 3191-3194.
74. Yousaf, M. N.; Houseman, B. T.; Mrksich, M. *Angew. Chem. Int. Ed.* **2001**, *40*, 1093-1096.
75. Yousaf, M. N.; Mrksich, M. *J. Am. Chem. Soc.* **1999**, *121*, 4286-4287.
76. Peto, C.; Batta, G.; Gyorgydeak, Z.; Sztaricskai, F. *J. Carbohydr. Chem.* **1996**, *15*, 465-483.
77. Wang, Q.; Chan, T. R.; Hilgraf, R.; Fokin, V. V.; Sharpless, K. B.; Finn, M. G. *J. Am. Chem. Soc.* **2003**, *125*, 3192-3193.
78. Fazio, F.; Bryan, M. C.; Blixt, O.; Paulson, J. C.; Wong, C. H. *J. Am. Chem. Soc.* **2002**, *124*, 14397-14402.
79. Saxon, E.; Armstrong, J. I.; Bertozzi, C. R. *Org. Lett.* **2000**, *2*, 2141-2143.
80. Nilsson, B. L.; Kiessling, L. L.; Raines, R. T. *Org. Lett.* **2001**, *3*, 9-12.

81. Grandjean, C.; Boutonnier, A.; Guerreiro, C.; Fournier, J. M.; Mulard, L. A. *J. Org. Chem.* **2005,** *70,* 7123-7132.
82. Takegawa, K.; Fan, J. Q. *Methods Enzymol.* **2003,** *362,* 64-74.
83. Fujita, K.; Takegawa, K. *Biochem. Bioph. Res. Co.* **2001,** *282,* 678-682.
84. Leeuwenburgh, M. A.; van der Marel, G. A.; Overcleeft, H. S. *Curr. Opin. Chem. Biol.* **2002,** *7,* 757-765.
85. McGarvey, G. J.; Benedum, T. E.; Schmidtmann, F. W. *Org. Lett.* **2002,** *4,* 3591-3594.
86. Biswas, K.; Coltart, D. M.; Danishefsky, S. J. *Tetrahedron Lett.* **2002,** *43,* 6107-6110.

Chapter 4

Haemophilus influenzae Type b Conjugate Vaccine with a Synthetic Capsular Polysaccharide Antigen: Chemical View

Vicente Verez Bencomo[1], René Roy[2], Maria C. Rodriguez[1], Annete Villar[1], Violeta Fernandez-Santana[1], Ernesto Garcia[1], Yury Valdes[1], Lázaro Heynngnezz[3], Ivan Sosa[3], and Ernesto Medina[3]

[1]Center for the Study of Synthetic Antigens, Faculty of Chemistry, University of Havana, Ciudad Habana, Cuba 10400
[2]Department of Chemistry, Université du Québec à Montréal, P.O. Box 8888, Succ. Centre-Ville, Montréal, Québec H3C 3P8, Canada
[3]Center for Genetic Engineering and Biotechnology, Apdo 6162, Cubanacan, Playa, Ciudad Habana, Cuba 10600

> Glycoconjugate vaccines containing capsular polysaccharides were developed in the last 20 years against *Haemophilus influenzae* type b. They provided effective prophylaxis against meningitis and pneumonia, especially amongst infants. Despite recent advancements in synthetic carbohydrate chemistry, the technology for using a synthetic antigen in a commercial vaccine remains a formidable challenge. Major chemical problems to be solved and the strategies employed to produce the first vaccine containing a synthetic carbohydrate antigen that becomes commercially available are described.

Introduction

Modern vaccines are increasingly composed of well-defined molecules isolated by costly fractionation/purification procedures from living organisms obtained by either growing the relevant pathogens or after cloning a non-pathogenic organism that over-expresses the relevant molecules (*1*).

Carbohydrates constitute important components of several relevant antigens. In bacteria, capsular or lipopolysaccharides represent the major antigens, while in parasites, virus, and cancer cells, glycoproteins or glycolipids comprise the other key antigens. Carbohydrate-based vaccines have long been the mainstream of vaccine development. In the past and at least in two different occasions, the advent of polysaccharide-based vaccines represented relevant contributions in the fight against group A and C meningococcal meningitis. In the seventies, important epidemics were controlled due to this achievement (*2*). The use of capsular polysaccharides for combating other infectious diseases, especially in infants, were not as successful (*3*). A typical example is found in infections caused by *Haemophilus influenzae* type b (Hib), one of the main causes of meningitis and pneumonia in infants worldwide.

Vaccinology made another tremendous step forward with the introduction of conjugate vaccines. The conjugation of polysaccharides to proteins confers the ability to induce protective antibody responses in infants otherwise unresponsive to capsular polysaccharide alone (*4*). Application of this technology to *Haemophilus influenzae* type b essentially eradicated this disease in industrialized countries wherein vaccination programs were inserted. However, several circumstances that include the high cost of conjugate vaccines complicated the global coverage of vaccination in other countries. As a consequence, poor countries still experience more than 600 000 infant's death every year from Hib-pneumoniae or meningitis because of the lack of more available vaccines (*5*).

The initial success of conjugate vaccines was an important stimulus for attempting the use of synthetic Hib-oligosaccharides in place of the natural polysaccharide. The chemical synthesis of antigens has many advantages over other modern approaches for the manufacturing of vaccines. It avoids the use of pathogenic cells and therefore the products are free of infectious materials or pathogen-derived molecules. They can be economically produced in large scale. The final products are usually better molecularly defined pharmaceuticals. However, despite these facts, none of the major human vaccines are actually made from synthetic carbohydrate antigens (*6*). Synthetic chemistry is usually less efficient than bacteria in producing the antigens needed in the manufacturing of vaccines. Thus, the development of suitable technologies for synthesizing the carbohydrate antigens represents a formidable challenge.

Anti-Hib conjugate vaccines

Actual anti-Hib vaccines are composed of either the capsular polysaccharide or its oligosaccharide fragments isolated from bacterial growth. Hence, the molecular weights of effective commercial vaccines cover a wide range of sizes, ranging from the intact capsular polysaccharide to relatively small oligosaccharide fragments (7) containing 5~15 ribosyl-ribitol-phosphate repeating units. Despite the enormous differences in composition, the mechanism of action is quite similar and therefore no major differences were found in clinical trials that could be associated to the size of the polysaccharide (8).

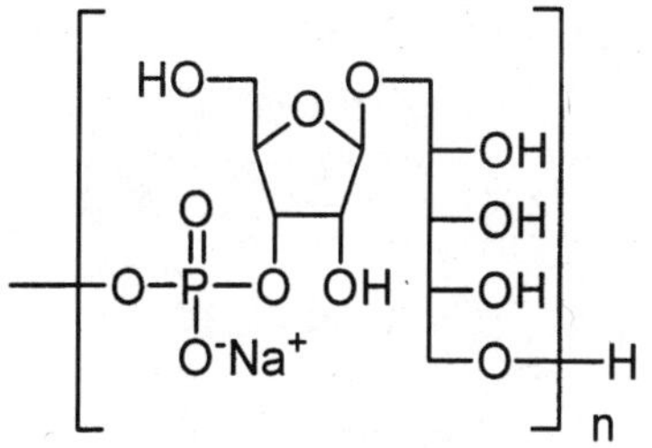

Figure 1. Structure of the repeating unit of the poly-(ribosyl-ribitol-phosphate) (PRP), the capsular polysaccharide of Haemophilus influenzae type b.

Studies conducted in laboratory animals have shown that smaller fragments containing 3-4 repeating units were sufficient for the induction of an efficacious anti-Hib immune response (9). Consequently, the initial synthetic strategies targeted antigens from 3 to 10 repeating units long.

Classical synthesis of PRP-fragments

As can be seen from the structure of the capsular polysaccharide (Figure 1), ribosyl-ribitol repeating units are connected by phosphodiester linkages. The retrosynthetic strategy for such structure is represented in Figure 2. The ideal combination of permanent and temporary protecting groups included benzyl and allyl ethers, respectively. The use of acetyl or other acyl protecting groups are not recommended because the required basic conditions employed for their removal could promote migrations of the phosphodiester linkages in the final product.

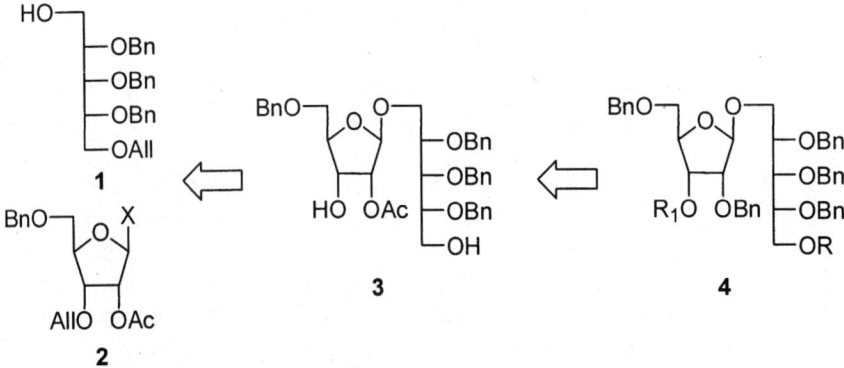

Figure 2. Retrosynthetic scheme for the construction of the repeating disaccharide unit from Hib capsular polysaccharide. R or R_1 is either phosphotriester, phosphoramidite, or H-phosphonate.

Ribitol Unit

Ribitol derivative **1** fitted with most application criteria for both laboratory scale and for larger industrial production. Usually **1** could be synthesized from D-ribose through three groups of operations (Figure 3): A) temporary protection of C-1 and transformation to free alcohol; B) permanent protection of O-2, 3, 4 and C) O-5 temporary protection (allyl). There are two different alternatives for selectively protecting position 1 in ribitol, either directly on the aldehyde carbonyl group of ribose as dithioacetal (*10*) or after reduction by its O-1 trityl ether (*11*). The synthesis through the dithioacetal is shorter and afforded better yields but face serious environmental problems. Our strategy was based on the combination of both alternatives.

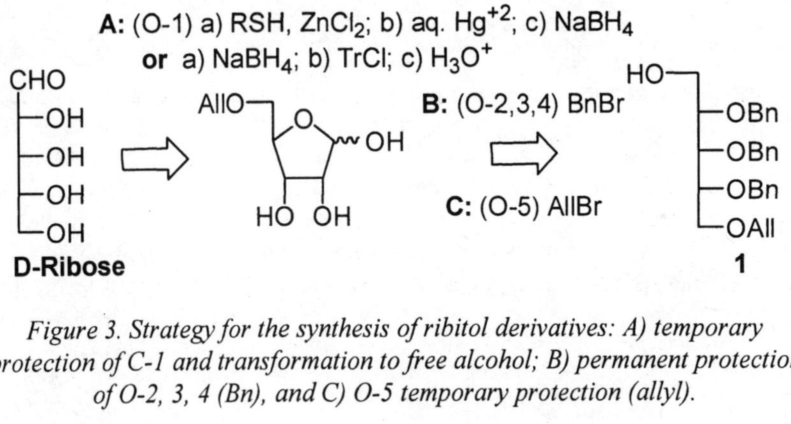

Figure 3. Strategy for the synthesis of ribitol derivatives: A) temporary protection of C-1 and transformation to free alcohol; B) permanent protection of O-2, 3, 4 (Bn), and C) O-5 temporary protection (allyl).

Ribosyl unit and disaccharide

For the selection of the ribosyl donor, the following constraints should be considered: a) the use of permanent benzyl protection at positions 2 and 5; b) use of an allyl or other temporary protective group at position 3; c) a suitable leaving group at the anomeric center; and d) a participating group at position 2. As constraint a and d are incompatible, ribosyl donor such as **2** are usually preferred (Fig. 2). A single acyl group at position 2 of the donor ensure the stereochemistry of β-D-linked disaccharide. It could be substituted later by a benzyl group. Structure **2** represents the generic formula of the majorities of donors employed in previous syntheses (*10,11*). The preparation of such donors are only possible through multistep synthesis. For this purpose, two major strategies were developed, one from D-ribose in 8 steps (*11*) and the second one from D-glucose through 1,2:5,6-di-*O*-isopropylidene-D-glucofuranose by the following reaction sequence (Figure 4) (*12*). First, inversion of the O-3 configuration and introduction of the temporary protective group (oxidation-reduction-allylation), followed by shortening of the side-chain through selective acid hydrolysis of the side-chain isopropylidene group, periodic acid oxidation, reduction, and benzylation. Finally, the desired donor **8** was obtained by selective acetal hydrolysis and acetylation.

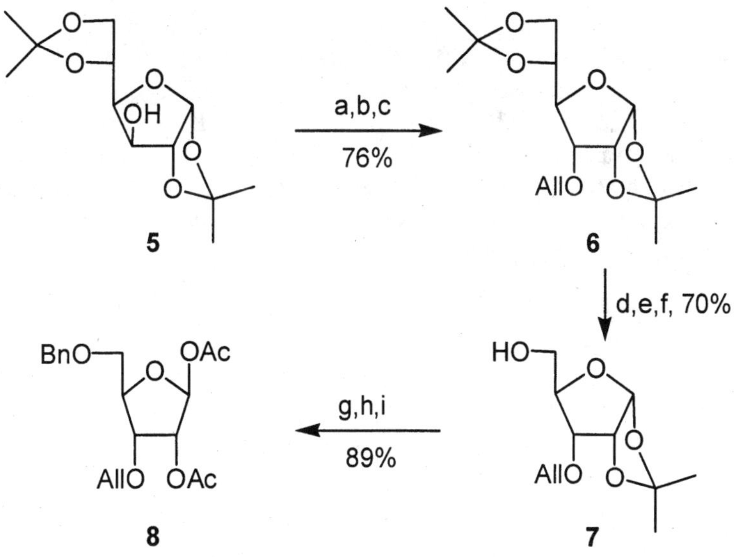

Figure 4. Synthesis of the ribose donor from D-glucose. Reaction conditions: a) RuO_2/IO_4^-; b) $NaBH_4$; c) AllBr/NaH; d) CH_3COOH; e) HIO_4/dioxane; f) $NaBH_4$; g) BnCl/NaH; h) Dowex H+, dioxane, 60°C; i) Ac_2O/Py.

The synthesis of the disaccharide intermediate was next performed by trimethylsilyltriflate-catalyzed coupling of **1** and **8** to afford the β-D-linked disaccharide in 79% yield (Figure 5). The introduction of the benzyl group at O-2 was then acomplished by deacetylation and benzylation by conventional procedures. Finally, deallylation with *t*-BuOK/DMSO afforded **10** in 70 % yield. Disaccharide **10** is a key intermediate that could be dimethoxytritylated at the ribitol O-5 position, thus allowing easy access to either the hemisuccinate **11**, designed for coupling to a solid support, or compound **12**, designed for further chain elongation (*12*).

Chain Elongation

Several strategies were available for chain elongation of the oligosaccharide using well-established nucleotide chemistry, including phosphoramidite, H-phosphonate, or phosphotriester active groups. All these alternatives were employed and provided satisfactorily results. The first synthesis of Hib fragments solving all the above problems was performed in van Boom's laboratory (*13*) using solution-phase chemistry. The examples that followed described the use of either solid phase chemistry (*14,15*) or polymer-bound solution techniques (*16*). The basic concepts developed in all these approaches, including our own early strategy were similar. For this reason, a description of our work for chain elongation will be discussed as an example.

The disaccharide **11** (Fig. 5) was coupled to a Merrifield resin and the DMTr group was removed by hydrolysis with trifluoroacetic acid. Next, four elongating cycles were performed with disaccharide **12** activated with 2,4,6-triisopropylbenzenesulfonyl chloride and 1-methylimidazole in acetonitrile-pyridine. The spacer **13** was further added and the resulting oligosaccharide was released from the resin together with the simultaneous removal of the O-chlorophenyl protecting groups (*17*). The overall yields were between 80-90%/cycle. Finally, the benzyl groups were removed by hydrogenolysis. The solid phase chemistry is generally a fast and high yielding process for the preparation of small quantities of pure compounds. H-Phosphonate gave similar yields/per coupling step and deprotection was usually easier (*15*).

However, for large scale production, solid phase chemistry has several drawbacks as it requires the use of very large excess of the elongating unit (5-10 equivalents). The recovery of the precious disaccharide is a complex process and, consequently the real yield based on disaccharide **12** wwas in the 10-15% range.

After deprotection, the pentamer was coupled to *Neiseria meningitidis* outer membrane protein (*18*) and the conjugate was used for immunization of mice. As

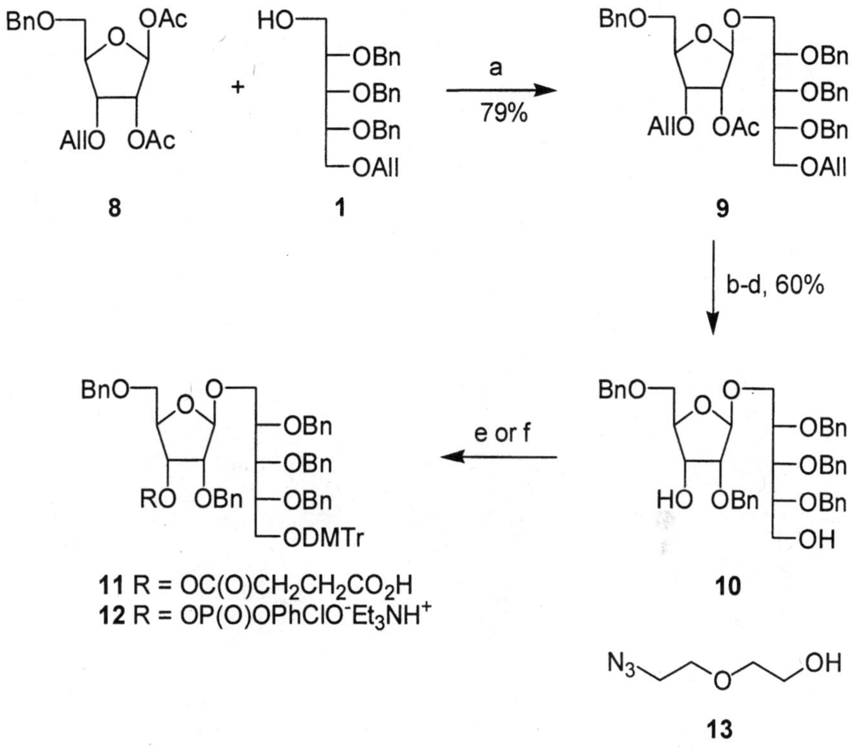

Figure 5. Example of the disaccharide intermediates synthesis. Reaction conditions: a) TMSTfO/DCE; b) NaOMe/MeOH; c) BnCl/NaH; d) tBuOK/DMSO; e) Succinic anhydride/Py, DMAP, 65% yield; f) TAMM reagent/CH$_3$CN, NMI, 85% yield.

can be seen from Figure 6, a strong anti-polysaccharide immune response was reached after three doses.

Fragments of the capsular polysaccharide developed by other synthetic procedures and having from 3 to 10 repeating units conjugated to either proteins or peptides were also immunogenic in laboratory animals. (*19,20*).

Unfortunately, for large scale production and eventual commercialization, none of the above processes were deemed acceptable, especially when the final cost of the vaccine was to be kept as low as possible. Moreover, the large volume of scientific literature published at the end of our immunological assays rendered patent applications more complicated, if not impossible. Hence, a totally different and cost effective synthetic strategy became critical.

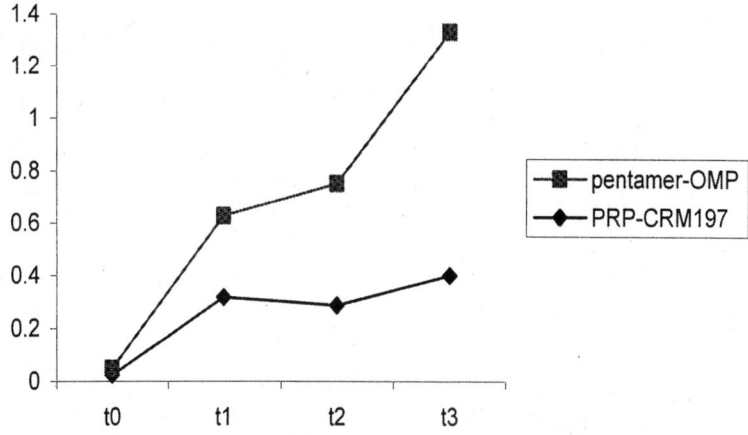

Figure 6. Mice immunogenicity of pentamer-OMP conjugate compared to a commercially available vaccine. Serum dilution 1/200.

New strategy for the synthesis of PRP fragments

The following criteria constitutes the key factors to be considered in the selection of the PRP-fragment size: a) the smaller fragments used in approved vaccines are mixtures having between 5-15 repeating units and b) some information was available concerning the minimal size of PRP recognized by the immune system of laboratory animals. However, the minimal size of PRP recognized by the immature infant immune system was still to be documented; c) if one considers the quantity of PRP per vaccine dose (~10 µg) independently from the size and the number of attachment sites in the protein is limited, then the quantity of protein carrier needed per dose will increase for shorter PRP fragments.

Although all the poly-(ribosyl-ribitol-phosphate) syntheses published at the end of the 1980's and during the 1990's paved the way for future improvements, all of them had the same drawback, *i. e.* they were rather complex and they all provided very low overall yields. After the analysis of published synthesis and using our previous experience, we came back to the synthetic scheme in our search for simplification and yields enhancement. The first modification was possible by a new selective benzylation of disaccharide **16** that was discovered. It allowed the shortening of a lengthy synthesis of D-ribofuranose derivative **2** which was substituted by the use of a very simple and commercially available ribofuranose peracetate **14** that readily glycosylated (*21*) ribitol **1** in the presence of BF_3-Et_2O, followed by deacetylation to afford disaccharide **16** having three free hydroxyl functions (Fig. 7). Initial attempts for the selective benzylation of triol **16** was performed by activation with dibutyltin oxide, Bu$_4$NI, and benzyl chloride. A mixture of only 2- and 3-O-monobenzylated compounds was obtained in a ratio of 1:1. Several additives were added in order to shift the possible equilibrium between the intermediate stannylidene derivatives and to improve the selectivity. The best of such additives was found to be sodium hydride as it not only promoted the initial benzylation predominantly at position 2, but also allowed further *in situ* benzylation at position 5. The successful synthesis of disaccharide **17** was one of the key achievements in the new synthesis of the required Hib antigen.

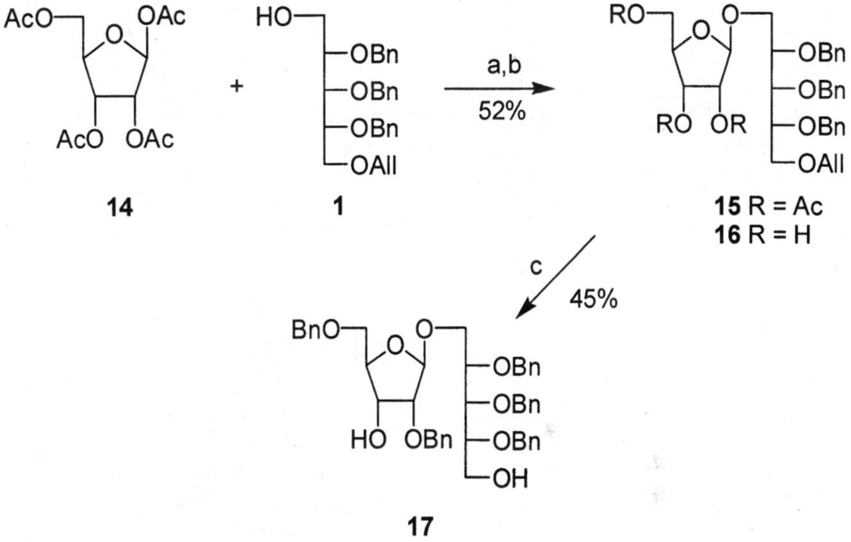

Figure 7. Synthesis of the key disaccharide 17. Reaction conditions: a) BF$_3$.Et$_2$O/DCE; b)NaOMe/MeOH; c) BnCl, Bu$_2$SnO, Bu$_4$NI, NaH, Tol.

After several attempts of constructing oligomers with precise sizes by solution, solid phase, or even by polymer-bound solution techniques, it was realized that, while all these procedures were excellent for the production of small quantities of compounds, they would fail in the production of large quantities of material in suitable yields. We next turned our attention to a single step polycondensation reaction that was previously shown (*22*) to provide short oligosaccharide mixtures in good yields using unrelated carbohydrates.

Thus, the H-phosphonate precursor **18** was prepared in excellent yield and purity (PCl$_3$, Imidazole, CH$_3$CN) and deprotected at the ribitol's primary hydroxyl group (Fig. 8). Extensive studies were then undertaken to evaluate its reaction with several quenching reagents. A fine tuning of the reactivity between the chain elongating **18** and chain terminating component **19** as well as conditions to avoid competing reaction such as acylation of free hydroxyl groups by the activating pivaloyl chloride were shown to be crucial. The position unto which was initially introduced the H-phosphonate functionality was also found to be critical in this respect (primary *versus* secondary hydroxyl groups).

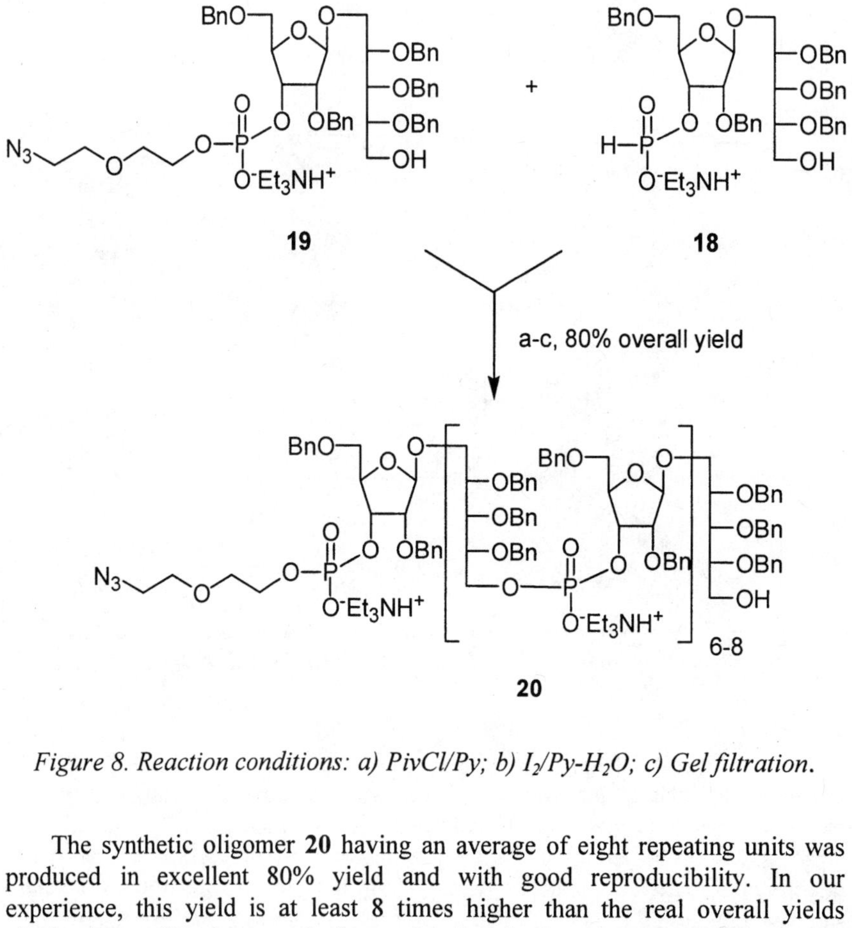

Figure 8. Reaction conditions: a) PivCl/Py; b) I$_2$/Py-H$_2$O; c) Gel filtration.

The synthetic oligomer **20** having an average of eight repeating units was produced in excellent 80% yield and with good reproducibility. In our experience, this yield is at least 8 times higher than the real overall yields obtained by solid-phase methods considering the large excess of disaccharide needed per cycle.

Another important issue was the complexity of removing efficiently all the protecting groups. Deprotection was performed by hydrogenolysis in the presence of 10 % Pd/C. The steric hindrance of triethylammonium salts made the complete removal of benzyl groups a difficult but imperative task for the production of a pharmaceutical product. Once removal of the entire benzyl groups and ion exchange of the triethylammonium salts were performed satisfactorily, the terminal linker functionality was treated with 3-maleimidopropionic acid N-hydroxysuccinimidate to provide **21**, the key synthetic fragment of *Haemophilus influenzae* type b ready for conjugation to protein carriers. This process was simple enough to be reproduced into 100 g scale (*23*). Figure 9 illustates the ^{1}H-NMR spectra of oligomer **21**.

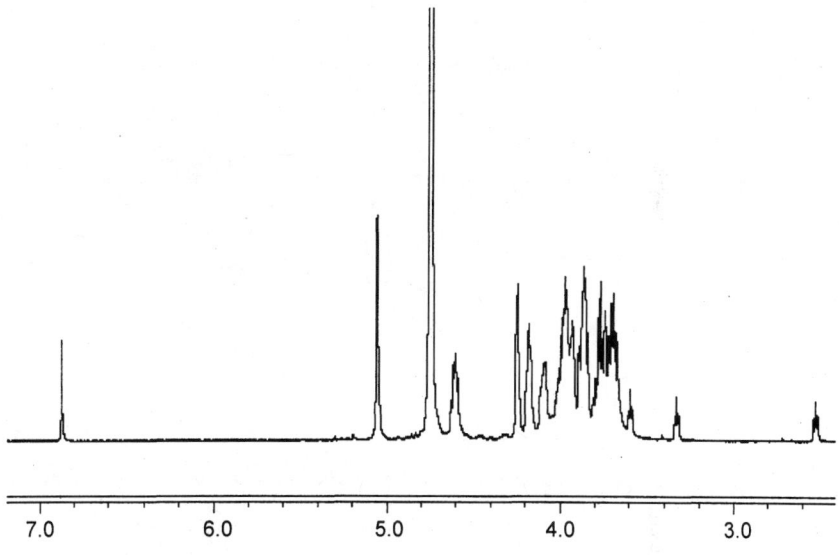

*Figure 9. 500 MHz ^{1}H-NMR spectra of compound **21** in D_2O.*

The conjugate vaccine

Several proteins from bacterial origin were tested as carrier candidates. Finally, tetanus toxoid was selected because of its many advantages: a) wide experience of its use alone or as a carrier for Hib conjugate vaccines; b) it is probably the cheapest bacterial protein available; c) it can be produced with an acceptable degree of purity; d) its solubility and the availability of its amino groups for conjugation are ideal. Several years ago (*24*) we adapted a procedure from peptide-protein chemistry for the conjugation of oligosaccharides to proteins. This process is distinguished by its simplicity and the high conjugation yield of the oligosaccharides. Tetanus toxoid (TT) was initially thiolated with N-hydroxysuccimide dithioproprionate to provide ~30 lysine amino groups modified (*25*). Compound **21** was then conjugated to give **22** in a yield of ~50% using very mild conditions (Fig. 10).

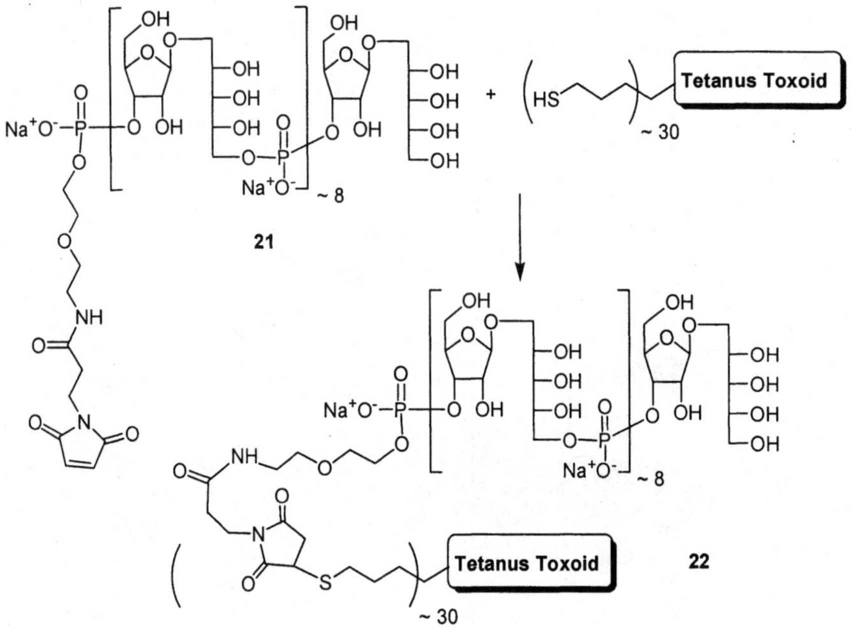

Figure 10. Reaction conditions: a) PBS solution, pH 7.4, 50% yield.

Compound **22** was further developed as a vaccine candidate. In a series of experiments (*25*), immunization of laboratory animals with **22** elicited antibodies similar to those elicited by commercially available vaccines in recognition and in functionality.

In the past, many vaccine candidates having synthetic carbohydrate antigens were obtained in the laboratory. However, in order to develop them as commercial vaccines, suitable and competitive technologies for production are required. Having demonstrated the similar behaviors of the synthetic PRP and its protein conjugate with its natural counterpart, coupled with a technology capable of producing gram quantities of the final vaccine, we were in a position to test the vaccine candidate in the clinic. Eight clinical assays starting in July 2001 were initiated. The population tested included adults, 4-5 years old children, and two-months old infants. Overall, the vaccine was demonstrated to be safe and induced a strong anti-PRP antibody response in the target population (*26*). The vaccine was registered in Cuba and starting in January 2004, it was included in the regular vaccination program.

In conclusion, this work represents the first time wherein a vaccine having a synthetic carbohydrate antigen became commercially available. It demonstrated the feasibility and opened new possibilities for vaccines against other human diseases (*27,28*).

Acknowledgments

We would like to thank the World Health Organization, the Pan-American Health Organization, and the following Cuban Institutions: State Council, Ministry of Science Technology and Environment, and Ministry of Health. We are also thankful to the many chemists, engineers, and laboratory assistants that were involved in the project over the years.

References

1. Ada, G. in *Vaccine Protocols*; Robinson, A.; Hudson, M.J.; Cranage, M.P., Eds.; Humana Press, Totowa, NJ, 2nd Edition, **2003**, pp 1-18.
2. Artenstein, M. S.; Gold, R.; Zimmerle, J. G.; Wyle, F. A.; Schnieder, H.; Harkins, C. *New England J. Med.* **1970**, *282*, 417-420.
3. Shapiro, E.D.; Murphy, T.V.; Wald, E.R.; Brady, C. A. *JAMA*, **1988**, *260*, 1419-1322.
4. Robbins, J. B.; Schneerson, R.; Szu, S. C.; Pozsgay, V. *Pure Appl. Chem.* **1999**, *71*, 745-754.
5. Peltola, H.; *Microbiology Rev.* **2000**, *13*, 302-317.
6. Francis, M. J. in *Vaccine Protocols*; Robinson, A.; Hudson, M.J.; Cranage Eds, M.P.; Humana Press, Totowa, NJ, 2nd Edition, **2003**, pp 115-131.
7. Constantino, P.; Norelli, F.; Giannozi, A.; D'Ascendi, S.; Bartoloni, A.; Kaur, S.; Tang, D.; Seid, R.; Viti, S.; Paffetti, R.; Bigio, M.; Pennatini, C.; Averani, G.; Guarnieri, V.; Gallo, E.; Ravenscropt, N.; Lazzeroni, C.; Rappuoli, R.; Ceccarini,C. *Vaccine* **1999**, *17*, 1251-1263.
8. Granoff, D.M.; Holmes, S. J. *Vaccin*, **1991**, *9*, S30-34.
9. Pillai, S.; Ciciriello, S.; Koster, M.; Eby, R. *Infect. Immun.* **1991**, *59*, 4371-4376.
10. Chan, L.; Just, G. *Tetrahedron*, **1990**, *46*, 151-162.
11. Hermans, J.P.G.; Poot, L.; Kloosterman, M.; van den Marel, G. A.; van Boeckel, C. A. A.; Evenberg, D.; Poolman, J.T.; Hoogerhout, P.; van Boom, J. H. *Rec. Trav. Chim. Pays-Bas*, **1987**, *106*, 498-504.
12. Chiu-Machado, I.; Madrazo-Alonso, O.; Verez-Bencomo, V. *J. Carbohydr. Chem.* **1994**, *13*, 464-474.

13. Hoogerhout, P.; Funke, C. W.; Mellema, J. R.; Wagenaars, G. N.; van Boeckel, C. A. A.; Evenberg, D.; Poolman, J. T.; Lefeber, A. W. M.; van der Marel, G. A.; van Boom, J. H. *J. Carbohydr. Chem*, **1988**, *7*, 399-416.
14. Elie,C. J. J.; Muntendam, H. J.; van den Elst, H.; van den Marel,G. A.; Hoogerhout,P.; van Boom, J. H. *Rec. Trav. Chim. Pays-Bas*, **1989**, *108*, 219-223.
15. Nilsson, S.; Bengtsson, M.; Norberg, T. *J. Carbohydr. Chem.* **1991**, *10*, 1-22.
16. Kandil, A. A.; Chan, N.; Chong, P.; Klein, M. *Synlett.*, **1992**, 555-557.
17. Lorenzo, Y. Diploma thesis, Fac. Chemistry, University of Havana, 1997.
18. Verez Bencomo, V.; Fernández Santana, V.; Figueroa Perez, I.; Adan Padron, A.; Lorenzo, Y.; Mondelo, A.; Castro-Palomino, J.; Sierra, G.; R.Barbera, XIX International Carbohydrate Symposium, San Diego, August 9-14, **1998**, BP 141
19. Peeters, C.A.A.; Evenberg, D.; Hoogerhout, P.; Kayhty, H.; Saarinen, L.; van Boeckel, C. A. A.; van der Marel,G. A.; van Boom, J. H.; Poolman, J. T. *Infect. Immun.* **1992**, *60*, 1826-1833.
20. Chong, P.; Chan , N.; Kandil, A.; Tripet, B.; James, O.; Yang, Y-P.; Shi, S-P.; Klein, M. *Infect. Immun.* **1997**, *65*, 4918-4925.
21. Chiu-Machado, I.; Castro-Palomino, J.C.; Madrazo-Alonso, O.; Lopetegui-Palacios, C.; Verez-Bencomo,V. *J. Carbohydr. Chem.* **1995**, *14*, 551-561.
22. Nikolaev, AV.; Chudek, J.A.; Fergusson, M.A.J. *Carbohydr. Res.*, **1995**, *272*, 179-189.
23. Verez-Bencomo, V.; Roy, R. WO Patent 01/16146, 2001; U.S. Patent 6,765,091, 2004.
24. Fernández-Santana, V.; Gonzalez-Lio, R.; Sarracent-Perez, J.; Verez-Bencomo, V. *Glycoconj J.* **1998**, *306*: 163-170.
25. Fernández-Santana, V.; Cardoso, F.; Rodriguez, A.; Carmenate, T.; Peña, L.; Valdés, Y.; Hardy, E.; Mawas, F.; Heynngnezz, L.; Rodríguez, M. C; Figueroa, I.; Chang, J.; Toledo, M. E.; Musacchio, A; Hernández, I.; Izquierdo, M.; Cosme, K.; Roy, R.; Verez-Bencomo, V. *Infect. Immun.* **2004**, *72*, 7115-7123.
26. Verez-Bencomo, V.; Fernández-Santana, V.; Hardy, E.; Toledo, M. E.; Rodríguez, M. C.; Heynngnezz, L.; Rodriguez, A.; Baly, A.; Herrera, L.; Izquierdo, M.; Villar, A.; Valdés, Y.; Cosme, K.; Deler, M. L.; Montane, M.; Garcia, E.; Ramos, A.; Aguilar, A.; Medina, E.; Toraño, G.; Sosa, I.; Hernandez, I.; Martínez, R.; Muzachio, A.; Carmenates, A.; Costa, L.; Cardoso, F.; Campa, C.; Diaz, M.; Roy, R. *Science* **2004**, *305*, 522-524.
27. Borman, S. *Chem. Eng. News*, **2004**, *82*, 31-35 .
28. Roy, R. *Drug Discovery Today: Technologies* **2004**, *1*, 327-336.

Chapter 5

Immunology of Experimental Synthetic Carbohydrate–Protein Conjugate Vaccines against *Streptococcus pneumoniae* Serotypes

Harm Snippe[1], Wouter T. M. Jansen[1], and Johannis P. Kamerling[2]

[1]Eijkman-Winkler Institute for Microbiology, Infectious Diseases and Inflammation, Vaccines Section, University Medical Center Utrecht, Heidelberglaan 100, 3584 CX Utrecht, The Netherlands
[2]Bijvoet Center, Department of Bio-Organic Chemistry, Utrecht University, Padualaan 8, 3584 CH Utrecht, The Netherlands

The design and development of synthetic routes for the preparation of oligosaccharide fragments of capsular polysaccharides of *Streptococcus pneumoniae* serotypes took 25 years of intensive research at the Bijvoet Center. During that time sets of oligosaccharides of type 3, 6A, 6B, 14, and 23F were synthesized. Meanwhile, the Eijkman-Winkler Institute performed immunological studies on the effects of these carbohydrates in mice and rabbits. Different carriers (including liposomes), dosages, and adjuvants were tested in order to obtain high antibody titers with opsonic activity. Opsonization promoted the uptake and killing of bacteria by phagocytes, resulting in protective immunity to a lethal challenge with the corresponding virulent *S. pneumoniae* strain. The immunogenicity of synthetic oligosaccharide-protein conjugates was compared with that of polysaccharide-protein conjugates for the same serotypes of *S. pneumoniae*.

History

Studies using synthetic carbohydrate antigens were started in the mid-1970s at the Eijkman-Winkler Institute (previously known as The Laboratory of Microbiology). They were the logical follow-up to studies being performed at the institute at that time on the immunogenicity of antigens composed of carrier proteins and synthetic hapten groups. Hapten-carrier complexes were first introduced into immunology by Karl Landsteiner in the early 1900s (*1*). He discovered that (i) small organic molecules with a simple structure, such as phenyl arsonates and nitrophenyls, do not provoke antibodies, and (ii) if the molecule is attached covalently, by simple chemical reactions, to a protein carrier, the antibodies are evoked. Since their introduction these hapten-carrier complexes have become excellent tools to elucidate the role of different antigen-reactive cells in the immune response (*2*). The key players in this immunological process are thymus-derived T cells and bone marrow-derived B cells. The former group of lymphoid cells is responsible for various phenomena of cell-mediated immunity, e.g., delayed hypersensitivity, allograft, and graft-versus-host reactions, and reacts with unknown determinants on the carrier protein. The latter group of cells gives rise to the precursors of antibody-secreting cells, and reacts with both the carrier protein and the synthetic haptenic determinants. This results in antibody formation to both the carrier and the hapten.

The goal of the research at the Eijkman-Winkler Institute was to perform basic studies in the field of immunology to gain more insight into T-cell/B-cell cooperation. To accomplish this, the immune system of animals was challenged with different hapten-carrier complexes, that varied with regard to the hapten/carrier ratio, the antigen dosage, the carrier types (proteins, polysaccharides (PS), and liposomes), the addition of adjuvants, and the route of immunization (intraperitoneally, intravenously, intramuscularly, or orally). While these investigations were in progress, contact was sought with the Department of Bio-Organic Chemistry (now part of the Bijvoet Center) to discuss possible collaborative investigations on synthetic carbohydrate antigens. The reason for the switch to carbohydrate antigens was to extend the studies to more relevant antigens and to address an immunological problem: Polysaccharides are so-called thymus-independent (TI) antigens that do not require T cells to induce an immune response. As a result the antibodies formed are mainly of the IgM type and have a low avidity. Moreover, no immunological memory is generated and the antigens are poorly immunogenic in infants. It was hypothesized that by linking small carbohydrates (oligosaccharides) to a carrier protein, the immunogenic behavior would change to that of a thymus-dependent (TD) antigen.

The two institutes tested this hypothesis using carbohydrate antigens from highly virulent serotype 3 pneumococci. It was decided to extend and combine the studies of both Goebel (*3, 4*) and Campbell and Pappenheimer (*5*), who first isolated the antigenic determinant of *Streptococcus pneumoniae* type 3. The hapten-inhibition studies by Mage and Kabat (*6*) demonstrated that the antibody-combining site of type 3 pneumococcal PS consists of two to three cellobiuronic acid units. In the dextran-anti-dextran system extensively studied by Kabat and colleagues (*7*), the upper size limit of the antibody-combining site appeared to be a hexa- or heptasaccharide and the lower limit was estimated to be somewhat larger than a monosaccharide. The immunologists of the Eijkman-Winkler Institute, therefore, required the synthesis of type 3 mono- to heptasaccharides for their antigen studies. This was a huge challenge for the chemists of the Bijvoet Center as the chemical synthesis of oligosaccharides was largely unexplored at that time. It was indeed proved that type 3 carbohydrates could be transformed into TD antigens by conjugation to a carrier protein. These initial studies formed the beginning of a long-lasting, successful collaboration between the two centers, exploring the synthesis and immunogenicity of numerous oligosaccharide-carrier protein conjugates of several pneumococcal serotypes.

Streptococcus pneumoniae

S. pneumoniae is a major cause of acute respiratory bacterial infections worldwide, leading to approximately 1 million childhood deaths each year (*8*). Antibiotic resistance is a growing problem in the treatment of pneumococcal infections (*9*). Despite the widespread use of antibiotics, the mortality and morbidity of pneumococcal disease remain high (*10*). In addition, resistant pneumococci are being increasingly observed (*10-12*). Host protection against *S. pneumoniae* is mainly mediated by complement- and antibody-dependent phagocytosis. Studies initiated in the previous century indicated that capsular PS are the most important virulence factors of pneumococci (*13*). They provide resistance to complement-mediated phagocytosis (*14*) and shield the inner structures of the pneumococcus from the host immune system (*15*). As a consequence, only antibodies targeted to capsular PS have been shown to fully protect humans against invasive pneumococcal disease.

Capsular polysaccharides

Pneumococci can be subdivided into more than 90 different serotypes, based on their capsular polysaccharide (*13*, *16*, *17*). These pneumococcal capsular types differ significantly in their virulence (*16*). Of the 90 serotypes known today, only a limited number (n=20) causes the majority (90%) of disease cases (*13*, *16*). The serotypes most commonly isolated from patients include serotypes 1 and 14 from the blood, serotypes 6, 10, and 23 from the cerebral spinal fluid, and serotypes 3, 19, and 23 from the middle ear fluid of infants (*18*). Serotypes

Table I. Survey of capsular polysaccharide structures of *Streptococcus pneumoniae* species, discussed in this overview

Serotype	Structure (Reference)
2	→4)-β-D-Glc$p$-(1→3)-[α-D-Glc$p$A-(1→6)-α-D-Glc$p$-(1→2)-]α-L-Rha$p$-(1→3)-α-L-Rha$p$-(1→3)-β-L-Rha$p$-(1→ (*19*)
3	→3)-β-D-Glc$p$A-(1→4)-β-D-Glc$p$-(1→ (*20*)
6A	→2)-α-D-Gal$p$-(1→3)-α-D-Glc$p$-(1→3)-α-L-Rha$p$-(1→3)-D-Rib-ol-(5→P→ (*21*)
6B	→2)-α-D-Gal$p$-(1→3)-α-D-Glc$p$-(1→3)-α-L-Rha$p$-(1→4)-D-Rib-ol-(5→P→ (*22*)
7F	→6)-[β-D-Gal$p$-(1→2)-]α-D-Gal$p$-(1→3)-β-L-Rha$p$2Ac-(1→4)-β-D-Glc$p$-(1→3)-[α-D-Glc$p$NAc-(1→2)-α-L-Rha$p$-(1→4)-]β-D-Gal$p$NAc-(1→ (*23*)
8	→4)-β-D-Glc$p$A-(1→4)-β-D-Glc$p$-(1→4)-α-D-Glc$p$-(1→4)-α-D-Gal$p$-(1→ (*24*)
9A	→4)-α-D-Glc$p$A-(1→3)-α-D-Gal$p$-(1→3)-β-D-Man$p$NAc-(1→4)-β-D-Glc$p$-(1→4)-α-D-Glc$p$-(1→ (*25*)
9N	→4)-α-D-Glc$p$A-(1→3)-α-D-Glc$p$-(1→3)-β-D-Man$p$NAc-(1→4)-β-D-Glc$p$-(1→4)-α-D-Glc$p$NAc-(1→ (*26*)
14	→6)-[β-D-Gal$p$-(1→4)-]β-D-Glc$p$NAc-(1→3)-β-D-Gal$p$-(1→4)-β-D-Glc$p$-(1→ (*27*)
17F	→3)-β-L-Rha$p$-(1→4)-β-D-Glc$p$-(1→3)-α-D-Gal$p$-(1→3)-[α-D-Gal$p$-(1→4)-]β-L-Rha$p$2Ac-(1→4)-α-L-Rha$p$-(1→2)-D-Ara-ol-(1→P→ (*28*)
18C	→4)-β-D-Glc$p$-(1→4)-[α-D-Glc$p$6Ac$_{0.3}$-(1→2)][Gro-(1→P→3)-]β-D-Gal$p$-(1→4)-α-D-Glc$p$-(1→3)-β-L-Rha$p$-(1→ (*29*)
19A	→4)-β-D-Man$p$NAc-(1→4)-α-D-Glc$p$-(1→3)-α-L-Rha$p$-(1→P→ (*30*)
19F	→4)-β-D-Man$p$NAc-(1→4)-α-D-Glc$p$-(1→2)-α-L-Rha$p$-(1→P→ (*31*)
22F	→4)-β-D-Glc$p$A-(1→4)-[α-D-Glc$p$-(1→3)-]β-L-Rha$p$2Ac$_{0.8}$-(1→4)-α-D-Glc$p$-(1→3)-α-D-Gal$f$-(1→2)-α-L-Rha$p$-(1→ (*32*)
23F	→4)-β-D-Glc$p$-(1→4)-[α-L-Rha$p$-(1→2)-][Gro-(2→P→3)-]β-D-Gal$p$-(1→4)-β-L-Rha$p$-(1→ (*33*)

6A, 6B, 14, 19F, and 23F are the main pediatric serotypes (*13*). These serotypes are poorly immunogenic, which might be the reason for their success as a pathogen in infants. Capsular PS are very long polymers of either linear or branched repeating units consisting of 2 (serotypes 3, 37) to 8 (serotype 17A) monosaccharides (*16, 17*). A survey of the capsular PS structures of *S. pneumoniae* discussed in this chapter is presented in Table I.

23-valent PS vaccine

The capsular PS of the 23 most prevalent serotypes have been included in the licenced 23-valent PS vaccine (*34, 35*). This vaccine provides protection against 73% of the strains and reduces the risk of systemic infection in the adult population by 83% (*36*). The efficacy of the vaccine, however, is debatable in groups at risk of pneumococcal infections (*37, 38*). For example, ineffective antibody responses after vaccination have been reported in the elderly (*39*) and the poor immunogenicity of the PS vaccine has been shown beyond doubt in infants (*40*).

Polysaccharide-protein conjugate vaccines

As noted above, the conventional 23-valent PS vaccine provides insufficient protection in groups at risk of pneumococcal infection. To overcome the thymus independency of PS antigens, PS are conjugated to carrier proteins according to the immunological concepts developed in the mid-1970s. Recently, clinical phase III trials were conducted on several conjugate vaccines (*41*), all of which included serotypes 4, 6B, 9V, 14, 18C, and 23F. Diphteria toxoid CRM197, meningococcus B outer membrane proteins, or tetanus toxoid were used as the carrier proteins (*42*). Several of these conjugate vaccines have since been licensed and seem effective against pneumococcal invasive disease, pneumoniae, and, to a lesser extent, otitis media (*43, 44*). It is, therefore, tempting to assume that the major obstacles in the prevention of pneumococcal infections have been overcome.

Despite the clinical and commercial potential of these vaccines, however, a number of major problems can still be envisioned. For example, the protective capacities of these vaccines have not yet been fully established since information on protective opsonic antibody levels in humans is still lacking; immunogenicity varies among the serotypes; and multivalent conjugate vaccines are very expensive and therefore not appropriate for vaccination programs in the developing world. The most important problem is the number of serotypes that must be included in the vaccine. Although a multivalent conjugate vaccine that comprises the nine most virulent serotypes would cover 80-90% of the pneumococcal serotypes isolated worldwide (*16*, *17*), this coverage may decrease dramatically in the future. It has been shown that pneumococcal serotypes not included in the current conjugate vaccine replace vaccine serotypes, and that pathogenic pneumococci have the propensity to switch their capsular types and thus evade vaccine-induced immunity (*45*, *46*).

Preparation of polysaccharide fragments via chemical degradation

Depending on the primary structure of the polysaccharide (Table I), specific degradation methodologies can generate oligosaccharide fragments suitable for conjugation with a carrier protein. A series of typical examples are presented in Table II. For an extension of this list, see Kamerling (*17*).

In the case of capsular serotype 3 PS, use can be made of its higher stability against mineral acids of the GlcA(β1→4)Glc glycosidic linkage compared with the Glc(β1→3)GlcA glycosidic linkage. Partial acid hydrolysis (TFA) mainly generates repeating units with GlcA at the non-reducing end (a series of cellobiuronic acid oligomers). A similar approach of partial acid hydrolysis can be followed for serotype 8 PS. A series of repeating units with Glc at the non-reducing end can be generated by the enzymatic depolymerization of serotype 3 PS with a *Bacillus palustris*-derived endo-β-glucuronidase.

Partial alkaline hydrolysis of serotype 6B PS causes cleavage of the phosphate linkage at one of the two sites, whereas acidic treatment (HF) affords repeating units of different length that are not phosphorylated at both ends.

Serotype 14 PS can be cleaved into repeating units of different lengths via partial alkaline de-*N*-acetylation followed by nitrous acid deamination to yield repeating units of different length, terminated at the "reducing site" with 2,5-anhydro-mannose (conversion of *N*-acetylglucosamine, via glucosamine, into 2,5-anhydro-mannose). An endo-β-galactosidase from *Cytophago keratolytica*, which catalyzes the hydrolysis of the (β1→4) linkage between galactose and glucose in lactose, can also be applied at this point.

In all cases the generated mixtures were separated into single entities using chromatographic procedures.

In principle, free oligosaccharides can also be obtained via organic synthetic routes. Nevertheless, when such an approach is chosen, the synthesis of oligosaccharides with an aglycon suitable for further conjugation with carrier protein is preferred (Table III).

Synthesis of amino-spacered oligosaccharide fragments from monosaccharides

Over the years series of oligosaccharide fragments, with an amino group-containing spacer, of capsular polysaccharides of *S. pneumoniae* have been prepared using multistep organic synthesis, sometimes in combination with glycosyltransferases. These oligosaccharide fragments can comprise part of a repeating unit or be as long as one or more repeating units. Table III presents a survey of reported products of serotypes 2, 3, 6B, 7F, 9A, 9N, 14, 17F, 19F, 22F, and 23F. For a more comprehensive list, see Kamerling (*17*). The amino group-terminated aglycon at the "reducing end" of the oligosaccharides enables a direct peptide coupling with aspartic acid or glutamic acid or, via an extra linker, with lysine residues.

Conjugation

Many conjugation protocols have been discussed over the years and a set of possibilities has been published by Kamerling (*17*). Only a few approaches will be considered in the context of this review, namely, those used in the preparation of the oligosaccharide-protein conjugates discussed in the immunology part below.

In the case of free oligosaccharides, conjugations with carrier proteins are mainly realized via reductive amination in the presence of sodium cyanoborohydride, usually including a linker (*17*). Typical examples comprise (i) coupling with 2-(4-aminophenyl)-ethylamine in the presence of sodium cyanoborohydride followed by reaction with thiophosgene, yielding a phenylisothiocyanate for conjugation, and (ii) treatment with ammonium chloride/sodium cyanoborohydride followed by reaction with the disuccinimidyl ester of adipic acid, yielding an active ester for conjugation. Oligosaccharides that have a terminal 2,5-anhydrohexose residue with a free aldehyde function can be coupled directly to the carrier protein via lysine residues using a reductive amination protocol with sodium cyanoborohydride.

Two procedures have been followed for oligosaccharides with an amino group-containing spacer coupled to the anomeric center of their "reducing end". In the first procedure, the aminoalkyl groups are *S*-acetylated with *N*-succinimidyl *S*-acetylthioacetate, followed by de-acetylation with hydroxylamine and coupling with a carrier protein of which the lysine residue has been bromoacetylated with *N*-succinimidyl bromoacetate. In the other, the aminoalkyl spacer is elongated with the linker diethyl squarate or *N*-hydroxysuccinimide-activated adipic acid diester and subsequently coupled with free lysine residues of the carrier protein.

Synthetic oligosaccharides in pneumococcal vaccines

The saccharides described in Tables II and III were used in immunological studies, e.g., as inhibitors to determine the epitope specificity of anti-PS

Table II. Free oligosaccharide fragments of capsular polysaccharides of *Streptococcus pneumoniae* species

Serotype	Product (Reference)
3	β-D-GlcpA-(1→4)-D-Glcp *(47, 48)*
	β-D-GlcpA-(1→4)-β-D-Glcp-(1→3)-β-D-GlcpA-(1→4)-D-Glcp *(47, 48)*
	β-D-GlcpA-(1→4)-[β-D-Glcp-(1→3)-β-D-GlcpA-(1→4)]$_2$-D-Glcp *(47, 48)* *
	β-D-GlcpA-(1→4)-[β-D-Glcp-(1→3)-β-D-GlcpA-(1→4)]$_3$-D-Glcp *(47, 48)*
	β-D-GlcpA-(1→4)-[β-D-Glcp-(1→3)-β-D-GlcpA-(1→4)]$_4$-D-Glcp *(48)*
	β-D-GlcpA-(1→4)-[β-D-Glcp-(1→3)-β-D-GlcpA-(1→4)]$_5$-D-Glcp *(48)*
	β-D-GlcpA-(1→4)-[β-D-Glcp-(1→3)-β-D-GlcpA-(1→4)]$_6$-D-Glcp *(48)*
	β-D-Glcp-(1→3)-D-GlcpA *(49)*
	β-D-Glcp-(1→3)-β-D-GlcpA-(1→4)-β-D-Glcp-(1→3)-D-GlcpA *(49)*
	β-D-Glcp-(1→3)-[β-D-GlcpA-(1→4)-β-D-Glcp-(1→3)]$_2$-D-GlcpA *(49)*
6A	α-D-Galp-(1→3)-α-D-Glcp-(1→3)-α-L-Rhap-(1→3)-D-Rib-ol *(50)*
	α-D-Galp-(1→3)-α-D-Glcp-(1→3)-α-L-Rhap-(1→3)-D-Rib-ol-(5→P→2)-α-D-Galp-(1→3)-α-D-Glcp-(1→3)-L-Rhap *(51)*
6B	α-D-Glcp-(1→3)-L-Rhap *(50)*
	α-D-Galp-(1→3)-α-D-Glcp-(1→3)-L-Rhap *(50)*
	α-D-Galp-(1→3)-α-D-Glcp-(1→3)-α-L-Rhap-(1→4)-D-Rib-ol-(5→P→2)-α-D-Galp-(1→3)-α-D-Glcp-(1→3)-L-Rhap *(51)*
	α-D-Galp-(1→3)-α-D-Glcp-(1→3)-α-L-Rhap-(1→4)-D-Rib-ol *(50, 52)*
	α-D-Galp-(1→3)-α-D-Glcp-(1→3)-α-L-Rhap-(1→4)-D-Rib-ol-(5→P→2)-α-D-Galp-(1→3)-α-D-Glcp-(1→3)-α-L-Rhap-(1→4)-D-Rib-ol *(52)*
	α-D-Galp2P-(1→3)-α-D-Glcp-(1→3)-α-L-Rhap-(1→4)-D-Rib-ol-5→P *(52)*
8	β-D-GlcpA-(1→4)-β-D-Glcp-(1→4)-D-Glcp *(53)*
	β-D-Glcp-(1→4)-α-D-Glcp-(1→4)-D-Galp *(53)*

Table II. *Continued.*

Serotype	Product (Reference)
14	β-D-Galp-(1→4)-β-D-Glcp-(1→6)-[β-D-Galp-(1→4)]-β-D-GlcpNAc-(1→3)-β-D-Galp-(1→4)-D-Glcp (54)
18C	α-D-Glcp-(1→2)-[β-D-Glcp-(1→4)-]β-D-Galp-(1→4)-α-D-Glcp-(1→3)-L-Rhap (55)
19F	β-D-ManpNAc-(1→4)-α-D-Glcp-(1→2)-L-Rhap (56)
19A	β-D-ManpNAc-(1→4)-α-D-Glcp-(1→3)-L-Rhap (57)

The marked oligosaccharide * refers to Table IV.

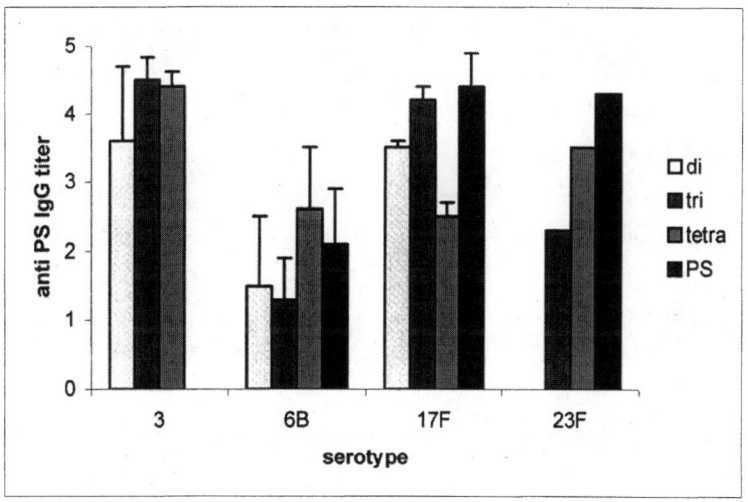

Figure 1. Induction of antibodies against S. pneumoniae *serotype 3, 6B, 17F, and 23F polysaccharide, by small oligosaccharide fragments representing one repeating unitor less of the corresponding polysaccharide. Anti-PS antibodies induced by the conjugated oligosaccharide fragments were compared with those evoked by the whole conjugated PS as determined by ELISA. For details, see Refs. given in Table IV.*

Table III. Synthesized amino-spacered fragments of capsular polysaccharides of *Streptococcus pneumoniae* species

Serotype	Product (Reference)
2	β-L-Rha*p*-(1→4)-β-D-Glc*p*-(1→3)-α-L-Rha*p*-(1→O(CH₂)₃NH₂ (*58*)
3	β-D-Glc*p*A-(1→4)-β-D-Glc*p*-(1→O(CH₂)₃NH₂ (*59*) *
	β-D-Glc*p*-(1→3)-β-D-Glc*p*A-(1→4)-β-D-Glc*p*-(1→O(CH₂)₃NH₂ (*59*) *
	β-D-Glc*p*A-(1→4)-β-D-Glc*p*-(1→3)-β-D-Glc*p*A-(1→4)-β-D-Glc*p*-(1→O(CH₂)₃NH₂ (*59*) *
	β-D-Glc*p*A-(1→4)-[β-D-Glc*p*-(1→3)-β-D-Glc*p*A-(1→4)]₂-β-D-Glc*p*-(1→O(CH₂)₃NH₂ (*60*)
6B	α-L-Rha*p*-(1→4)-D-Rib-ol-(5→*P*→(CH₂)₃NH₂ (*61*) †
	D-Rib-ol-(5→*P*→2)-α-D-Gal*p*-(1→O(CH₂)₃NH₂ (*61*)
	α-L-Rha*p*-(1→4)-D-Rib-ol-(5→*P*→2)-α-D-Gal*p*-(1→O(CH₂)₃NH₂ (*61*)
	D-Rib-ol-(5→*P*→2)-α-D-Gal*p*-(1→3)-α-D-Glc*p*-(1→O(CH₂)₃NH₂ (*61*) †
	D-Rib-ol-(5→*P*→2)-α-D-Gal*p*-(1→3)-α-D-Glc*p*-(1→3)-α-L-Rha*p*-(1→O(CH₂)₃NH₂ (*62*)
	α-L-Rha*p*-(1→4)-D-Rib-ol-(5→*P*→2)-α-D-Gal*p*-(1→3)-α-D-Glc*p*-(1→O(CH₂)₃NH₂ (*62*) †
	α-D-Glc*p*-(1→3)-α-L-Rha*p*-(1→4)-D-Rib-ol-(5→*P*→2)-α-D-Gal*p*-(1→O(CH₂)₃NH₂ (*62*)
	α-D-Gal*p*-(1→3)-α-D-Glc*p*-(1→3)-α-L-Rha*p*-(1→4)-D-Rib-ol-(5→*P*→(CH₂)₃NH₂ (*62, 63*)
7F	β-L-Rha*p*-(1→4)-β-D-Glc*p*-(1→3)-β-D-Gal*p*NAc-(1→O(CH₂)₃NH₂ (*58*)
9A	α-D-Gal*p*-(1→3)-β-D-Man*p*NAc-(1→4)-β-D-Glc*p*-(1→O(CH₂)₂NH₂ (*64*)
	α-D-Glc*p*A6Me-(1→3)-α-D-Gal*p*-(1→3)-β-D-Man*p*NAc-(1→4)-β-D-Glc*p*-(1→O(CH₂)₂NH₂ (*65*)
	α-D-Glc*p*-(1→4)-α-D-Glc*p*A6Me-(1→3)-α-D-Gal*p*-(1→3)-β-D-Man*p*NAc-(1→4)-β-D-Glc*p*-(1→O(CH₂)₂NH₂ (*65*)
9N	α-D-Glc*p*-(1→3)-β-D-Man*p*NAc-(1→4)-β-D-Glc*p*-(1→O(CH₂)₂NH₂ (*64*)

Table III. *Continued.*

Serotype	Product (Reference)
14	β-D-Gal$p$-(1→4)-β-D-Glc$p$-(1→6)-β-D-Glc$p$NAc-(1→O(CH$_2$)$_3$S(CH$_2$)$_2$NH$_2$ (*66*)
	β-D-Gal$p$-(1→4)-β-D-Glc$p$-(1→6)-[β-D-Gal$p$-(1→4)]-β-D-Glc$p$NAc-(1→O(CH$_2$)$_3$S(CH$_2$)$_2$NH$_2$ (*66*) **#**
	β-D-Gal$p$-(1→4)-β-D-Glc$p$-(1→6)-[β-D-Gal$p$-(1→4)]-β-D-Glc$p$NAc-(1→3)-β-D-Gal$p$-(1→O(CH$_2$)$_6$NH$_2$ (*67*)
	β-D-Gal$p$-(1→4)-β-D-Glc$p$-(1→6)-[β-D-Gal$p$-(1→4)]-β-D-Glc$p$NAc-(1→3)-β-D-Gal$p$-(1→4)-β-D-Glc$p$-(1→O(CH$_2$)$_6$NH$_2$ (*67*)
	β-D-Gal$p$-(1→4)-β-D-Glc$p$-(1→6)-[β-D-Gal$p$-(1→4)]-β-D-Glc$p$NAc-(1→O(CH$_2$)$_6$NH$_2$ (*68*)
	β-D-Gal$p$-(1→4)-β-D-Glc$p$-(1→6)-[β-D-Gal$p$-(1→4)]-β-D-Glc$p$NAc-(1→3)-β-D-Gal$p$-(1→4)-β-D-Glc$p$-(1→6)-[β-D-Gal$p$-(1→4)]-β-D-Glc$p$NAc-(1→O(CH$_2$)$_6$NH$_2$ (*68*)
	β-D-Glc$p$-(1→6)-[β-D-Gal$p$-(1→4)]-β-D-Glc$p$NAc-(1→3)-β-D-Gal$p$-(1→O(CH$_2$)$_6$NH$_2$ (*69*)
	β-D-Glc$p$-(1→6)-[β-D-Gal$p$-(1→4)]-β-D-Glc$p$NAc-(1→3)-β-D-Gal$p$-(1→4)-β-D-Glc$p$-(1→O(CH$_2$)$_6$NH$_2$ (*69*)
	β-D-Gal$p$-(1→4)-β-D-Glc$p$NAc-(1→3)-β-D-Gal$p$-(1→4)-β-D-Glc$p$-(1→O(CH$_2$)$_6$NH$_2$ (*69*)
	β-D-Gal$p$-(1→4)-β-D-Glc$p$NAc-(1→3)-β-D-Gal$p$-(1→4)-β-D-Glc$p$-(1→6)-[β-D-Gal$p$-(1→4)]-β-D-Glc$p$NAc-(1→O(CH$_2$)$_6$NH$_2$ (*69*)
	β-D-Gal$p$-(1→4)-{β-D-Glc$p$-(1→6)-[β-D-Gal$p$-(1→4)]-β-D-Glc$p$NAc-(1→3)-β-D-Gal$p$-(1→4)}$_2$-β-D-Glc$p$-(1→6)-[β-D-Gal$p$-(1→4)]-β-D-Glc$p$NAc-(1→O(CH$_2$)$_6$NH$_2$ (*70*)
17F	α-L-Rha$p$-(1→2)-D-Ara-ol-(1→P→(CH$_2$)$_3$NH$_2$ (*71*) ¶
	α-L-Rha$p$-(1→2)-D-Ara-ol-(1→P→4)-α-L-Rha$p$-(1→O(CH$_2$)$_3$NH$_2$ (*71*) ¶
	α-L-Rha$p$-(1→2)-D-Ara-ol-(1→P→4)-α-L-Rha$p$-(1→4)-β-D-Glc$p$-(1→O(CH$_2$)$_3$NH$_2$ (*71*) ¶

Continued on next page.

Table III. Continued.

Serotype	Product (Reference)
19F	β-D-ManpNAc-(1→4)-α-D-Glcp-(1→2)-α-L-Rhap-(1→P→4)-β-D-ManpNAc-(1→4)-α-D-Glcp-(1→2)-α-L-Rhap-(1→P→4)-β-D-ManpNAc-(1→4)-α-D-Glcp-(1→2)-α-L-Rhap-(1→P→(CH$_2$)$_2$NH$_2$ (72)
22F	β-L-Rhap-(1→4)-α-D-Glcp-(1→3)-α-D-Gal$f$-(1→O(CH$_2$)$_3$NH$_2$ (58)
23F	β-L-Rhap-(1→4)-β-D-Glcp-(1→4)-β-D-Galp-(1→O(CH$_2$)$_3$NH$_2$ (58) §
	β-D-Glcp-(1→4)-[α-L-Rhap-(1→2)][Gro-(2→P→3)]-β-D-Galp-(1→4)-β-L-Rhap-(1→O(CH$_2$)$_3$NH$_2$ (73) §

The marked oligosacchrides *, †, #, ¶, and § refer to Table IV.

antibodies (74-78), to inhibit pneumococcal adhesion to epithelial cells (79-83), and as an experimental conjugate vaccine in murine models (84-98). Table IV summarizes the results gathered over the years using different carrier molecules. The liposomal presentation of oligosaccharides resulted only in the formation of short-lived IgM antibodies in the mice strains tested (47, 49, 85).

Moreover, no memory formation was induced. This carrier system, therefore, is currently unsuitable for further vaccination studies (49). In contrast, all of the protein carriers tested so far have proved suitable for vaccination studies because long-lived IgG antibodies and memory formation were induced in all cases (Table IV). The application of adjuvants resulted in long-lasting antibody levels, which provided full protection against a pneumococcal challenge as demonstrated for serotypes 3, 6B, and 17F. The adjuvants also increased the antibody levels, antibody avidity, subclass distribution, and opsonic capacities of the sera (92).

As shown in Figure 1, anti-PS responses could be induced for all serotypes tested so far using very small oligosaccharide fragments comprising only one repeating unit or less of the corresponding PS. The serotypes listed in Table IV belong to the most common serotypes isolated from adults and infants. The immune responses we found against the oligosaccharides were highly specific and, interestingly, some of the oligosaccharide-protein conjugate vaccines induced even higher levels of anti-PS antibodies than the conventional PS-protein conjugate counterpart (94).

Discussion

Although the vaccine potential of small synthetic oligosaccharides has been proven in murine models, their potential as vaccine candidates in humans is still

Table IV. Minimal oligosaccharide fragments conjugated to carrier proteins can induce (protective) anti-PS responses in mice and rabbits

Sero-type	Carrier-protein	Oligosaccharide chain length[a] inducing (protective) anti-PS responses				Refs.
		Mice		Rabbits		
		Anti-PS response[b]	Protection[c]	Anti-PS response	Protection[d]	
3	liposomes	penta	penta	ND[e]	ND	49, 85, 91
3	CRM 197[f]	di, tri, tetra	di, tri, tetra	ND	ND	85, 95, 97
6A	KLH[g]	tri, tetra	ND	tri, tetra	tri, tetra	61-63, 96
6B	KLH	di, tri, tetra	tetra	di, tri, tetra	di, tri, tetra	61-63, 96
14	CRM 197	tetra	ND	ND	ND	66, 78
17F	KLH	di, tri, tetra	di, tri, tetra	di, tri, tetra	ND	92, 93
23F	KLH	ND	ND	tri, tetra	ND	58, 73, 76

a: Number of monosaccharides within 1 oligosaccharide fragment, coupled to a carrier protein; b: As measured by ELISA using PS as coating antigen (96); c: As measured in a mouse challenge model (96); d: Passive protection of mice by anti- oligosaccharide antibodies raised in rabbits (96); e: ND, not done; f: CRM 197, a nontoxic variant of diphtheria toxin; g: KLH, keyhole limpet hemocyanin.

Type 3: For the structure of the hexa-derived pentasaccharide, marked *, see Table II.

Type 3: For the structures of the di-, tri-, and tetrasaccharides, marked *, see Table III.

Types 6A/6B: For the structures of the di-, tri-, and tetrasaccharides, marked †, see Table III.

Type 14: For the structure of the tetrasaccharide, marked #, see Table III.

Type 17F: For the structures of the di, tri-, and tetrasaccharides, marked ¶, see Table III.

Type 23F: For the structures of the tri- and tetrasaccharides, marked §, see Table III.

uncertain. A breakthrough for this type of vaccine was made in 2004 by Verez Bencomo et al. (99) when they introduced a synthetic oligosaccharide vaccine for *Haemophilus influenza* type b in Cuba. This vaccine is currently being marketed in South America and India. Numerous questions must still be answered, however, before a synthetic pneumococcal vaccine is available for the general public. It is known that epitope sizes on pneumococcal PS can vary between serotypes (59, 92), between species (59), and with age (100). Moreover, humans may require longer saccharides than mice for an optimal immune response, presumably because of the better presentation of conformational epitopes (100). Again, this may vary per serotype since 6A and 6B (96), but not 23F (76) oligosaccharide fragments are recognized by human vaccination antisera, indicating the presence of epitopes for human anti-PS antibodies on the former, but not the latter oligosaccharide fragments. A potential drawback of minimal synthetic oligosaccharide-conjugate vaccines is the absence of conformational epitopes, which are present on the entire PS. On the other hand, the thymus dependency of an oligosaccharide conjugate increases with the shortening of the oligosaccharide chain length. Therefore, an optimum oligosaccharide length has to be defined for PS, containing conformational epitopes: one long enough to express conformational epitopes, yet short enough to induce an optimal Th-cell response. Three methods can be applied to obtain oligosaccharides that contain several repeating units of the PS: partial hydrolysis of the PS, chemical synthesis, and biosynthesis (17). Several oligosaccharides have been successfully prepared by hydrolysis. One disadvantage of this method is that the production of these oligosaccharides is less well controlled, which leads to less-defined products i.e., varying in size and exact linkage types. The chemical synthesis of oligosaccharides with several repeating units is possible, but is complex and requires a large number of synthesis steps. The combination of chemical synthesis and biosynthesis might be a promising alternative in the near future to produce large oligosaccharides in sufficient amounts. Information on the biosynthetic pathways of capsular PS of serotypes 3, 14, and 19F has become available (17). Moreover, several glycosyltransferases have been identified and cloned (101-103) and one of them has already been produced on a large scale (104). In the future, the conjugation of certain saccharides to a carrier protein or peptide might be performed in part enzymatically since several enzymes have now been identified that can glycosylate specific amino acid residues (102-105).

Our findings and the work of others on synthetic oligosaccharide-protein conjugates should encourage the exploration of synthetic oligosaccharide-conjugate vaccines. Establishing whether animals and humans can produce antibodies against minimal synthetic oligosaccharides, however, is just the first step in the development of a synthetic pneumococcal vaccine. The next steps are to define the most immunogenic protective epitopes on the PS and to optimize the presentation of these epitopes to the immune system.

In conclusion, the generation of new synthetic saccharides should be stimulated considering the large number of *S. pneumoniae* serotypes. Then, depending on the serotype, natural PS can, when needed and when possible, be replaced sequentially by their (bio)synthetic oligosaccharide counterparts in future semi-synthetic pneumococcal vaccines.

Acknowledgements

We thank all former PhD students and post-docs who contributed to this project: Drs. Jan van Dam, Ted Slaghek, Frans Koeman, Guusta van Steijn, Mark Thijssen, Dirk Lefeber, Jutta Niggemann, Dirk Michalik, and John Joosten for the synthesis and conjugation of the oligosaccharides, and Arend Jan van Houte, Jan van Dam, André Verheul, Guy Zigterman, Enrique Alonso de Velasco, and Barry Benaissa-Trouw for the immunology research. The continuous support of Prof. Dr. Jan Willers, Prof. Dr. Jan Verhoef, and Prof. Dr. Hans Vliegenthart is highly acknowledged. The work was financially supported by the Faculties of Medicine and Chemistry of Utrecht University, the former Institute of Molecular Biology and Medical Biotechnology of Utrecht University, Preventiefonds, Medigon, the Dutch Innovation Oriented Program Biotechnology, and the European Union (grants ERB CHRX CT940442, BIO CT95-0138, ERB BIO 4CT960158).

References

1. Landsteiner, K.; Jacobs, J. *J. Exp. Med.* **1935**, *61*, 643-656.
2. Mitchison, N. A.; Rajewsky, K.; Taylor, R. B. In *Developmental aspects of antibody formation and structure;* Sterzl, J.; Riha, I., Eds.; Academia Publishing House: Praha, 1970.
3. Goebel, W. F. *J. Exp. Med.* **1939**, 69, 353-364.
4. Goebel, W. F. *J. Exp. Med.* **1940**, *72*, 33-48.
5. Campbell, J. H.; Pappenheimer, A. M. *Immunochemistry* **1966**, *3*, 195-212.
6. Mage, R. G.; Kabat, E. A. *J. Immunol.* **1963**, *91*, 633-640.
7. Kabat, E. A.; Mayer, M. M. *Experimental immunochemistry*; Thomas: Springfield, IL, USA, 1948.
8. Pneumococcal vaccines. WHO position paper. *Wkly. Epidemiol. Rec.* **1999**, *74*, 177-183.
9. Tan, T. Q. *Curr. Opin. Microbiol.* **2000**, *3*, 502-507.
10. Kellner, J. D. *Semin. Respir. Infect.* **2001**, *16*, 186-195.
11. Whitney, C. G.; Farley, M. M.; Hadler, J. L.; Harrison, H.; Lexau, C.; Reingold, A.; Lefkowitz, L.; Cieslak, P. R.; Cetron, M.; Zell, E. R.; Jorgensen, J. H.; Schuchat, A. *N. Engl. J. Med.* **2000**, *343*, 1917-1924.

12. Schreiber, J. R.; Jacobs, M. R. *Pediatr. Clin. North Am.* **1995**, *42*, 519-537.
13. Alonso de Velasco, E.; Verheul, A. F.; Verhoef, J.; Snippe, H. *Microbiol. Rev.* **1995**, *59*, 591-603.
14. Brown, E. J.; Joiner, K. A.; Cole, R. M.; Berger, M. *Infect. Immun.* **1983**, *39*, 403-409.
15. Nielsen, S. V.; Sørensen, U. B. S.; Henrichsen, J. *Microb. Pathog.* **1993**, *14*, 299-305.
16. Van Dam, J. E. G.; Fleer, A.; Snippe, H. *Antonie Van Leeuwenhoek* **1990**, *58*, 1-47.
17. Kamerling, J. P. In *Streptococcus pneumoniae*; Tomasz, A., Ed.; Mary Ann Liebert Inc.: Larchmont, NY, 2000; pp 81-114.
18. Hausdorff, W. P.; Yothers, G.; Dagan, R.; Kilpi, T.; Pelton, S. I.; Cohen, R.; Jacobs, M. R.; Kaplan, S. L.; Levy, C.; Lopez, E. L.; Mason, E. O. Jr.; Syriopoulou, V.; Wynne, B.; Bryant, J. *Pediatr. Infect. Dis. J.* **2002**, *21*, 1008-1016.
19. Jansson, P.-E.; Lindberg, B.; Andersson, M.; Lindquist, U.; Henrichsen, J. *Carbohydr. Res.* **1988**, *182*, 111-117.
20. Reeves, R. E.; Goebel, W. F. *J. Biol. Chem.* **1941**, *139*, 511-519.
21. Larm, O.; Lindberg, B. *Adv. Carbohydr. Chem. Biochem.* **1976**, *33*, 295-322.
22. Kenne, L.; Lindberg, B.; Madden, J. K. *Carbohydr. Res.* **1979**, *73*, 175-182.
23. Moreau, M.; Richards, J. C.; Perry, M. B.; Kniskern, P. J. *Carbohydr. Res.* **1988**, *182*, 79-99.
24. Jones, J. K. N.; Perry, M. B. *J. Am. Chem. Soc.* **1957**, *79*, 2787-2793.
25. Richards, J. C.; Perry, M. B. In *The molecular immunology of complex carbohydrates*; Wu, A. M., Ed.; Plenum: New York, NY, 1988; pp 593-594.
26. Jones, C.; Mulloy, B.; Wilson, A.; Dell, A.; Oates, J. E. *J. Chem. Soc., Perkin Trans. 1* **1985**, 1665-1673.
27. Lindberg, B.; Lönngren, J.; Powell, D. A. *Carbohydr. Res.* **1977**, *58*, 177-186.
28. Jennings, H. J.; Pon, R. A. In *Polysaccharides in medicinal applications*; Dumitriu, S., Ed.; Marcel Dekker: New York, NY, 1996; pp 443-479.
29. Lindberg, J. Thesis Stockholm University 1990.
30. Jennings, H. J.; Rosell, K.-G.; Carlo, D. J. *Can. J. Chem.* **1980**, *58*, 1069-1074.
31. Katzenellenbogen, E.; Jennings, H. J. *Carbohydr. Res.* **1983**, *124*, 235-245.
32. Richards, J. C.; Perry, M. B.; Kniskern, P. J. *Can. J. Chem.* **1989**, *67*, 1038-1050.
33. Richards, J. C.; Perry, M. B. *Biochem. Cell Biol.* **1988**, *66*, 758-771.
34. Requejo, H. I. *Rev. Hosp. Clin. Fac. Med. Sao Paulo* **1993**, *48*, 130-138.
35. Robbins, J. B.; Austrian, R.; Lee, C.-J.; Rastogi, S. C.; Schiffman, G.; Henrichsen, J.; Mäkelä, P. H.; Broome, C. V.; Facklam, R. R.; Tiesjema, R. H.; Parke, J. C. *J. Infect. Dis.* **1983**, *148*, 1136-1159.

36. Hutchison, B. G.; Oxman, A. D.; Shannon, H. S.; Lloyd, S.; Altmayer, C. A.; Thomas, K. *Can. Fam. Physician* **1999**, *45*, 2381-2393.
37. Avanzini, M. A.; Carra, A. M.; Maccario, R.; Zecca, M.; Pignatti, P.; Marconi, M.; Comoli, P.; Bonetti, F.; De Stefano, P.; Locatelli, F. *J. Clin. Immunol.* **1995**, *15*, 137-144.
38. King, J. C., Jr.; Vink, P. E.; Farley, J. J.; Parks, M.; Smilie, M.; Madore, D.; Lichenstein, R.; Malinoski, F. *Pediatr. Infect. Dis. J.* **1996**, *15*, 192-196.
39. Romero-Steiner, S.; Musher, D. M.; Cetron, M. S.; Pais, L. B.; Groover, J. E.; Fiore, A. E.; Plikaytis, B. D.; Carlone, G. M. *Clin. Infect. Dis.* **1999**, *29*, 281-288.
40. Meyer, M.; Gahr, M. *Monatsschr. Kinderheilkd.* **1993**, *141*, 770-776.
41. Eskola, J. *Vaccine* **2000**, *19* (Suppl 1), S78-S82.
42. Pomat, W. S.; Lehmann, D.; Sanders, R. C.; Lewis, D. J.; Wilson, J.; Rogers, S.; Dyke, T.; Alpers, M. P. *Infect. Immun.* **1994**, *62*, 1848-1853.
43. Briles, D. E.; Ades, E.; Paton, J. C.; Sampson, J. S.; Carlone, G. M.; Huebner, R. C.; Virolainen, A.; Swiatlo, E.; Hollingshead, S. K. *Infect. Immun.* **2000**, *68*, 796-800.
44. Lee, L. H.; Lee, C. J.; Frasch, C. E. *Crit. Rev. Microbiol.* **2000**, *28*, 27-41.
45. Toltzis, P.; Jacobs, M. R. *Infect. Dis. Clin. North. Am.* **2005**, *19*, 629-645.
46. Whithney, C. G. *Pediatr. Infect. Dis. J.* **2005**, *24*, 729-730.
47. Snippe, H.; van Houte, A.-J.; van Dam, J. E. G.; de Reuver, M. J.; Jansze, M.; Willers, J. M. N. *Infect. Immun.* **1983**, *40*, 856-861.
48. Lefeber, D. J.; Gutiérrez Gallego, R.; Grün, C. H.; Proietti, D.; D'Ascenzi, S.; Costantino, J.; Kamerling, J. P.; Vliegenthart, J. F. G. *Carbohydr. Res.* **2002**, *337*, 819-825.
49. Snippe, H.; Zigterman, G. J. W. J.; van Dam, J. E. G.; Kamerling, J. P. In *Liposomes as drug carriers*; Gregoriadis, G., Ed.; John Wiley & Sons Ltd.: Chichester, 1988; pp 183-196.
50. Slaghek, T. M.; van Oijen, A. H.; Maas, A. A. M.; Kamerling, J. P.; Vliegenthart, J. F. G. *Carbohydr. Res.* **1990**, *207*, 237-248.
51. Slaghek, T. M.; Maas, A. A. M.; Kamerling, J. P.; Vliegenthart, J. F. G. *Carbohydr. Res.* **1991**, *211*, 25-39.
52. Van Dam, J. E. G.; Breg, J.; Komen, R.; Kamerling, J. P.; Vliegenthart, J. F. G. *Carbohydr. Res.* **1989**, *187*, 267-286.
53. Koeman, F. A. W.; Kamerling, J. P.; Vliegenthart, J. F. G. *Tetrahedron* **1993**, *49*, 5291-5304.
54. Koeman, F. A. W.; Meissner, J. W. G.; van Ritter, H. R. P.; Kamerling, J. P.; Vliegenthart, J. F. G. *J. Carbohydr. Chem.* **1994**, *13*, 1-25.
55. Van Steijn, A. M. P.; van der Ven, J. G. M.; van Seeventer, P.; Kamerling, J. P.; Vliegenthart, J. F. G. *Carbohydr. Res.* **1992**, *229*, 155-160.
56. Panza, L.; Ronchetti, F.; Russo, G.; Toma, L. *J. Chem. Soc., Perkin Trans. 1*, **1987**, 2745-2747.
57. Panza, L.; Ronchetti, F.; Toma, L. *Carbohydr. Res.* **1988**, *180*, 242-245.

58. Van Steijn, A. M. P.; Kamerling, J. P.; Vliegenthart, J. F. G. *J. Carbohydr. Chem.* **1992**, *11*, 665-689.
59. Lefeber, D. J.; Kamerling, J. P.; Vliegenthart, J. F. G. *Chem. Eur. J.* **2001**, *20*, 4411-4421.
60. Lefeber, D. J.; Aldaba Arévalo, E.; Kamerling, J. P.; Vliegenthart, J. F. G. *Can. J. Chem.* **2003**, *80*, 76-81.
61. Thijssen, M.-J. L.; van Rijswijk, M. N.; Kamerling, J. P.; Vliegenthart, J. F. G. *Carbohydr. Res.* **1998**, *306*, 93-109.
62. Thijssen, M.-J. L.; Bijkerk, M. H. G.; Kamerling, J. P.; Vliegenthart, J. F. G. *Carbohydr. Res.* **1998**, *306*, 111-125.
63. Thijssen, M.-J. L.; Halkes, K. M.; Kamerling, J. P.; Vliegenthart, J. F. G. *Bioorg. Med. Chem.* **1994**, *2*, 1309-1317.
64. Alpe, M.; Oscarson, S. *Carbohydr. Res.* **2002**, *337*, 1715-1722.
65. Alpe, M.; Oscarson, S. *Carbohydr. Res.* **2003**, *338*, 2605-2609.
66. Niggemann, J.; Kamerling, J. P.; Vliegenthart, J. F. G. *Bioorg. Med. Chem.* **1998**, *6*, 1605-1612.
67. Michalik, D.; Vliegenthart, J. F. G.; Kamerling, J. P. *J. Chem. Soc. Perkin Trans. 1* **2002**, 1973-1981.
68. Joosten, J. A. F.; Kamerling, J. P.; Vliegenthart, J. F. G. *Carbohydr. Res.* **2003**, *338*, 2611-2627.
69. Joosten, J. A. F.; Lazet, B. J.; Kamerling, J. P.; Vliegenthart, J. F. G. *Carbohydr. Res.* **2003**, *338*, 2629-2651.
70. Sundgren, A.; Lahmann, M.; Oscarson, S. *J. Carbohydr. Chem.* **2005**, *24*, 379-391.
71. Veeneman, G. H.; Gomes, L. J. F.; van Boom, J. H. *Tetrahedron* **1989**, *45*, 7433-7448.
72. Nilsson, M.; Norberg, T. *J. Chem. Soc., Perkin Trans. 1*, **1998**, 1699-1704.
73. Van Steijn, A. M. P.; Kamerling, J. P.; Vliegenthart, J. F. G. *Carbohydr. Res.* **1991**, *211*, 261-277.
74. Van Dam, G. J.; Verheul, A. F. M.; Zigterman, G. J. W. J.; de Reuver, M. J.; Snippe, H. *Mol. Immunol.* **1989**, *26*, 269-274.
75. Van Dam, G. J.; Zigterman, G. J. W. J.; Verheul, A. F. M.; de Reuver, M. J.; Snippe, H. *J. Immunol.* **1989**, *143*, 3049-3053.
76. Alonso de Velasco, E.; Verheul, A. F. M.; van Steijn, A. M. P.; Dekker, H. A. Th.; Feldman, R. G.; Fernández, I. M.; Kamerling, J. P.; Vliegenthart, J. F. G.; Verhoef, J.; Snippe, H. *Infect. Immun.* **1994**, *62*, 799-808.
77. Snippe, H.; Jansen, W. T. M.; Kamerling, J. P.; Lefeber, D. J. *Clin. Diagn. Lab. Immunol.* **2000**, *7*, 325.
78. Mawas, F.; Niggemann, J.; Jones, C.; Corbel, M. J.; Kamerling, J. P.; Vliegenthart, J. F. G. *Infect. Immun.* **2002**, *70*, 5107-5114.
79. Sharon, N.; Ofek, I. *Glycoconjugate J.* **2000**, *17*, 659-664.
80. Ukkonen, P.; Varis, K.; Jernfors, M.; Herva, E.; Jokinen, J.; Ruokokoski, E.; Zopf, D.; Kilpi, T. *Lancet* **2000**, *356*, 1398-1402.

81. Tong, H. H.; McIver, M. A.; Fisher, L. M.; DeMaria, T. F. *Microb. Pathog.* **1999**, *26*, 111-119.
82. Barthelson, R.; Mobasseri, A.; Zopf, D.; Simon, P. *Infect. Immun.* **1998**, *66*, 1439-1444.
83. Idanpaan-Heikkila, I.; Simon, P. M.; Zopf, D.; Vullo, T.; Cahill, P.; Sokol, K.; Tuomanen, E. *J. Infect. Dis.* **1997**, *176*, 704-712.
84. Snippe, H.; van Houte, A. J.; van Dam, J. E. G.; de Reuver, M. J.; Jansze, M.; Willers, J. M. N. *Infect. Immun.* **1983**, *40*, 856-861.
85. Snippe, H.; van Dam, J. E. G.; van Houte, A. J.; Willers, J. M. N.; Kamerling, J. P.; Vliegenthart, J. F. G. *Infect. Immun.* **1983**, *42*, 842-844.
86. Zigterman, G. J. W. J.; van Dam, J. E. G.; Snippe, H.; Rotteveel, F. T. M.; Jansze, M.; Willers, J. M. N.; Kamerling, J. P.; Vliegenthart, J. F. G. *Infect. Immun.* **1985**, *47*, 421-428.
87. Van Dam, J. E. G.; van Halbeek, H.; Kamerling, J. P.; Vliegenthart, J. F. G.; Snippe, H.; Jansze, M.; Willers, J. M. N. *Carbohydr. Res.* **1985**, *142*, 338-343.
88. Zigterman, G. J. W. J.; Snippe, H.; Jansze, M.; Ernste, E.; de Reuver, M. J.; Willers, J. M. N. *Adv. Biosci.* **1988**, *68*, 301-308.
89. Van Dam, G. J.; Zigterman, G. J. W. J.; Verheul, A. F. M.; de Reuver, M. J.; Snippe, H. *J. Immunol.* **1988**, *143*, 3049-3053.
90. Zigterman, G. J. W. J.; Schotanus, K.; Ernste, E. B. H. W.; van Dam, G. J.; Jansze, M.; Snippe, H.; Willers, J. M. N. *Infect. Immun.* **1989**, *57*, 2712-2718.
91. Snippe, H.; Verheul, A. F. M.; van Dam, J. E. G. In *Immunological adjuvants and vaccines;* Gregoriadis, G.; Allison, A. C.; Poste, G., Eds.; NATO ASI series, Life Sciences Vol. 179, Plenum Press: London, 1998; pp 107-122.
92. Alonso de Velasco, E.; Verheul, A. F. M.; Veeneman, G. H.; Gomes, L. J. F.; van Boom, J. H.; Verhoef, J.; Snippe, H. *Vaccine* **1993**, *11*, 1429-1436.
93. Jansen, W. T. M.; Verheul, A. F. M.; Veeneman, G. H.; van Boom, J. H. *Vaccine*, **2000**, *20*, 19-21.
94. Alonso de Velasco, E., Dekker, H. A. Th.; Antal, P.; Jalink, K. P., van Strijp, J. A. G.; Verheul, A. F. M.; Verhoef, J.; Snippe, H. *Vaccine* **1994**, *12*, 1419-1422.
95. Benaissa-Trouw, B.; Lefeber, D. J.; Kamerling, J. P.; Vliegenthart, J. F. G.; Kraaijeveld, K.; Snippe, H. *Infect. Immun.* **2001**, *69*, 4698-4701.
96. Jansen, W. T. M.; Hogenboom, S.; Thijssen, M.-J. L.; Kamerling, J. P.; Vliegenthart, J. F. G.; Verhoef, J.; Snippe, H.; Verheul, A. F. M. *Infect. Immun.* **2001**, *69*, 787-793.
97. Lefeber, D. J.; Benaissa-Trouw, B.; Vliegenthart, J. F. G.; Kamerling, J. P.; Jansen, W. T. M.; Kraaijeveld, K.; Snippe, H. *Infect. Immun.* **2003**, *71*, 6915-6920.
98. Jansen, W. T. M.; Snippe, H. *Indian J. Med. Res.* **2004**, *119 (Suppl)*, 7-12.

99. Verez Bencomo, V.; Fernandez Santana, V.; Hardy, E.; Toledo, M. E.; Rodriguez, M. C.; Heynngnezz, L.; Rodriguez, A.; Baly, A.; Herrera, L.; Izquierdo, M.; Villar, A.; Valdes, Y.; Cosme, K.; Deller, M. L.; Montane, M.; Garcia, E.; Ramos, A.; Aguillar, A.; Medina, E.; Torano, G.; Sosa, I.; Hernandez, I.; Martinez, R.; Muzachio, A.; Carmentates, A.; Costa, L.; Cardoso, F.; Campa, C.; Diaz, M.; Roy, R. *Science* **2004**, *305*, 522-525.
100. Pichichero, M. E.; Porcelli, S.; Treanor, J.; Anderson, P. *Vaccine* **1998**, *16*, 83-91.
101. Shpak, E.; Leykam, J. F.; Kieliszewski, M. J. *Proc. Natl. Acad. Sci. U. S. A.* **1999**, 96, 14736-14741.
102. Di Virgilio, S.; Glushka, J.; Moremen, K.; Pierce, M. *Glycobiology* **1999**, *9*, 353-364.
103. Elling, L. *Adv. Biochem. Eng. Biotechnol.* **1997**, *58*, 89-144.
104. Yamamoto, T.; Nagae, H.; Kajihara, Y.; Terada, I. *Biosci. Biotechnol. Biochem.* **1998**, *62*, 210-214.
105. Imperiali, B.; O'Connor, S. E. *Curr. Opin. Chem. Biol.* **1999**, *3*, 643-649.

Chapter 6

From Epitope Characterization to the Design of Semisynthetic Glycoconjugate Vaccines against *Shigella flexneri* 2a Infection

Laurence A. Mulard[1,*] and Armelle Phalipon[2]

[1]Unité de Chimie Organique, CNRS URA 2128 and [2]Unité de Pathogénie Microbienne Moléculaire, INSERM U389, Institut Pasteur, 28 rue du Dr Roux, 75724 Paris Cedex 15, France

The four-step conception of a potential *Shigella flexneri* 2a glycoconjugate vaccine exposing synthetic oligosaccharides mimicking the serotype-specific protective determinants is described. (i) Study of the recognition of synthetic O-antigen fragments by protective murine mIgGs showed that the O-antigen exhibits a serotype-specific immunodominant epitope and that chain elongation improves binding. (ii) Five epitope-related tri- to pentadecasaccharides were synthesized and coupled via single point-attachment to tetanus toxoid or PADRE. (iii) The immunogenicity of the conjugates was assessed in mice. (iv) The protective efficacy of sera induced by the most immunogenic conjugates was evaluated in a murine model of pulmonary infection. A pentadecasaccharide was identified as a good candidate for further development of a *S. flexneri* 2a semi-synthetic glycoconjugate vaccine.

Introduction

Shigella is a Gram negative enterobacterium known as the causative agent of shigellosis, a human dysenteric syndrome characterized by a spectrum of symptoms varying from watery diarrhoea to severe dysentery [1]. These symptoms largely reflect bacterial invasion into the colonic and rectal mucosa, which provokes the induction of an acute inflammation responsible for massive tissue destruction. Shigellosis represents about one third of the total death due to diarrheal diseases. More than 99% of the estimated 164.7 annual episodes, resulting in some 1.1 million death, occur in developing countries, especially in areas where sanitary conditions are insufficient. Of utmost concern is the observation that close to 70% of all episodes and 60% of all death involve children under 5 years of age [2]. *Shigella*, which is transmitted by the feco-oral route, has a high propagation rate in areas of poor sanitary conditions, especially since the infectious dose is as low as about 100 bacteria. *Shigella flexneri* is the major strain responsible for the endemic form of the disease, particularly in developing countries. Of all the known serotypes, *S. flexneri* 2a is the most prevalent one in a large number of areas [2]. Considering (i) the increased number of isolated antibiotic-multiresistant *Shigella* strains, (ii) the poor benefit of oral rehydratation therapy, (iii) acute complications leading often to death, and (iv) the absence of significant improvement of sanitary conditions in most developing countries, vaccine-based prevention remains the only way out. There is a crucial need for a safe and efficacious vaccine against the most common serotypes, among which *S. flexneri* 2a comes first in line together with *S. flexneri* 1b, 3a, 6, *S. sonnei*, and *S. dysenteriae* 1 which is responsible for devastating epidemics [3]. Various strategies were undertaken to reach this goal. The two major ones, orally administered live attenuated vaccine strains and lipopolysaccharide (LPS)-derived vaccines, have gone through various stages of clinical trial successfully [1,4]. However, there is yet no licensed vaccine for shigellosis.

As for several other Gram negative bacteria, *S. flexneri* 2a LPS is an essential virulence factor for the bacterium. This major bacterial surface antigen is also the target of the host adaptive immunity. Anti-LPS antibodies (Abs) are actually elicited upon infection, both as secretory IgA (SIgA) locally at the intestinal level and as serum IgG systemically [1]. Upon natural infection or following vaccine trials, the Ab-mediated protection has been shown to be mostly species- and serotype-specific [5-7], pointing to the O-specific polysaccharide (O-SP) moiety of the LPS, also termed O-antigen (O-Ag), as the target of the protective immune response. Indeed, *Shigella* species and serotypes are defined by the structure of their O-Ag repeating unit (RU) [8]. As whole LPS could not be used as immunogen due to its highly toxic lipid A, the non toxic acid-detoxified LPS (pmLPS) has been used instead for the development of

Shigella vaccines. Analogously to the successful conversion of bacterial capsular polysaccharides from T-cell independent antigens (Ag) to T-cell dependent ones [9-11], *S. flexneri* 2a pmLPS was turned into a potent T-cell dependent immunogen through its covalent coupling onto a protein carrier. Several *S. flexneri* 2a pmLPS-protein conjugates were shown to be safe and immunogenic in adults[12,13] as well as in young children [14]. Noteworthy, encouraging results were obtained with a *S. sonnei* pmLPS-rEPA conjugate vaccine administered parenterally, showing protection in about 75% of the vaccinees during a *S. sonnei* outbreak [15]. Even though encouraging data are available, pmLPS-protein conjugate vaccines remain complex constructs especially when obtained from randomly activated pmLPS. Since potential loss of antigenicity may occur upon the detoxification and/or carrier-conjugation steps, accurate controls of these two crucial issues are required. In addition, appropriate consideration should be given to the increasing requirements from regulatory agencies for always better-defined molecules to be used in human.

In line with progress in the field dealing with other bacterial pathogens [16,17] or closely related ones such as *S. dysenteriae* type 1 [18] or *S. flexneri* Y, which was extensively studied although it is not pathogenic [19], we have considered alternatives to conventional pmLPS-protein conjugates, such as the use of accurate synthetic mimics of the bacterial O-Ag, or the use of universal T helper peptides as carrier [20,21]. The former approach has been mostly developed along two lines including the use of either synthetic oligosaccharides or peptides [22] mimicking the carbohydrate determinants recognized by anti-O-Ag protective monoclonal antibodies (mAbs), that is mAbs conferring protection in experimental models of infection. Such mimics are expected to induce an anti-LPS Ab response when appropriately presented to the immune system. We present here an overview of our contribution to the development of synthetic oligosaccharide-based conjugates as potential vaccines targeting *S. flexneri* 2a infection.

We chose the following strategy: (i) the use of carbohydrate haptens suitable for single-point attachment onto a carrier to overcome the limitations due to LPS-random chemical modifications and/or detoxification, (ii) the control of various parameters such as the length and nature of the carbohydrate hapten, its loading onto the carrier, as well as the choice of the carrier, to allow the design of glycoconjugates with optimal immunogenicity. A four-step process was thus developed, encompassing (i) the identification of the protective *S. flexneri* 2a epitopes, (ii) the conception of the candidate glycoconjugates, (iii) the study of the immunogenicity of the glycoconjugates in mice, and (iv) when appropriate, the analysis of the protective efficacy of the anti-*S. flexneri* 2a LPS Abs induced by the glycoconjugates.

Results and Discussion

1. Identification of the protective carbohydrate epitopes

1.A. Protective serotype-specific mAbs

As mentioned above, anti-*S. flexneri* 2a LPS SIgA and IgG responses are induced upon *Shigella* infection. Several studies suggest that both kinds of Ab confer protection against homologous challenge [1]. Therefore, theoretically, both monoclonal IgA (mIgA) and IgG (mIgG) are suitable tools to identify the protective oligosaccharide determinants carried by the O-Ag, that is fragments best recognized by protective mAbs. However, attempts in getting mouse mIgA specific for *S. flexneri* 5a have revealed that, since mice are resistant to oral *Shigella* infection, the priming of local intestinal IgA-mediated response is highly limited (A. Phalipon, personal communication). Consequently, only a weak number of mIgAs can be obtained [23-25]. Besides, using two anti-*S. flexneri* 5a LPS mIgAs, we showed that the IC_{50} of fragments of the O-Ag, including the pentasaccharide RU, was at best in the 25 millimolar range [22], preventing any precise investigation of the protective epitopes. Therefore, we decided to essentially use mIgGs to characterize the protective *S. flexneri* 2a epitopes.

In order to mimic the diversity of anti-LPS IgG subclasses induced following natural infection with *Shigella* [26], murine mIgG specific for serotype 2a O-Ag, and representative of each of the four murine IgG subclasses were successfully identified upon immunization of mice with killed bacteria. Interestingly, during screening for *S. flexneri* LPS recognition, most of the hybridoma cells tested (about 90%) were shown to secrete serotype-specific mAbs. This result differs slightly from previous reports on the isolation of mAbs directed to determinants common to several *S. flexneri* serotypes including 2a and 5a [27,28]. Five mIgGs were selected: F22-4 (IgG1), D15-7 (IgG1), A2-1 (IgG2a), E4-1 (IgG2b) and C1-7 (IgG3). Each LPS/mIgG interaction was characterized by measuring the IC_{50} that was shown to range from 2 to 20 ng/mL and the protective capacity of the selected mIgGs was ascertained using the above mentioned murine model of pulmonary infection [23]. As anticipated, protection was serotype-specific (data not shown) and dependent on the amount of mIgG passively administered intranasally locally prior to intranasal bacterial challenge [29]. Upon passive transfer using 20 µg of Ab, all mIgGs were shown to induce a reduction of the lung-bacterial load (Figure 1B), which was accompanied by a reduction of inflammation, and therefore of subsequent tissue destruction (Figure 1A). In contrast, using 2 µg of mIgG, only mIgG D15-7, A2-1 and E4-1 were shown to significantly reduce the lung-bacterial load in comparison to control mice, but with much less efficiency than that observed using 20 µg (Figure 1B).

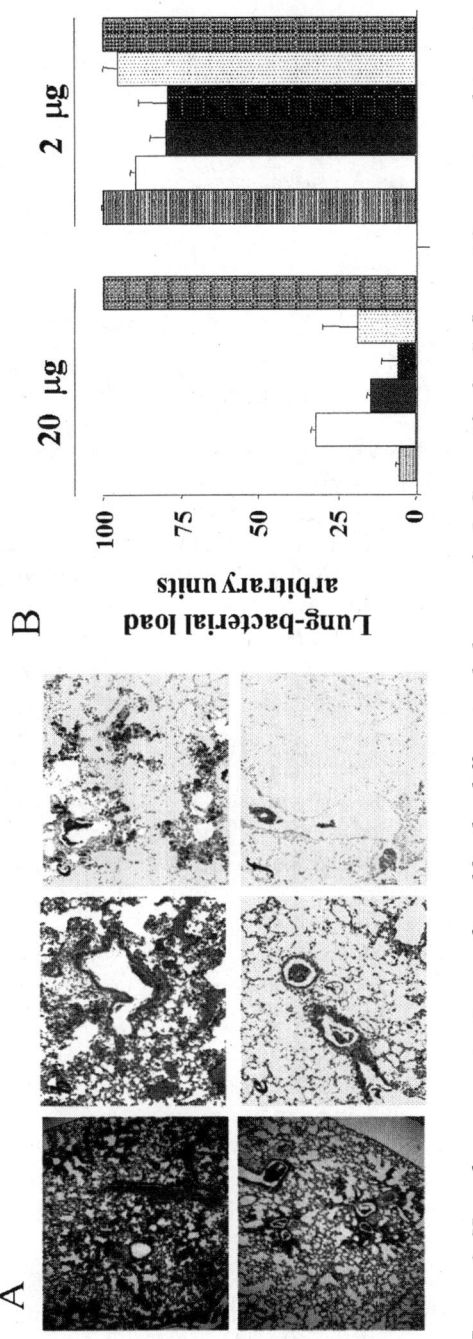

Figure 1. Homologous protection conferred by the different subclasses of mIgG specific for S. flexneri 2a serotype determinants. A: Histopathological study of mouse lungs. Upper row: control mice. Lower row: mice receiving mIgG. Hemalun Eosin staining: a and d magnification x 40 ; b and e magnification x100. Immunostaining using an anti-LPS Ab specific for S. flexneri serotype 2a: c and f magnification x 100. B: mice receiving intranasally (i.n.) 20 µg or 2 µg of purified.mIgG, respectively, 1h prior to i.n. challenge with a sublethal dose of virulent S. flexneri 2a bacteria. Lung-bacterial load was expressed using arbitrary units with 100 corresponding to the bacterial count in lungs of control mice. Standard deviations are represented (n = 10 mice per group, 3 independant experiments).

1.B. Synthetic methyl glycosides representative of fragments of the O-Ag

For the predominant *S. flexneri* serotype 2a, the biological RU is the branched pentasaccharide **AB(E)CD** (**I**) [8]. It is composed of a linear tetrasaccharide backbone made of three L-rhamnose residues **A**, **B**, and **C** and a *N*-acetyl-D-glucosamine residue **D**, that is common to all *S. flexneri* except serotype 6, and represents the RU of the non prevalent *S. flexneri* serotype Y. The RU of serotype 2a is characterized by the α-D-glucose residue **E** branched at position 4 of rhamnose **C**.

```
          A                 B                 C                 D
2)-α-L-Rhap-(1→2)-α-L-Rhap-(1→3)-α-L-Rhap-(1→3)-β-D-GlcNAcp(1→
                                    (1↑4)
                                 [α-D-Glcp]        E            (I)
```

In addition to mono-, di- and trisaccharides representative of fragments of the linear backbone of *S. flexneri* 2a O-Ag, all **E**-bearing mono-, di-, tri-, tetra- and pentasaccharides obtained by circular permutation of the four monosaccharides defining the linear tetrasaccharide backbone of the O-Ag biological RU, thus corresponding to all possible serotype-specific pentasaccharides and shorter sequences encountered along the bacterial polysaccharide, were synthesized. They were used as probes in order to study the contribution of each O-Ag sugar residue in LPS recognition by the selected protective mIgGs. An octa- and a decasaccharide fragments of the *S. flexneri* 2a O-Ag, chosen arbitrarily for their easier synthetic accessibility, were prepared as additional probes allowing to investigate the impact of hapten length on binding. All targets were obtained through multi-step chemical synthesis as their methyl glycoside so as to mimic the ring and anomeric forms that the reducing residue adopts in the natural polymer (Table 1). Key features of the syntheses which were developed are highlighted below.

Disaccharide EC. Focus was put on the synthesis of the serotype-specific oligosaccharides, thus bearing residue **E**. Retrosynthetic analysis of fragments of the O-Ag shows that all glycosidic linkages involved in the linear tetrasaccharide backbone are 1,2-*trans* glycosidic linkages. On the contrary, the **EC** glycosidic linkage is of the 1,2-*cis* type. It was thus anticipated to be the most challenging to build. We reasoned that this linkage should be constructed independently, and subsequently elongated selectively either at its reducing end or at its non-reducing end, and eventually in both directions. Thus, an **EC** disaccharide building block acting as an acceptor was needed. Besides, for most targets, except those bearing a reducing **C**, the **EC** disaccharide should act as a donor or a potential donor.

A suitable precursor to residue **E** should act as a donor bearing permanent protecting groups at all positions and a non participating protecting group at

Table 1. Synthetic methyl glycosides representative of *S. flexneri* 2a O-Ag

Mono-	Di-	Tri-	Tetra-	Pentasaccharides and larger
A=B=C D	AB * BC * CD * DA *	DAB * BCD * CDA * ABC *		
E	EC [30]	ECD [30] B(E)C [30]	B(E)CD [31] ECDA' [32] AB(E)C [33]	AB(E)CD [31] B(E)CDA' [32] D"AB(E)C [33] ECDA'B' [31] B(E)CDA'B'(E')C' [34] D"AB(E)CDA'B'(E')C [35]

* Di- and trisaccharides devoid of the **E** residue were either previously described or synthesized according to known procedures [36-40].

position 2. Thus, 2,3,4,6-tetra-*O*-benzyl-D-glucopyranose (**1**) was selected as a commercially available precursor to residue **E**. Using conventional procedures, hemiacetal **1** was turned into the bromide, the fluoride, and the trichloroacetimidate donor **2**, **3**, and **4**, respectively. All donors were condensed to either methyl α-L-rhamnopyranoside (**5**) or the corresponding allyl glycoside **6**. Use of the fluoride **3**, in combination with the silylated acceptor **7** and triflic anhydride as the promoter, gave the α**EC** glycosylation product **8** in 55% yield which was initially found satisfactory [30]. However in the long run, the trichloroacetimidate **4** was the preferred donor [32]. Indeed, following the inverse procedure in a 1/10 mixture of dichloromethane (DCM) and Et$_2$O, condensation of the latter to acceptor **6** in the presence a catalytic amount of TMSOTf, resulted in a mixture of **8** and the β**EC** glycosidation products from which **8** was isolated in 73% yield (Scheme 1). As a consequence, all the synthetic strategies developed towards the targets shown in Table 1 relied on the use of the trichloroacetimidate chemistry for the construction of the glycosidic linkages. Conversion of **8** into acceptor **9** suitable for elongation at position 3$_C$ was a two step process involving acidic hydrolysis of the isopropylidene and subsequent benzoylation at position 2$_C$ of the resulting diol (87%).

Monosaccharides A and B. When they do not act as chain terminators or as reducing end terminus, residues **A** and **B** are involved in the same substitution pattern. For that reason, the trichloroacetimidate donor **11** [41], which is easily accessible from the crystalline orthoester intermediate **10** [42], was selected as their common precursor. Compound **10** was obtained in five steps from L-rhamnose. Besides permanent benzyl groups at position 3 and 4, donor **11** has an acetyl moiety at position 2 which acts as a participating group upon glycosidation to

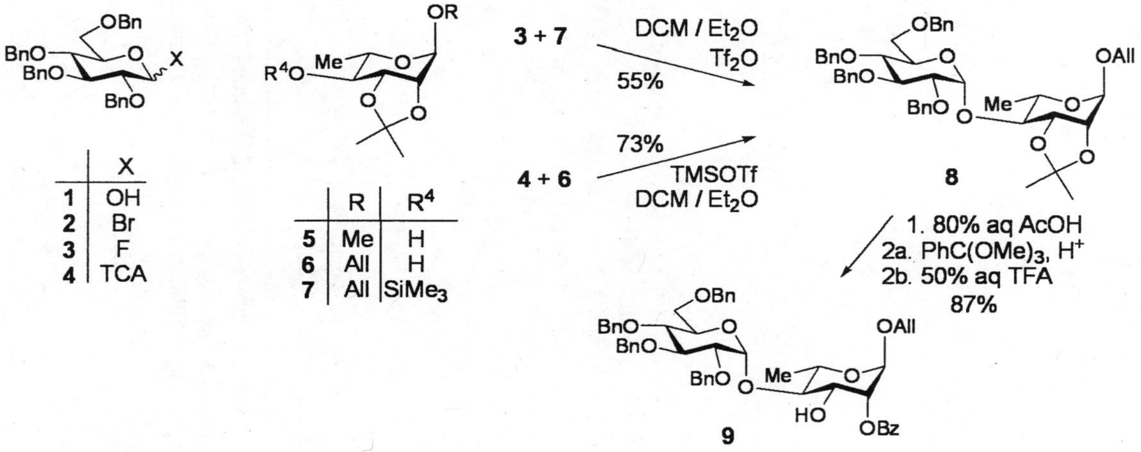

Scheme 1. Synthesis of the EC acceptor and potential donor.

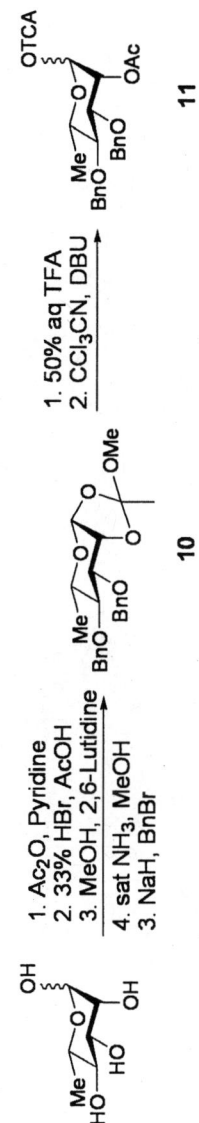

Scheme 2. Synthesis of the trichloroacetimidate precursor to residues A and B.

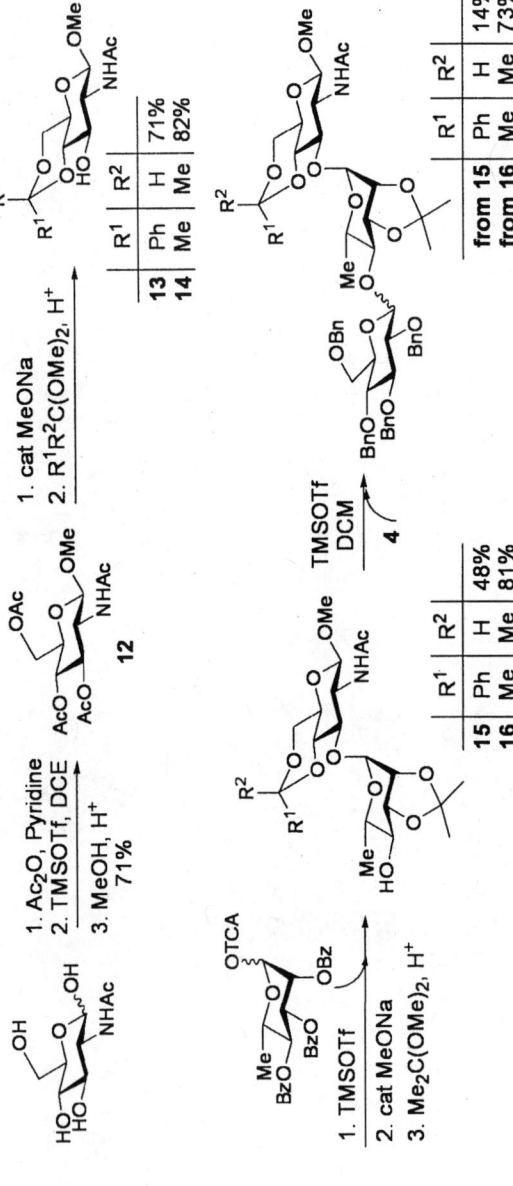

Scheme 3. The isopropylidene as a selective protecting group at position 4 and 6 of residue D.

favour the required 1,2-*trans* glycosidic linkage and may be selectively removed in the presence of both benzyl ethers and benzoyl esters to allow subsequent elongation at position 2_B and 2_A, if necessary (Scheme 2).

Monosaccharide **D**. Study on acceptor **D** was first performed in the methyl glycoside series (Scheme 3). Peracetylated **12** was obtained by opening, under strong acidic conditions, the oxazoline intermediate easily obtained from *N*-acetyl-D-glucosamine (71%). Transesterification of **12**, and subsequent regioselective acetalation gave benzylidene **13** or isopropylidene **14**, which were next converted to the **CD** acceptors **15** and **16**, respectively [30]. A benzylidene acetal protecting positions 4 and 6 of the *N*-acetyl-D-glucosamine residue **D** has been used successfully in the construction of large fragments of the O-Ag of *S. flexneri* serotype Y [43]. However, attempts to condense acceptor **15** with donors **3** and **4** gave at best a mixture of the αE and βE condensation products in 14% yield [30]. These results were tentatively correlated to the structure of **15**, resulting in poor solubility. Moreover, earlier observations [44] and NMR data [30] indicated that **15** adopted a preferred conformation derived from the *exo*-anomeric effect [45], resulting in anisotropic shielding of proton 6_C by the phenyl ring of the benzylidene acetal, and possibly explaining the poor outcome of the various glycosidation attempts. As anticipated, the isopropylidene acceptor **16** could be condensed to donor **4** to give a mixture of αE and βE condensation products in an acceptable 73% yield. Being compatible with the trichloroacetimidate-based glycosidation conditions in use, the isopropylidene acetal was thus adopted as a protecting group for all precursors to residue **D** when regioselective protection at position OH-4 and OH-6 was needed.

The *N*-protecting group to be introduced at position 2 of D-glucosaminyl donors used as precursors to residue **D** was investigated closely, with particular emphasis on the use of tetrachlorophtaloyl **17** or trichloroacetamidyl **18** (Figure 2) [46]. The donor properties of **18** were found slightly better. Besides, building blocks having a 2-*N*-trichloroacetyl-glucosaminyl residue at their non reducing end, retained their good acceptor properties, allowing chain elongation at position 3_D [34]. In addition, it was found that the trichloroacetamide could be easily converted to the required acetamide moiety by palladium-activated hydro-dechlorination [34] as an alternative to tributyltin hydride-mediated radical dechlorination [47]. The *N*-trichloracetyl donor **18** was adopted as a convenient precursor to residue **D**.

Convergent strategies to pentasaccharides and larger fragments

Interestingly, for fragments up to pentasaccharides, a convergent synthesis was developed in the case of pentasaccharide **B(E)CDA** only [32]. Indeed, the easiest linkage to build on the linear backbone was thought to be the **C-D** linkage since (i) it is a 1,2-*trans* linkage and (ii) introduction of a participating group at

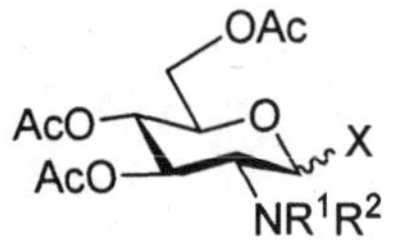

	R^1, R^2	X
17	Cl$_4$Pht	Br
18	C(O)CCl$_3$, H	OTCA

*Figure 2. Glucosaminyl donors as precursors to residue **D**.*

position 2_C is feasible. As anticipated, condensation of a **DA** methyl glycoside acceptor (**19**) to a **B(E)C** trichloroacetimidate donor (**20**) gave the fully protected pentasaccharide **21** in an acceptable 81% yield (Scheme 4). Others showed that disconnection at the **D-A** linkage was not suitable even when involving di- or trisaccharide building blocks [48,49], and the poor outcome of disconnections at the **A-B** and **B-C** linkages was reported in the course of this study [35]. Accordingly, convergent syntheses of **B(E)CDA'B'(E')C'** [34] and **D"AB(E)CDA'B'(E')C'** were developed successfully using a common **DA'B'(E')C'** acceptor. Interestingly in the latter example, the use of two closely related pentasaccharide building blocks, a **D"AB(E)C** donor and a **DA'B'(E')C'** acceptor, was seen as a significant drawback for purification purposes [35], suggesting that building blocks involved in glycosylation processes should be differentiated.

1.C. Epitope mapping

Synthetic oligosaccharides listed in Table 1 were tested for their recognition by the 5 protective mIgGs using inhibition ELISA assays to define an IC$_{50}$ (Table 2) [29]. **ECD** was the only trisaccharide recognized, underlining the crucial contribution of both the branched glucosyl residue (**E**) and the neighbouring *N*-acetyl-glucosaminyl residue (**D**) to Ab recognition. Larger oligosaccharides lacking the **ECD** sequence, such as **AB(E)C** or **D"AB(E)C** did not show any binding whereas **ECDA'** did. Interestingly, F22-4 was the only mAb showing measurable affinity for **ECD**, **ECDA'**, and **ECDA'B'**. Binding of F22-4 to these three ligands was similar suggesting a minor, if any, contribution of residues **A'** and **B'** to recognition. Rhamnose **B** was also shown to be a key element in Ab recognition. Indeed, **B(E)CD** was recognized by all the protective mIgG, except A2-1 and C1-7 for which the minimal sequences necessary for recognition were pentasaccharides **AB(E)CD** or **B(E)CDA'**. Interestingly, non reducing **A** in **AB(E)CD** had a somewhat variable impact on mAb recognition with a positive effect in the case of A2-1, and C 7-1, and a negative one when considering the other mAbs. Two families of mIgGs were thus identified: the first one represented by F22-4 recognizing the **ECD** trisaccharide, and the second one

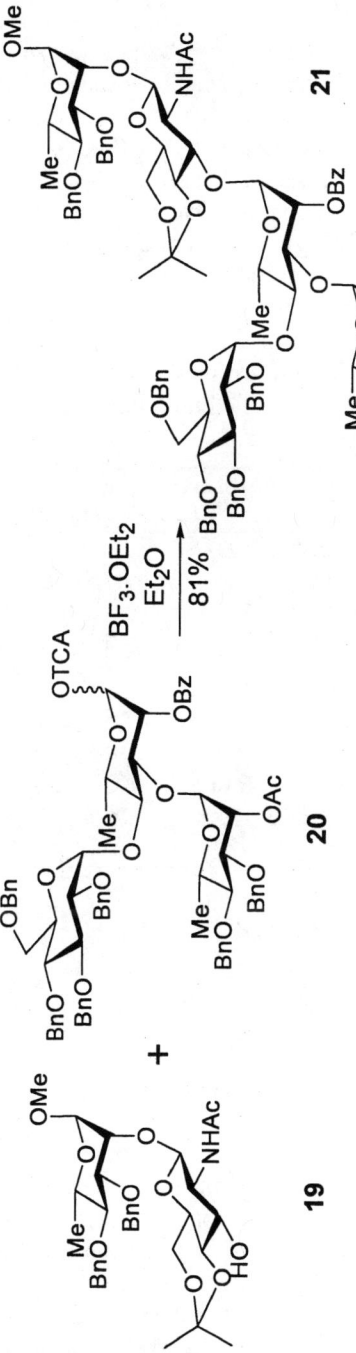

Scheme 4. Block synthesis of fragments of *S. flexneri 2a* O-Ag.

comprising the remaining four mIgGs, that recognized the same common B(E)CD sequence flanked or not by residue A/A' at either end. This observation was confirmed when studying the length-dependent oligosaccharide-Ab recognition using extended oligosaccharides such as the octa-B(E)CDA'B'(E)'C' and decasaccharide D"AB(E)CDA'B'(E)'C'. Interestingly, the decasaccharide showed the highest affinity for all mAbs except F22-4, which best recognized the octasaccharide. Overall, the branched tetrasaccharide B(E)CD was seen as an immunodominant "protective" determinant, with elongation of the oligosaccharide sequence increasing mAb binding.

Table 2. IC_{50} (µM) of the binding of protective mIgGs to *S. flexneri* 2a LPS by synthetic methyl glycosides (the maximum ligand concentration used is 1 mM).

Methyl glycosides	Anti-*S. flexneri* 2a serotype specific protective mIgGs				
	F22-4 (IgG1)	D15-7 (IgG1)	A2 (IgG2a)	E4-1 (IgG2b)	C1-7 (IgG3)
ECD	179 ± 93	> 1000	> 1000	> 1000	> 1000
ECDA'	18 ± 102	> 1000	> 1000	> 1000	> 1000
ECDA'B'	354 ± 40	> 1000	> 1000	> 1000	> 1000
B(E)CD	5.0 ± 0.9	198 ± 79	> 1000	87 ± 17	> 1000
AB(E)CD	21 ± 9	490 ± 100	378 ± 24	287 ± 66	734 ± 200
B(E)CDA'	2.5 ± 0.4	240 ± 65	340 ± 80	75 ± 9	400 ± 65
B(E)CDA'B'(E')C'	0.22 ± 0.02	60 ± 23	15 ± 5	12 ± 4	242 ± 124
D"AB(E)CDA'B'(E')C'	5.0 ± 1.4	12 ± 4	3.0 ± 1.8	4.4 ± 1.7	19 ± 5

Standard deviation is indicated (±).
> 1000: no inhibition of the oligosaccharide tested at this concentration.

2. Design and conception of the glycoconjugates

Our goal was to design conjugates exposing the carbohydrate haptens at the surface of the carrier for optimal interaction with the immune system. Among the different parameters influencing the immunogenicity of the glycoconjugates are the length and nature of the hapten [18], the hapten loading on the carrier [18,50], the nature of the linker between the hapten and the carrier [51,52], and the carrier itself [53]. In order to provide access to a broad diversity of constructs and consequently, to the optimisation of the above mentioned parameters, glycoconjugates were synthesized according to a modular approach involving three partners. Basically, it relied on (i) the use of appropriate oligosaccharide haptens functionalized with an aminoethyl spacer at their reducing end to allow site-selective modification, and thus to eventually serve as a suitable anchoring point; (ii) the incorporation of a thioacetyl acetamido linker as a masked thiol functionality for chemoselective ligation, and (iii) the use of a carrier activated at the ε amino groups of its lysine residues by maleimido groups complementary to thiols in the ligation process (Figure 3).

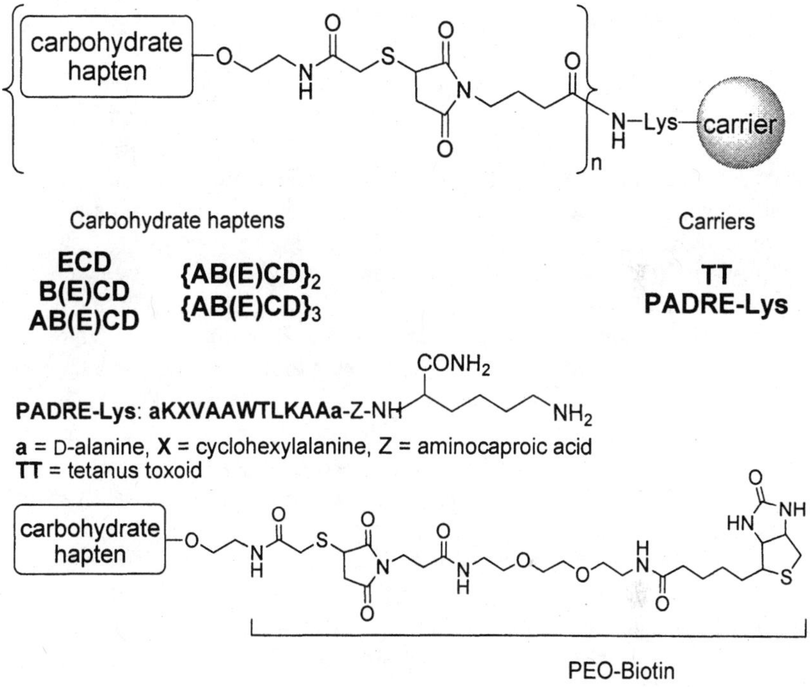
carbohydrate hapten
N-Lys-carrier
Carbohydrate haptens
ECD
B(E)CD
AB(E)CD
{AB(E)CD}2
{AB(E)CD}3
Carriers
TT
PADRE-Lys
CONH2
PADRE-Lys: aKXVAAWTLKAAa-Z-NH
NH2
a = D-alanine, X = cyclohexylalanine, Z = aminocaproic acid
TT = tetanus toxoid
carbohydrate hapten
PEO-Biotin

Figure 3. Representation of tri-, tetra-, penta-, deca-, and pentadecasaccharide conjugates. Tri- and tetrasaccharides are overlapping fragments of the pentasaccharide, which represents the biological RU of S. flexneri 2a O-Ag. Deca- and pentadecasaccharides are a dimer and a trimer of the RU, respectively. TT and PADRE conjugates were used as immunogens. Biotinylated derivatives were used as coating agents for immunogenicity analysis.

2.A. The carbohydrate haptens

Basis for the selection: A set of oligosaccharides was next selected based on the above mentioned antigenicity data: (i) **ECD** since it was the shortest sequence recognized at least by F22-4, (ii) **B(E)CD** since it was the tetrasaccharide recognized by 3 out of 5 mIgGs in contrast to **ECDA'** recognized by F22-4 only and **AB(E)C** not recognized at all, (iii) **AB(E)CD** since it represents the biological O-Ag RU and was almost as well recognized by the 5 mIgGs as **B(E)CDA** (Table 2). Since antigenicity studies outlined the positive impact of chain elongation on the synthetic *S. flexneri* 2a O-SP fragments/Ab recognition process, the decasaccharide **{AB(E)CD}$_2$** representing 2 biological RU and the pentadecasaccharide **{AB(E)CD}$_3$** representing 3 biological RU were also selected, although they differ from the haptens used for the antigenicity study and they may not be the easiest synthetic targets. In fact, our choice derived from two observations: (i) the crucial input of reducing **D** in mIgG recognition of short fragments which suggests that a given number of repeats should comprise the corresponding number of **ECD** sequences, (ii) the potential critical impact of non-reducing **A** on the specificity of sera induced by glycoconjugates since terminal non reducing residues of carbohydrate haptens may be immunodominant, leading to diversion of the Ab response [54].

Synthesis: the strategic points: All aminoethyl glycosides were synthesized according to a common convergent strategy taking into account the various synthetic observations made during the synthesis of the methyl glycosides. Basically, the strategy developed for constructing the short aminoethyl haptens [55] **22**, **23**, and **24** (Figure 4) relied on the condensation of a **D** acceptor (**31**) functionalized at the anomeric position with an azidoethyl spacer (Scheme 5) to a trichloroacetimidate **EC** [33], **B(E)C** [34], or **AB(E)C** donor [55], respectively. In the case of the larger haptens, such as the deca- **36** and the pentadecasaccharide **37**, a convergent iterative strategy was designed [56]. Thus in addition to **31**, the syntheses developed involved two building blocks, namely a **DAB(E)C** potential acceptor at position 3_D acting as a trichloroacetimidate donor (**32**), and an **AB(E)C** tetrasaccharide trichloroacetimidate donor (**34**) (Scheme 5).

Condensation of **31** and **32** was best performed in 1,2-dichloroethane (DCE), at 75°C, in the presence of trifluoromethanesulfonic acid as the promoter. Selective transesterification at the non reducing C-3_D gave the corresponding hexasaccharide acceptor (**33**), which was converted to the target decasaccharide **36** upon condensation with the tetrasaccharide donor **34** followed by a final three-step deprotection and concomitant Pd$_C$-mediated reduction of the azide to give the required aminoethyl aglycon. Alternatively, acceptor **33** was glycosylated with **32** to give the undecasaccharide acceptor **35** after selective 3_D-*O*-deacetylation. The latter was converted to the target pentadecasaccharide **37** as described for the preparation of **36** from **33**. Interestingly, building blocks **32** and **34** were obtained from the common tetrasaccharide intermediate **30** [35] (Figure 4). Noteworthy, donor **32** which is the key "repeating building block" is used twice in the synthesis of **37**. Besides, should further elongation of the hapten be necessary, the condensation/3_D-de-*O*-acetylation sequence involving **32** could be reiterated.

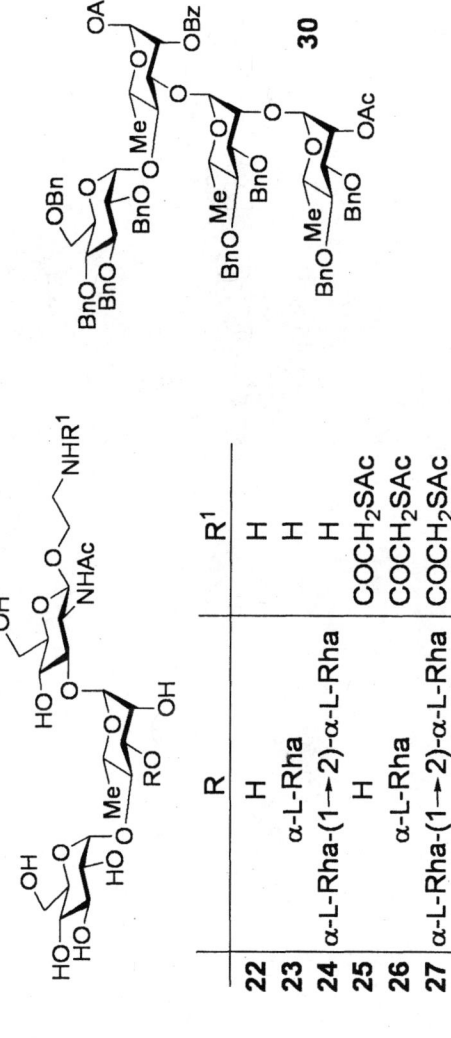

Figure 4. The aminoethyl tri- (**22**), tetra- (**23**), and pentasaccharide (**24**), and corresponding thioacetyl haptens, respectively. The key **AB(E)C** building block (**30**) for the synthesis of **25** and **26**.

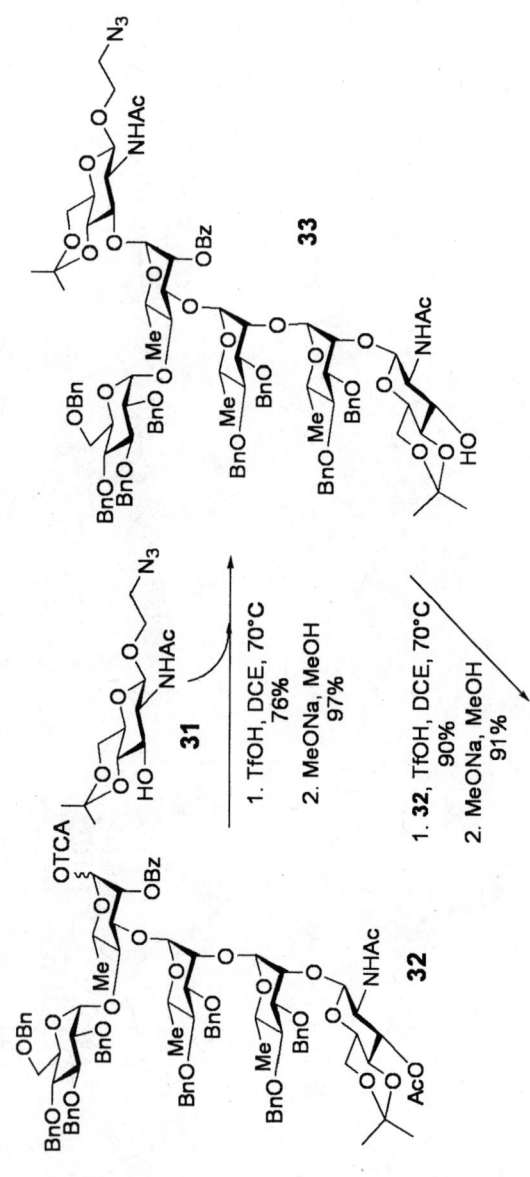

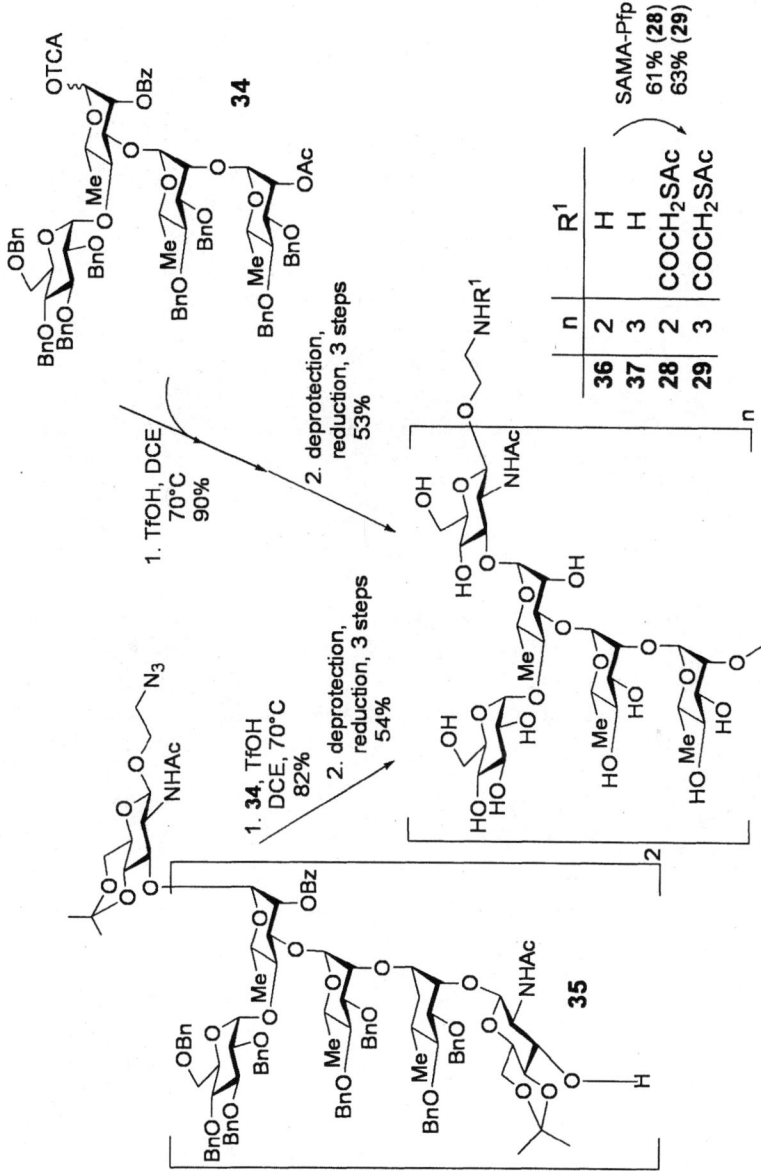

Scheme 5. Synthesis of the aminoethyl deca- 36 and pentadecasaccharide 37, and conversion to the thioacetyl glycosides 28 and 29, respectively.

2.B. The conjugation chemistry

We took advantage of the specific and high-yielding nucleophilic addition of a thiol group to the double bond of a maleimide in the presence of other nucleophiles [57]. Among the various maleimide-based cross-linking agents which are commercially available as their activated esters, several were reported to generate a significant anti-linker immune response [58,59]. In some instances, aromatic-based linkers were shown to produce total abrogation of the anti-hapten Ab response [60]. More recently, cyclohexyl-based linkers were demonstrated to act as highly antigenic linkers suppressing the Ab response to weak antigens [52]. Since induction of high anti-linker Ab levels appeared to be somewhat related to the use of constrained linkers, 4-(N-maleimido)-n-butanoyl, which was believed to be relatively flexible, was the linker of choice in this study.

The aminoethyl haptens **22-24**, **36** and **37** were converted to the corresponding masked thiol precursors **25-29** by reaction with S-acetylthioglycolic acid pentafluorophenyl ester (SAMA-Pfp) in yields ranging from 53 to 75%. Derivatization was monitored by RP-HPLC with detection at 215 nm and structures were ascertained based on MS and NMR analysis.

2.C. The carriers and glycoconjugates thereof

Two strategies were undertaken, that involving a protein carrier which allows the design of a set of semi-synthetic glycoproteins, also termed neoglycoproteins exposing the carbohydrate haptens in a multivalent fashion, and that relying on the use of an universal T-helper peptide, which results in the conception of a set of fully synthetic monovalent glycoconjugates, also termed neoglycopeptides.

Oligosaccharide-protein conjugates: neoglycoproteins. Several immunogenic proteins have been evaluated as potential carriers, but to our knowledge only four are components of currently licensed vaccines, namely Tetanus Toxoid (TT), Diphteria Toxoid (DT), CRM_{197} a non toxic mutant of the DT, and OMP, an outer membrane protein of *Neisseria meningitidis* [61]. In order to get as close as possible to a product compatible with use in humans, we chose from the early beginning to work with TT (gift from Sanofi Pasteur), a well-known carrier of 150 kDa whose immunogenicity in human has been extensively studied. TT was thus derivatized by reaction of the accessible side chain amino group of its lysine residues with a large excess of N-(γ-maleimidobutiryloxy) sulfosuccinimide ester. *In situ* unmasking of **25-29** and concomitant covalent addition of the resulting thiol bearing haptens to maleimido-linked TT next gave five sets of chemically defined semi-synthetic conjugates differing in hapten length [29]. The conjugation yields, in terms of protein recovery in the conjugates, were found to range from 52% to 95% based on a colorimetric assay using BSA as a standard [62]. The hapten loading was estimated using the SELDI-TOF-MS-based ProteinChip® System (Surface-Enhanced Laser Desorption Ionization-Time Of Flight) developed by Ciphergen [63,64] showing an average value of carbohydrate haptens per molecule of protein close to 13-14 (Figure 5).

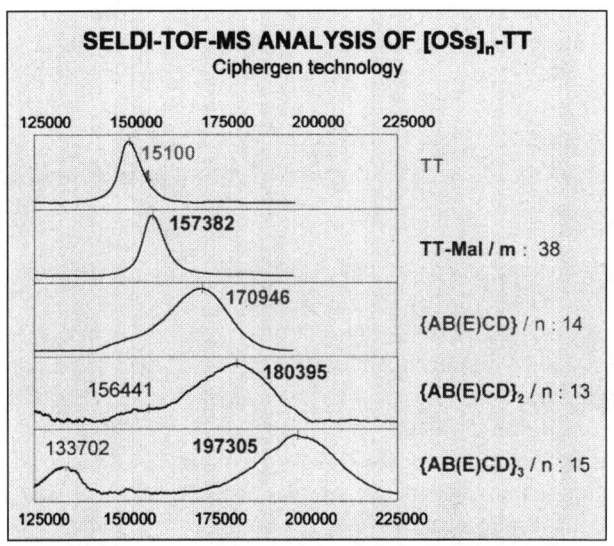

Figure 5. Hapten loading quantification of the oligosaccharide-protein conjugates ([OSs]$_n$-TT) using the SELDI-TOF-MS-based ProteinChip System (chip NP20, matrix EAM1)

Oligosaccharide-peptide conjugates: neoglycopeptides. The increasing development of new conjugate vaccines is in part associated to the availability of additional carriers providing appropriate T cell help. In parallel to the search for new protein carriers, the use of universal T helper epitopes as carriers for T cell independent antigens, including polysaccharides or short carbohydrates has been undertaken [21]. Universal T helper epitopes are peptides able to bypass the broad polymorphism of human leukocyte-associated antigens (HLA), and therefore to induce a T-dependent Ab response in a large human population exhibiting diverse haplotypes. In addition to the use of universal T helper epitopes of bacterial and viral origin [20,65] or polyepitopes corresponding to strings of the former [65,66], synthetic peptides were engineered based on the knowledge of peptide sequences preferentially bound to the most prevalent HLA antigens of a given human population. Among those, the non natural pan HLA DR-binding

Epitope (PADRE), shown to provide effective T cell help for both short oligosaccharides [67] and polysaccharides [68], was selected as a potential carrier to construct a fully synthetic neoglycopeptide vaccine to *S. flexneri* 2a infection (Figure 3). PADRE bearing an additional Lysine residue at its C-terminus (PADRE-Lys) was thus synthesized and covalently linked to 4-(*N*-maleimido)-*n*-butanoyl at the resulting side chain amino group. Hydroxylamine mediated chemoselective ligation of haptens **25-29** to the maleimido activated PADRE-Lys was monitored by RP-HPLC [69]. The target neoglycopeptides were isolated as single products whose identity was assessed by MS analysis, in yields ranging from 44% to 67% [55,56].

3. Immunogenicity of the *S. flexneri* 2a oligosaccharide-based conjugates in mice

3.A. Fully synthetic glycoconjugates incorporating PADRE as carrier.

All neoglycopeptides were shown to be antigenic (data not shown). To start, the tri-, tetra-, and pentasaccharide conjugates were used to immunize C57Bl/6 adult mice, intraperitoneally (i.p.), three times at three week intervals followed by a boost one month later, using an equivalent of 10 or 100 µg of oligosaccharide per dose in the presence of alun [67] (R. Lo-Man, personal communication). Immunogenicity of the neoglycopeptides was assessed by measuring the anti-oligosaccharide IgG response induced upon immunization in ELISA using the corresponding biotinylated oligosaccharides as coated antigens (Figure 3). Whatever the immunizing dose and the conjugate, no anti-oligosaccharide Ab response was detected. As OF1 mice were shown to be more potent in responding to glycopeptides than C57Bl/6 mice (R. Lo Man, personal communication), they were used for immunization with the penta-, deca- or pentadecasaccharide (20 µg/dose plus alun) for which preliminary immunogenicity data when coupled to TT were encouraging (see below). No anti-oligosaccharide Ab response was detected whatever the neoglycopeptide. We did not even succeed in inducing an Ab response when modifying the PADRE sequence as described [67] (data not shown). Despite the successful results using monovalent PADRE-neoglycopeptides reported in the literature [67], we may hypothesize than in our case, the selected haptens are non immunogenic when presented in the monovalent form. Interestingly, using a non immunogenic tetravalent lysine core known as the MAP (Multiple Antigen Peptide system) [70], others have shown that when working with fully synthetic constructs, a multimeric presentation in a form of clusters of short bacterial carbohydrate haptens covalently linked to an universal T helper peptide may be more efficient than the monomeric strategy [21].

3.B. Semi-synthetic glycoconjugates incorporating TT as carrier.

All semi-synthetic conjugates, namely **ECD-TT, B(E)CD-TT, AB(E)CD-TT, {AB(E)CD}$_2$-TT** and **{AB(E)CD}$_3$-TT**, were shown to be antigenic (data not shown). Seven week-old BALB/c mice were immunized i.p., three times at three week intervals followed by a boost one month later, using an equivalent of 10 µg of oligosaccharide per dose in the absence of any adjuvant. Immunogenicity of the different glycoconjugates was assessed by measuring the anti-oligosaccharide and anti-LPS IgG titer in ELISA by using the corresponding biotinylated oligosaccharides (Figure 3) and purified LPS as coated antigens, respectively.

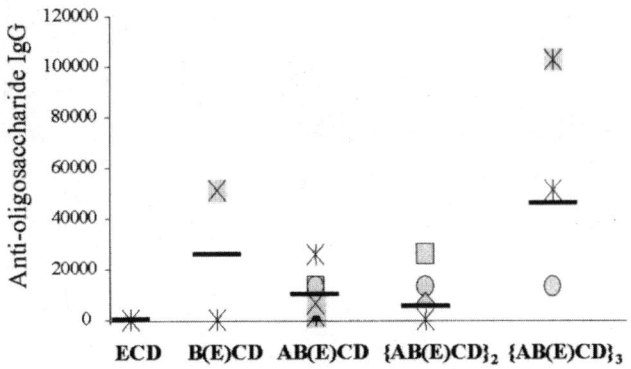

TT-glycoconjugates differing in hapten length

Figure 6: Anti-oligosaccharide Ab titer induced by the selected glycoconjugates after the boost. The Ab titer was defined as the last dilution of sera giving an OD value of at least twice that obtained with pre-immune sera. Each group comprises 14 mice except for the tri- and tetrasaccharide including 7 mice. Data shown here are representative of three independent experiments.

No cross-reactivity towards *S. flexneri* 2a LPS or the selected oligosaccharides was detected in sera of control mice immunized with TT alone (140 µg) and eliciting an high anti-TT Ab titer (10^6). **ECD-TT** did not elicit any anti-oligosaccharide IgG response (Figure 6), and as expected, any anti-*S. flexneri* 2a LPS IgG response (Figure 7). Anti-oligosaccharide Abs were induced by **B(E)CD-TT**, but no anti-*S. flexneri* 2a LPS IgG response was detected (Figures 6 and 7). In contrast, glycoconjugates incorporating 1 (**AB(E)CD-TT**), 2 (**{AB(E)CD}$_2$-TT**), or 3 RU (**{AB(E)CD}$_3$-TT**) raised both an anti-oligosaccharide and an anti-*S. flexneri* 2a LPS IgG response (Figures 6 and 7) that was shown to increase by a factor 2 to 2.5 between the third and fourth immunizations (data not shown). Interestingly, we observed that the anti-*S. flexneri* 2a LPS IgG titer elicited (Figure 7) as well as the number of mice responding was highly dependent on the hapten length. Whereas only 28.5% mice responded to **AB(E)CD-TT**, 85% responded to **{AB(E)CD}$_2$-TT** and 100% to **{AB(E)CD}$_3$-TT**. In addition, mice immunized with **AB(E)CD-TT** elicited an anti-*S. flexneri* 2a LPS IgG titer significantly different from that induced by **{AB(E)CD}$_3$-TT** but not from that induced by **{AB(E)CD}$_2$-TT** (p= 0.005 and 0.2, respectively). Similarly, mice immunized with **{AB(E)CD}$_2$-TT** elicited an anti-*S. flexneri* 2a LPS IgG titer significantly lower than that elicited with **{AB(E)CD}$_3$-TT** (p= 0.0002).

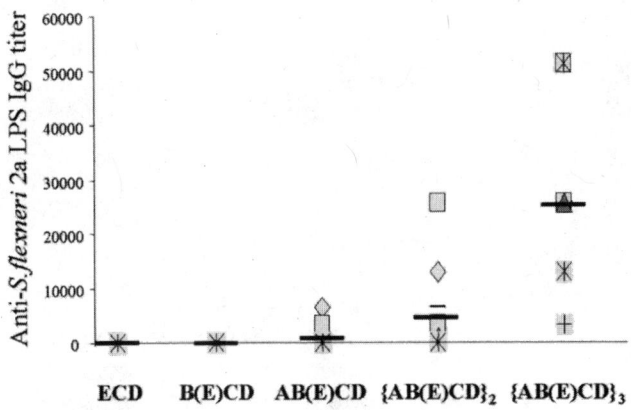

Figure 7: Anti-S. flexneri 2a LPS Ab titer induced by the selected glycoconjugates after the boost. Same groups of mice as in Figure 6. See legend of Figure 6.

Moreover, as observed for other bacterial diseases earlier on [18], a **pmLPS-TT** conjugate prepared in house via CNBr-mediated random activation of the *S. flexneri* 2a pmLPS [71] was shown to be less effective in inducing an anti-*S. flexneri* 2a LPS Ab response than **{AB(E)CD}$_3$-TT**. Indeed, the induced response was similar to that induced by **AB(E)CD-TT** (data not shown).

Taken together, these results demonstrate that the pentadecasaccharide **{AB(E)CD}$_3$** is the most accurate mimic of the O-Ag. These data add to previous immunogenicity studies reported for other bacterial pathogens, suggesting a high diversity of immunogenic behaviours for bacterial polysaccharide Ags. On one hand, neoglycoproteins incorporating oligosaccharides comprising one RU or smaller fragments were found immunogenic in mice, and shown to induce fully protective Ab in mice and/or in rabbits [54,72-75]. On the other end, it was suggested that bacterial carbohydrate haptens should comprise more than one RU to be turned to efficient immunogens upon coupling to a carrier [53,76,77]. Data shown above clearly suggest that *S. flexneri* 2a O-Ag belongs to the later group of bacterial polysaccharides as it is also the case for the O-Ag of another pathogenic *Shigella*, namely *S. dysenteriae* 1 [77].

4. Protective capacity of anti-S. flexneri 2a LPS Ab induced by the oligosaccharide-TT glycoconjugates.

To assess the potential of the glycoconjugates in inducing protective immunity, naive mice were administered with glycoconjugate-induced polyclonal sera incubated with virulent *S. flexneri* 2a prior to intranasal administration. Protection was assessed at 24 h post-infection by measuring the lung-bacterial load in comparison to control naive mice receiving *S. flexneri* 2a bacteria incubated with non immune serum. In order to compare the intrinsic capacity of anti-*S. flexneri* 2a LPS 2a Abs induced by the different oligosaccharides, independently of the induced Ab titer, we selected for each group sera exhibiting the same anti-*S. flexneri* 2a LPS Ab titer (2.5 x10^4). Serum of mice immunized with **B(E)CD-TT** was used as a negative control since no anti-*S. flexneri* 2a LPS IgG response was induced by this hapten. As shown in Figure 9, the best protection was conferred by **{AB(E)CD}$_3$-TT** (3 RU)-induced anti-*S. flexneri* 2a LPS Abs. This protection was, as expected, depending on the amount of anti-*S. flexneri* 2a LPS 2a Abs, since dilution of the polyclonal sera (1/100) prior to incubation with the virulent bacteria and subsequent intranasal administration, led to a significant decrease of the protection factor. Although exhibiting a significantly lower protection factor, the anti-*S. flexneri* 2a LPS Abs induced by the **{AB(E)CD}-TT** (1 RU) and **{AB(E)CD}$_2$-TT** (2 RU), respectively, also conferred significant protection in comparison to the negative control group.

Conclusion

In view of developing a chemically defined glycoconjugate vaccine to *S. flexneri* serotype 2a, we have characterized the "protective" serotype-specific determinants carried by the O-Ag, and analyzed their potential as functional mimics of the native antigen. In terms of Ab recognition, the importance of the branched glucosyl **E** and its surrounding residues has been emphasized. However, immunogenicity analysis of conjugates incorporating selected oligosaccharides based on available antigenicity data have demonstrated that not

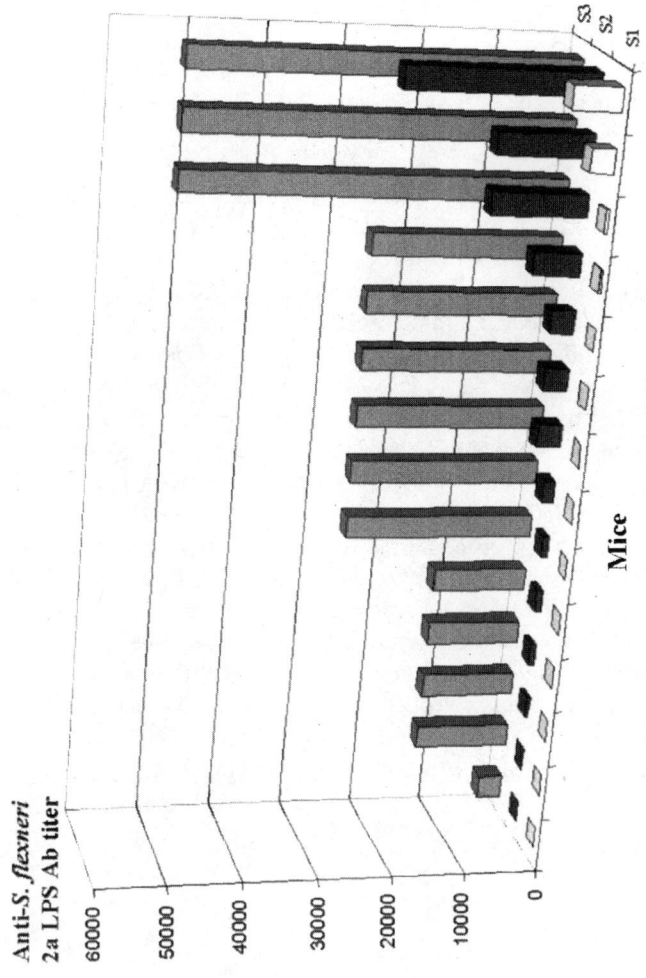

Figure 8. Immunogenicity of the selected oligosaccharides used as TT-glycoconjugates. The results presented for each mouse individually are representative of 3 independent experiments [29].

***Glycoconjugates:* S1: AB(E)CD; S2: {AB(E)CD}$_2$; S3: {AB(E)CD}$_3$.**

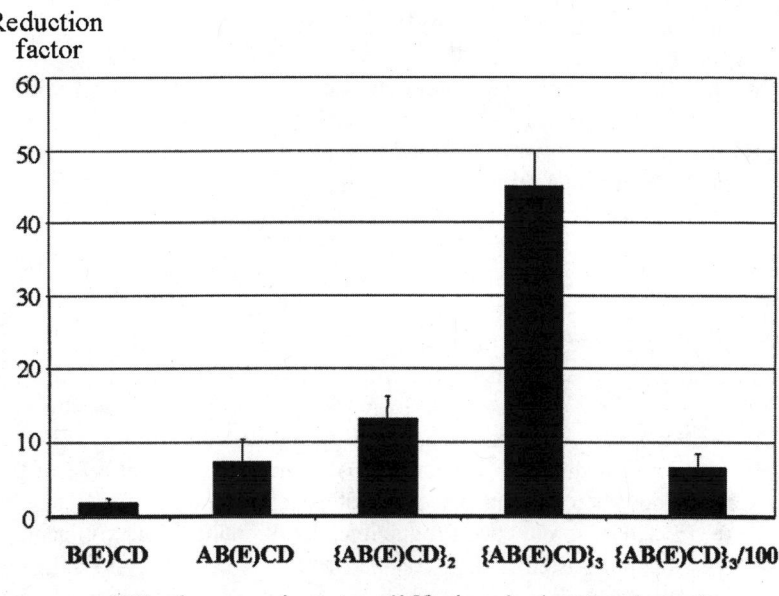

Figure 9. Protection conferred by the glycoconjugate-induced anti-S. flexneri 2a LPS Abs. The protection factor was defined as the factor of reduction of the lung-bacterial load in mice receiving immune sera in comparison to control mice receiving non immune sera. Polyclonal sera from mice infected with the different glycoconjugates are indicated. {AB(E)CD}$_3$/100 means that the serum from mice immunized with the {AB(E)CD}$_3$-TT has been diluted 100 fold prior incubation with the bacteria and subsequent intranasal administration.

all antigenic sequences accurately mimic the natural O-Ag. The most striking observation revealed by our study is the strong enhancement of the anti-*S. flexneri* 2a LPS IgG titer resulting from elongation of the hapten length from the penta- **AB(E)CD** to the deca- **{AB(E)CD}$_2$**, and subsequently, the pentadecasaccharide **{AB(E)CD}$_3$** corresponding to 1, 2 and 3 biological RU, respectively. These results demonstrate that the pentadecasaccharide **{AB(E)CD}$_3$** is, among the oligosaccharides that have been tested, the best functional mimic of the O-Ag. Therefore, it is a good candidate for further investigation in order to develop a chemically defined glycoconjugate vaccine against *S. flexneri* 2a infection.

Aknowledgements

We are extremely grateful to all our former and present collaborators who participated in this work, and whose contribution is acknowledged in the references listed. We warmly thank Philippe J. Sansonetti for his unfailing support and Marie-Aline Bloch for her trust in the project. We would like to emphasize in particular the contribution from Corina Costachel and Véronique Marcel-Peyre (sharing their work time between Unité de Chimie Organique and Unité de Pathogénie Microbienne Moléculaire, Institut Pasteur, Paris), Fabienne Segat-Dioury, Karen Wright, Frédéric Bélot, Catherine Guerreiro Inverno, Cyrille Grandjean, and Françoise Baleux (Unité de Chimie Organique, Institut Pasteur, Paris), Audrey Thuizat and Myriam Tanguy (Unité de Pathogénie Microbienne Moléculaire, Institut Pasteur, Paris) to the synthetic and biological aspects of this work. We are also grateful to Joël Ughetto-Monfrin (Unité de Chimie Organique, Institut Pasteur, Paris), Jacques d'Alayer (Plate-forme d'Analyse et de Microséquençage des Protéines, Institut Pasteur, Paris), and Farida Nato (Plate-forme de Production de Protéines Recombinantes et d'Anticorps, Institut Pasteur, Paris) for their technical help. We are indebted to Aventis Pasteur for their gracious gift of TT, and we thank Monique Moreau in particular.

P. J. S. is a Howard Hughes Medical Institute scholar. This work has been supported by the "Ministère Français de la Recherche", the "Direction Générale des Armées", Fondation pour la Recherche Médicale (to C. C.) the C.A.N.A.M. fellowship (to C. C. and F. S-D.), the Ms Frank Howard fellowship (to K. W.), the Roux fellowship (to F. B.), and the Transversal Research Program from the Pasteur Institute (PTR 99 which included a fellowship to C. G.).

References

1. Phalipon, A.; Sansonetti, P. J. *Crit. Rev. Immunol.* **2003**, *23*, 371-401.
2. Kotloff, K. L.; Winickoff, J. P.; Ivanoff, B.; Clemens, J. D.; Swerdlow, D. L.; Sansonetti, P. J.; Adak, G. K.; Levine, M. M. *Bull. WHO* **1999**, *77*, 651-666.
3. WHO *Weekly Epidemiol. Rec.* **1997**, *72*, 73-79.
4. Jennison, A. V.; Verma, N. K. *FEMS Microbiol. Rev.* **2004**, *28*, 43-58.
5. Mel, D. M.; Terzin, A. L.; Vuksic, L. *Bull. WHO* **1965**, *32*, 647-677.
6. Mel, D. M.; Arsic, B. L.; Nikolic, B. D.; Radovanic, M. L. *Bull. WHO* **1968**, *39*, 375-380.
7. Rasolofo-Razanamparany, V.; Cassel-Beraud, A.-M.; Roux, J.; Sansonetti, P. J.; Phalipon, A. *Infect. Immun.* **2001**, *69*, 5230-5234.
8. Lindberg, A. A.; Karnell, A.; Weintraub, A. *Rev. Infect. Dis.* **1991**, *13*, S279-S284.

9. Ada, G.; Isaacs, D. *Clin. Microbiol. Infect.* **2003**, *9*, 79-85.
10. Lockhart, S. *Expert Rev. Vaccines* **2003**, *2*, 633-648.
11. Roy, R. *Drug Discovery Today: Technologies* **2004**, *1*, 327-336.
12. Taylor, D. N.; Trofa, A. C.; Sadoff, J.; Chu, C.; Bryla, D.; Shiloach, J.; Cohen, D.; Ashkenazi, S.; Lerman, Y.; Egan, W.; Schneerson, R.; Robbins, J. B. *Infect. Immun.* **1993**, *61*, 3678-3687.
13. Passwell, J. H.; Harlev, E.; Ashkenazi, S.; Chu, C.; Miron, D.; Ramon, R.; Farzan, N.; Shiloach, J.; Bryla, D. A.; Majadly, F.; Roberson, R.; Robbins, J. B.; Schneerson, R. *Infect. Immun.* **2001**, *69*, 1351-1357.
14. Ashkenazi, S.; Passwell, J. H.; Harlev, E.; Miron, D.; Dagan, R.; Farzan, N.; Ramon, R.; Majadly, F.; Bryla, D. A.; Karpas, A. B.; Robbins, J. B.; Schneerson, R. *J. Infect. Dis.* **1999**, *179*, 1565-1568.
15. Cohen, D.; Ashkenazi, S.; Green, M. S.; Gdalevich, M.; Robin, G.; Slepon, R.; Yavzori, M.; Orr, N.; Block, C.; Ashkenazi, I.; Shemer, J.; Taylor, D. N.; Hale, T. L.; Sadoff, J. C.; Pavliovka, D.; Schneerson, R.; Robbins, J. B. *The Lancet* **1997**, *349*, 155-159.
16. Peeters, C. C. A. M.; Evenberg, D.; Hoogerhout, P.; Kayhty, H.; Saarinen, L.; van Boeckel, C. A. A.; van der Marel, G. A.; van Boom, J. H.; Poolman, J. T. *Infect. Immun.* **1992**, *60*, 1826-1833.
17. Verez-Bencomo, V.; Fernández-Santana, V.; Hardy, E.; Toledo, M. E.; Rodríguez, M. C.; Heynngnezz, L.; Rodriguez, A.; Baly, A.; Herrera, L.; Izquierdo, M.; Villar, A.; Valdés, Y.; Cosme, K.; Deler, M. L.; Montane, M.; Garcia, E.; Ramos, A.; Aguilar, A.; Medina, E.; Toraño, G.; Sosa, I.; Hernandez, I.; Martínez, R.; Muzachio, A.; Carmenates, A.; Costa, L.; Cardoso, F.; Campa, C.; Diaz, M.; Roy, R. *Science* **2004**, *305*, 522-525.
18. Pozsgay, V.; Chu, C.; Panell, L.; Wolfe, J.; Robbins, J. B.; Schneerson, R. *Proc. Natl. Acad. Sci. USA* **1999**, *96*, 5194-5197.
19. Hossany, R. B.; Johnson, M. A.; Eniade, A. A.; Pinto, B. M. *Bioorg. Med. Chem.* **2004**, *12*, 3743-3754.
20. Alonso de Velasco, E.; Merkus, D.; Anderton, S.; Verheul, A. F.; Lizzio, E. F.; Van der Zee, R.; Van Eden, W.; Hoffman, T.; Verhoef, J.; Snippe, H. *Infect. Immun.* **1995**, *63*, 961-968.
21. Chong, P. E.; Chan, N.; Kandil, A.; Tripet, B.; James, O.; Yang, Y.-P.; Shi, S.-P.; Klein, M. *Infect. Immun.* **1997**, *65*, 4918-4925.
22. Clément, M.-J.; Fortuné, A.; Phalipon, A.; Marcel-Peyre, V.; Simenel, C.; Imberty, A.; Delepierre, M.; Mulard, L. A. *J. Biol. Chem.* **2006**, *281*, 2317-2332.
23. Phalipon, A.; Kauffmann, M.; Michetti, P.; Cavaillon, J.-M.; Huerre, M.; Sansonetti, P.; Krahenbuhl, J.-P. *J. Exp. Med.* **1995**, *182*, 769-778.
24. Phalipon, A.; Cardona, A.; Kraehenbuhl, J.-P.; Edelman, L.; Sansonetti, P. J.; Corthesy, B. *Immunity* **2002**, *17*, 107-115.
25. Fernandez, M. I.; Pedron, T.; Tournebize, R.; Olivo-Marin, J.-C.; Sansonetti, P. J.; Phalipon, A. *Immunity* **2003**, *18*, 739-749.

26. Islam, D.; Wretlin, B.; Ryd, M.; Lindberg, A. A.; Christensson, B. *Infect. Immun.* **1995**, *63*, 2045-2061.
27. Carlin, N. I.; Lindberg, A. A. *Infect. Immun.* **1987**, *55*, 1412-1420.
28. Hartman, A. B.; Verg, L. L. V. d.; Mainhart, C. R.; Tall, B. D.; Smith-Gill, S. J. *Clin. Diagn. Lab. Immunol.* **1996**, *3*, 584-589.
29. Phalipon, A.; Costachel, C.; Grandjean, C.; Thuizat, A.; Guerreiro, C.; M.Tanguy; Nato, F.; Normand, B. V.-L.; Marcel-Peyre, V.; Sansonetti, P. J.; Mulard, L. A. *J. Immunol.* **2006**, *176*, 1686-1694.
30. Mulard, L. A.; Costachel, C.; Sansonetti, P. J. *J. Carbohydr. Chem.* **2000**, *19*, 849-877.
31. Mulard, L. A.; Guerreiro, C. *Tetrahedron* **2004**, *60*, 2475-2488.
32. Segat, F.; Mulard, L. A. *Tetrahedron: Asymmetry* **2002**, *13*, 2211-2222.
33. Costachel, C.; Sansonetti, P. J.; Mulard, L. A. *J. Carbohydr. Chem.* **2000**, *19*, 1131-1150.
34. Bélot, F.; Costachel, C.; Wright, K.; Phalipon, A.; Mulard, L. A. *Tetrahedron. Lett.* **2002**, *43*, 8215-8218.
35. Bélot, F.; Wright, K.; Costachel, C.; Phalipon, A.; Mulard, L. A. *J. Org. Chem.* **2004**, *69*, 1060-1074.
36. Auzanneau, F.-I.; Bundle, D. R. *Can. J. Chem.* **1993**, *71*, 534-548.
37. Auzanneau, F.-I.; Hanna, H. R.; Bundle, D. R. *Carbohydr. Res.* **1993**, *240*, 161-181.
38. Pozsgay, V.; Brisson, J.-R.; Jennings, H. J. *Can. J. Chem.* **1987**, *65*, 2764-2769.
39. Hanna, H. R.; Bundle, D. R. *Can. J. Chem.* **1993**, *71*, 125-134.
40. Nifant'ev, N. E.; Shashkov, A. S.; Khatuntseva, E. A.; Tsvetkov, Y. E.; Sherman, A. A.; Kotchetkov, N. K. *Russ. J. Bioorg. Chem.* **1994**, *20*, 556-565.
41. Zhang, J.; Mao, J. M.; Chen, H. M.; Cai, M. S. *Tetrahedron: Asymmetry* **1994**, *5*, 2283-2290.
42. Castro-Palomino, J. C.; Rensoli, M. H.; Verez-Bencomo, V. *J. Carbohydr. Chem.* **1996**, *15*, 137-146.
43. Pinto, B. M.; Reimer, K. B.; Morissette, D. G.; Bundle, D. R. *J. Chem. Soc. Perkin Trans. 1* **1990**, 293-299.
44. Bundle, D. R.; Josephson, S. *J. Chem. Soc. Perkin Trans. 1* **1979**, 2736-2739.
45. Lemieux, R. U.; Koto, S. *Tetrahedron* **1974**, *30*, 1933-1944.
46. Mulard, L. A.; Clément, M.-J.; Segat-Dioury, F.; Delepierre, M. *Tetrahedron* **2002**, *58*, 2593-2604.
47. Blatter, G.; Beau, J.-M.; Jacquinet, J.-C. *Carbohydr. Res.* **1994**, *260*, 189-202.
48. Pinto, B. M.; Reimer, K. B.; Morissette, D. G.; Bundle, D. R. *J. Org. Chem.* **1989**, *54*, 2650-2656.
49. Pinto, B. M.; Reimer, K. B.; Morissette, D. G.; Bundle, D. R. *Carbohydr. Res.* **1990**, *196*, 156-166.

50. Anderson, P. W.; Pichichero, M. E.; Stein, E. C.; Porcelli, S.; Betts, R. F.; Connuck, D. M.; Korones, D.; Insel, R. A.; Zahradnik, J. M.; Eby, R. *J. Immunol.* **1989**, *142*, 2462-2468.
51. Mawas, F.; Niggemann, J.; Jones, C.; Corbet, M. J.; Kamerling, J. P.; Vliegenthart, J. F. G. *Infect. Immun.* **2002**, *70*, 5107-5114.
52. Buskas, T.; Li, Y.; Boons, G-J. *Chem. Eur. J.* **2004**, *10*, 3517-3524.
53. Peeters, C. C. A. M.; Lagerman, P. R.; der Weers, O.; Oomen, L. A.; Hoogerhout, P.; Beurret, M.; Poolman, J. T. In *Vaccine Protocols*; Robinson, A., Farrar, G., Wiblin, C., Eds.; Humana Press Inc.: Totowa N. J., 1996; Vol. 87, pp 111-133.
54. Svenson, S. B.; Lindberg, A. A. *Infect. Immun.* **1981**, *32*, 490-496.
55. Wright, K.; Guerreiro, C.; Laurent, I.; Baleux, F.; Mulard, L. A. *Org. Biomol. Chem.* **2004**, *2*, 1518-1527.
56. Bélot, F.; Guerreiro, C.; Baleux, F.; Mulard, L. A. *Chem. Eur. J.* **2005**, *11*, 1625-1635.
57. Hermanson, G. T. *Bioconjugate techniques*; Academic Press: New York, 1996.
58. Peeters, J. M.; Hazendonk, T. G.; Beuvery, E. C.; Tesser, G. I. *J. Immunol. Methods* **1989**, *120*, 133-143.
59. Boeckler, C.; Frisch, B.; Muller, S.; Schuber, F. *J. Immunol. Methods* **1996**, *191*, 1-10.
60. Cruz, L. J.; Iglesias, E.; Aguilar, J. C.; Quintana, D.; Garay, H. E.; Duarte, C.; Reyes, O. *J. Peptide Sci.* **2001**, *7*, 511-518.
61. Ward, J. I.; Zangwill, K. M. In *Vaccines*; Plotkin, S. A., Orenstein, W. A., Eds.; W. B. Saunders: Philadelphia, 1999, pp 183-221.
62. Lowry, O. H.; Rosebrough, N. J.; Farr, A. L.; Randall, R. J. *J. Biol. Chem.* **1951**, *193*, 265-275.
63. Worderwülbecke, S.; Cleverley, S.; Weinberger, S. R.; Wiesner, A. *Nature Methods* **2005**, *2*, 393-395.
64. Saksena, R.; Chernyak, A.; Karavanov, A.; Kovac, P. In *Recognition of carbohydrates in biological systems*; Lee, Y. C., Lee, R. T., Eds.; Academic Press: San Diego (USA), 2003; Vol. 632, pp 125-139.
65. Paradiso, P. R.; Dermody, K.; Pillai, S. *Vaccine Res.* **1993**, *2*, 239-248.
66. Falugi, F.; Petracca, R.; Mariani, M.; Luzzi, E.; Mancianti, S.; Carinci, V.; Melli, M. L.; Finco, O.; Wack, A.; Tommasco, A. D.; Magistris, M. T. D.; Costantino, P.; Giudice, G. D.; Abrignani, S.; Rappuoli, R.; Grandi, G. *Eur. J. Immunol.* **2001**, *31*, 3816-3824.
67. Alexander, J.; del Guercio, A.-F.; Maewal, A.; Qiao, L.; Fikes, J.; Chesnut, R. W.; Paulson, J.; Bundle, D. R.; DeFrees, S.; Sette, A. *J. Immunol.* **2000**, *164*, 1625-1633.
68. Alexander, J.; del Guercio, M.-F.; Frame, B.; Maewal, A.; Sette, A.; Nahm, M. H.; Newman, M. J. *Vaccine* **2004**, *22*, 2362-2367.

69. Brugghe, H. F.; Timmermans, H. A. M.; van Unen, L. M. A.; Hove, G. J. T.; der Werken, G. W.; Poolman, J. T.; Hoogerhout, P. *Int. J. Peptide Protein Res.* **1994**, *43*, 166-172.
70. Tam, J. P.; Spetzler, J. C., *Methods Enzymol.* **1997**, *289*, 612-637.
71. Pavliakova, D.; Chu, C.; Bystricky, S.; Tolson, N. W.; Shiloach, J.; Kaufman, J. B.; Bryla, D.; Robbins, J. B.; Schneerson, R. *Infect. Immun.* **1999**, *67*, 5526-5529.
72. Goebel, W. F. *J. Exp. Med.* **1939**, *69*, 353-364.
73. Benaissa-Trouw, B.; Lefeber, D. J.; Kamerling, J. P.; Vliegenthart, J. F. G.; Kraaijeveld, K.; Snippe, H. *Infect. Immun.* **2001**, *69*, 4698-4701.
74. Jansen, W. T. M.; Hogenboom, S.; Thijssen, M. J. L.; Kamerling, J. P.; Vliegenthart, J. F. G.; Verhoef, J.; Snippe, H.; Verheul, A. F. M. *Infect. Immun.* **2001**, *69*, 787-793.
75. Jansen, W. T. M.; Verheul, A. F. M.; Veeneman, G. H.; van Boom, J. H.; Snippe, H. *Vaccine* **2002**, *20*, 19-21.
76. Evenberg, D.; Hoogerhout, P.; van Boeckel, C. A. A.; Rijkers, G. T.; Beuvery, E. C.; van Boom, J. H.; Poolman, J. T. *J. Infect. Dis.* **1992**, *165*, S152-155.
77. Pozsgay, V. In *Adv. Carbohydr. Chem. Biochem.*; Horton, D., Ed.; Academic Press: San Diego, 2000; Vol. 56, pp 153-199.

Chapter 7

Automated Oligosaccharide Synthesis to Create Vaccines for Malaria and Other Parasites

Bridget L. Stocker, Alexandra Hölemann, and Peter H. Seeberger

Laboratory for Organic Chemistry, Swiss Federal Institute of Chemistry, Zurich, Switzerland

Malaria and leishmaniasis are two of the most devastating parasitic infections. Although much progress has been made towards the understanding and treatment of these diseases, an effective vaccine remains elusive. To this end, our attention focused on the synthesis of malaria and leishmaniasis glycoconjugates for use as vaccine candidates. The synthesis of the malarial vaccine involved developing an "anti-disease" vaccine. Synthesis of a key portion of the malarial toxin's mannan cap was accelerated using automated solid-phase oligosaccharide synthesis methodology and, following conjugation to the protein carrier, the toxin was tested as a vaccine candidate. A series of second generation anti-disease vaccine candidates were subsequently synthesized. The key tetrasaccharide cap of the lipophosphoglycans found on the surface of leishmaniasis parasites was also synthesized, using both solution and automated solid-phase methodology, and conjugated to a protein carrier. Initial vaccination studies for both glycoproteins were encouraging.

Introduction

Although automated oligopeptide *(1)* and oligonucleotide *(2)* synthesis is well established, only in recent years has an automated approach to oligosaccharide synthesis been introduced *(3-5)*. Solid-phase automated synthesis significantly accelerates carbohydrate assembly allowing for the assembly of complex structures within hours, rather than the weeks or months required for more traditional assembly. The carbohydrates are attached to functionalized Merrifield's resin through the reducing end of the sugar, a strategy commonly referred to as the acceptor-bound approach, and the oligosaccharide is constructed in a linear fashion using either glycosyl phosphates *(6,7)* or glycosyl trichloroacetimidates *(8)*. A modified automated peptide synthesizer allows for the automation of the process and by using an excess of reagents, that are subsequently removed by washing the resin, high coupling yields and minimal purification steps are required. Oligosaccharides prepared by automated methodology include those of the phytoalexin elicitor family of glucans *(9)*, oligorhamnosides *(10)*, N-linked complex-type glycoproteins *(10, 11)*, HIV-1 viral surface envelope glycoproteins *(12)*, the Lewis blood group oligosaccharides *(13)*, and the parasitic vaccine candidates against malaria and leishmaniasis discussed here.

Malaria is the most devastating tropical parasitic disease in the world and plagues many developing countries, particularly those in sub-Saharan Africa. Each year malaria infects 5-10% of humanity causing more than 300 million clinical cases *(14,15)*. The annual mortality rate is estimated to be between one and two million. Young children are particularly susceptible to the disease with 3000 children dying of malaria each day. Although the source of malarial infection, a unicellular (protozoal) parasite of the *Plasmodium* genus, has been known since the 1880's, the complexity of the malaria life cycle, the wide variety of immune responses induced by the malaria parasite, and an incomplete knowledge of protective immunity, have hindered vaccine development.

Four different *Plasmodium* species are known to infect humans and cause malaria: the more common *P. falciparum* (malaria tropica) and *P. vivax*, and the less common *P. malariae* (malaria tertiana) and *P. ovale* (malaria quartana). Clinical symptoms of malaria, including fever, anemia, vomiting, headache and flu-like symptoms appear 9 to 14 days after the initial infection. If untreated malaria rapidly progresses to severe anemia, convulsions and coma, eventually causing death by either destroying red blood cells (leading to fatal anemia) and/or by clogging the prevenous capillaries that carry blood to the brain (cerebral malaria) or other vital organs. *P. falciparum* causes most lethal infections, while parasites of the species *P.vivax* and *P. ovale* are able to persist in dormant stages (hypnozoites) in the liver for years - resulting in clinical relapses at regular intervals.

A limited number of drugs are available for the treatment of malaria. Quinine (an alkaloid initially isolated in the 17th century from the bark of cinchona trees) in combination with other antimalarial agents remains the therapy of choice despite some regional decreases in the clinical response of *P. falciparum* to quinine alone *(16,17)*. In 1934 the synthetic derivative chloroquine became the mainstay treatment for malaria, however nowadays, chloroquine-resistant strains of *P. faciparum* are common in all endemic areas. More recently, artemisinin and its derivatives have exhibited potent antimalarial activity with no observed resistance *(18)*. Nevertheless, due to the increasing prevalence of malaria, the emerging resistance against conventional drugs, and the expense of traditional treatments, the development of new antimalarial agents, particularly vaccine candidates, remains paramount.

Leishmaniasis is another widespread tropical disease and is endemic in 88 countries on four continents *(14,15)*. Three different forms of leishmaniasis occur in humans. Visceral leishmaniasis, also known as *kala azar*, causes swelling of the spleen and liver, and is often lethal when left untreated. The most common form of the disease, cutaneous leishmaniasis, results in debilitating skin lesions, while mucocutaneous leishmaniasis causes lesions in the mucous membranes, leading to facial disfigurement. The impact of leishmaniasis on public health has been greatly underestimated. Each year about 1.5 to 2 million new infections are reported, culminating in 6000 deaths. Although leishmaniasis is most prevalent in tropical settings, the disease has recently been diagnosed in overseas travelers and U.S Gulf War veterans *(19)*, and has emerged as an opportunistic infection of HIV patients *(20)*. It is feared that the disease is currently becoming endemic in the U.S.A. *(21)*.

Treatment of leishmaniasis requires lengthy therapy using expensive pentavalent antimony medications that are also known to exhibit toxic side effects. Moreover, drug resistance is emerging. Spread by the bite of infected sandflies, the parasite inhabits the macrophages, the part of the immune response designed to kill the invading organisms, thus presenting a challenge for therapeutic strategies. Several vaccine candidates are being explored, including whole killed antigens as well as surface antigens, however to date, no effective vaccine has been developed *(15)*.

Development of a Carbohydrate-based Vaccine against Malaria

Mechanisms of Malaria Pathogenesis

In 1886 Camillo Golgi developed the first truly modern proposition of the toxin theory of malarial pathogenesis by demonstrating that malarial fever is synchronous with the developmental cycle of the blood-stage parasite *(22, 23)*.

He and other turn-of-the-century 'Italian school' malariologists hypothesized that the proximal cause of the fever was a released toxin of parasite origin *(24)*. During the early years of the 20th century it was proposed that not only the fever, but also other pathological processes in malaria, were the result of a parasite toxin and it was suggested that this agent exerted systematic effects through the induction of endogenous mediators of host origin *(25)*. Later, the tumor necrosis factor (TNF) was identified as a major host mediator of disease *(26-29)*. Nowadays it is appreciated that a wide range of pathology in clinically severe malaria infections result from a systematic inflammatory cascade initiated by the action on host tissue of toxin-induced TNF and the production of related pyrogenic cytokines, such as interleukin-1 (IL-1) and interlekin-6 (IL-6) *(30)*.

Anti-Malarial Carbohydrate-Based Vaccines

In contrast to other diseases, humans are only able to develop partial immunity to malaria *(31)*. In areas of high transmission, immunity to malaria is acquired in two stages: an initial phase of clinical immunity occurs despite persistent high parasitemias, followed some years later by an antiparasite immunity which limits parasite numbers, replication and burden within the host. Vaccines designed to provide protective immunity against malaria by lowering the parasite burden have several disadvantages. Evidence for considerable redundancy in invasion pathways, immune evasion strategies, and problems of major histocompatibility complex (MHC)-linked genetic restrictions in the immune response to the parasiticidal antigens, complicate vaccine development *(30)*. Antigenic diversity and variation may also enhance evasion of immunity and breakthrough parasites. For these reasons, multi-component vaccines are desirable, and a great deal of effort is currently devoted to the expression of recombinant protein or DNA vaccine constructs, synthetic peptides, dimeric proteins, and viral-vectored constructs. Many have been tested in clinical or preclinical trials *(15)*, however to date, no effective anti-malarial vaccine or vaccine candidate is available *(32)*.

A conceptually different approach thus focuses on the development of 'anti-disease' vaccines *(30)*. Here patients are immunized against those parasite products known to cause host pathology with the goal of reducing morbidity and mortality rather than providing for complete immunity. Anti-disease vaccines are currently used for immunization against tetanus and diphtheria, hence providing plausibility of the concept and further encouraging the search for the pathogenic malarial toxin.

Identification of the Malarial Toxin

As TNF is believed to play a central role in the etiology of severe malaria, the production of this cytokine *in vitro* is often taken as a surrogate marker of parasite toxicity, and as an endpoint for tracking the purification of toxin activity. In 1992 it was proposed that the toxin should be a phospholipid *(33)*, and a few years later a toxin of *Plasmodium falciparum* origin, shown to induce macrophages to secrete TNF, was isolated and assigned *(30, 34-40)*. The toxin (Figure 1), which belongs to the class of glycolipids known as glycolsylphosphatidylinositols (GPIs), represents at least 95% (conservative estimate) of the total carbohydrate modification of *P. falciparum* schizonts *(41-43)*. Though not yet proven, loss of GPI is likely to be a lethal mutation *(30)*. Recent evidence indicates the existence of an unsaturated fatty acyl chain (*cis*-vaccenic acid) in the *sn*-2 position. Further exploration revealed lipid chains with high structural diversity, in turn leading to remarkable differences in the activity of the GPIs *(44)*. Deacetylation using chemical or enzymatic hydrolysis makes the carbohydrate moiety non-toxic, highlighting the importance of both the lipid and carbohydrate portions *(34-40)*.

Automated Oligosaccharide Synthesis to Create a Vaccine against Malaria

The first chemical synthesis of the malarial GPI glycan **1** (Scheme 1), without lipid residues, was accomplished using a linear solution-phase approach *(45)*. Later a more rapid assembly of **1** was achieved by combining automated solid-phase synthesis and solution phase fragment coupling *(46)*. Here the GPI was dissected into two fragments: the tetra-mannosyl fragment **2** containing a pentenyl **2a** or imidate **2b** at the reducing end, rapidly prepared using solid-phase methodology, and disaccharide **3** *(45)*, synthesized in solution. While it would have been ideal to prepare the entire carbohydrate skeleton on solid-phase, the challenging α-linkage between the inositol and glucosamine prevented a fully automated approach.

The automated synthesis of tetrasaccharide **2** used the octenediol functionalized Merrifields resin **4** and the four readily available mannosyl trichloroacetimidate building blocks, **5 – 8** (Scheme 2). To ensure high coupling efficiencies a double glycosylation strategy was employed with the imidate being used in a 5-fold excess for each glycosylation. Mannosyl trichloroacetimidate **5** was first coupled to the resin - with each glycosylation taking 20 minutes using 0.5 equivalents of TMSOTf - then the temporary acetate protecting group was removed using two times eight equivalents of sodium methoxide (NaOMe). This protocol was repeated with each of the successive imidates, **6, 7** and **8**, to give the resin-bound tetrasaccharide in a total assembly time of less than six hours. Cleavage of the octenediol linker, using Grubbs' catalyst in an atmosphere of ethylene, furnished the *n*-pentenyl tetrasaccharide **2a**.

Analysis of the crude reaction product by HPLC revealed the presence of two major products: the desired tetrasaccharide **2a** (44% relative area), and deletion sequences (15% relative area). Purification by HPLC afforded **2a** as the expected α/β mixture at the reducing-end mannose with all other linkages being

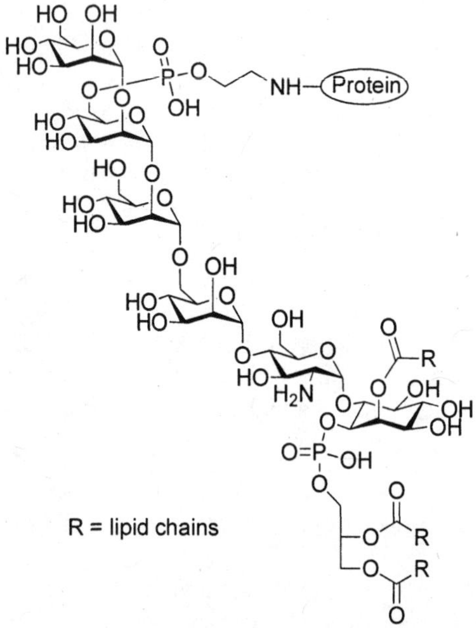

Figure 1. The malarial toxin

exclusively α. Initially it was anticipated that **2a** could be coupled directly to **3** using NIS, TESOTf activated glycosylations *(47)*, however an exploratory reaction between **3** and a model pentenyl mannoside monomer gave no coupling product. Consequently, **2a** was converted to imidate **2b** before being coupled with disaccharide **3** to give hexasaccharide **9** in modest yield (Scheme 3) *(45)*.

Conversion of hexasaccharide **9** to the malaria toxin **1** commenced with cleavage of both the triisopropyl and the acetonide substituents using 0.5 M HCl in methanol (Scheme 4). Temporary protection of the primary hydroxyl with *tert*-butyldimethylsilyl (TBS) allowed for the regioselective installation of the cyclic phosphate on the inositol core. Removal of the TBS group, followed by phosphorylation with phosphoramidite **10** and oxidation gave the corresponding *bis*-phosphate **11** as a mixture of diastereomers. Removal of the β-cyanoethoxy group with DBU and the simultaneous global deprotection of the benzyl ethers, the carbamate, and the azide using Birch conditions, gave the desired toxin **1**.

To prepare an immunogen, the synthetic pseudosaccharide **1** was treated with 2-iminothiolane **12**, to introduce a sulfhydryl at the ethanolamine, and then conjugated to maleimide-activated ovalbumin (OVA), in a molar ratio of 3.2:1, or keyhole limpet haemocyanin (KLH), in a molar ratio of 191:1 (Scheme 5). These adducts were then used in mouse immunization studies.

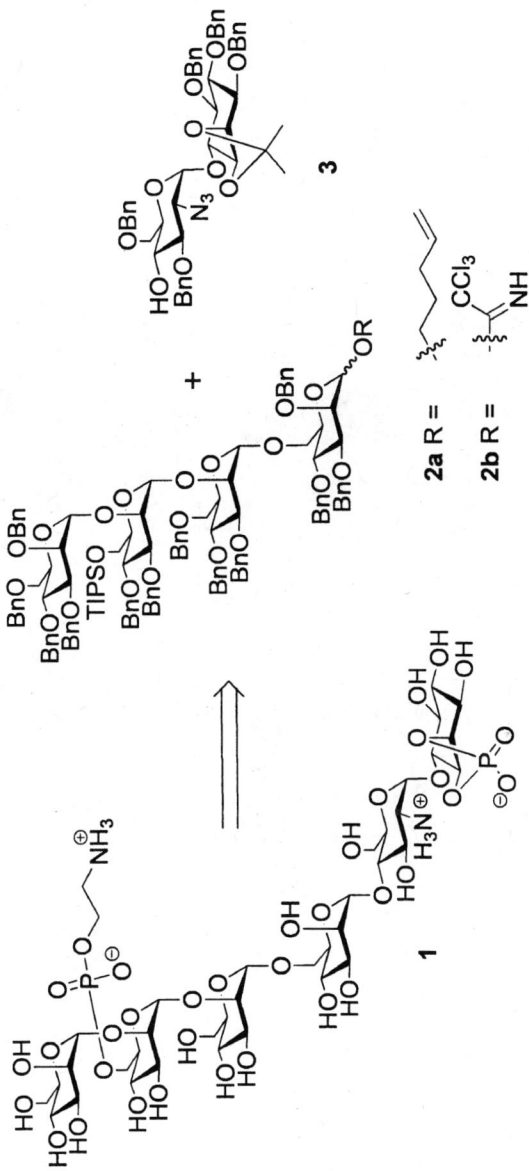

Scheme 1. Retrosynthesis of the GPI malaria toxin

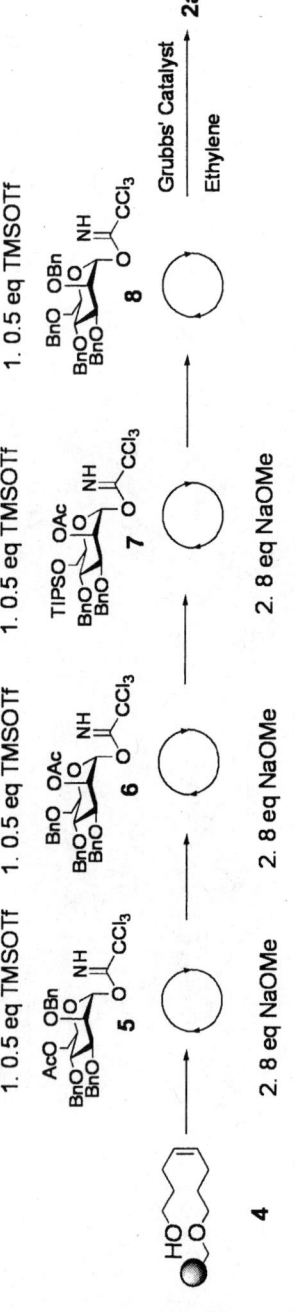

Scheme 2. The automated synthesis of the tetra-mannan cap

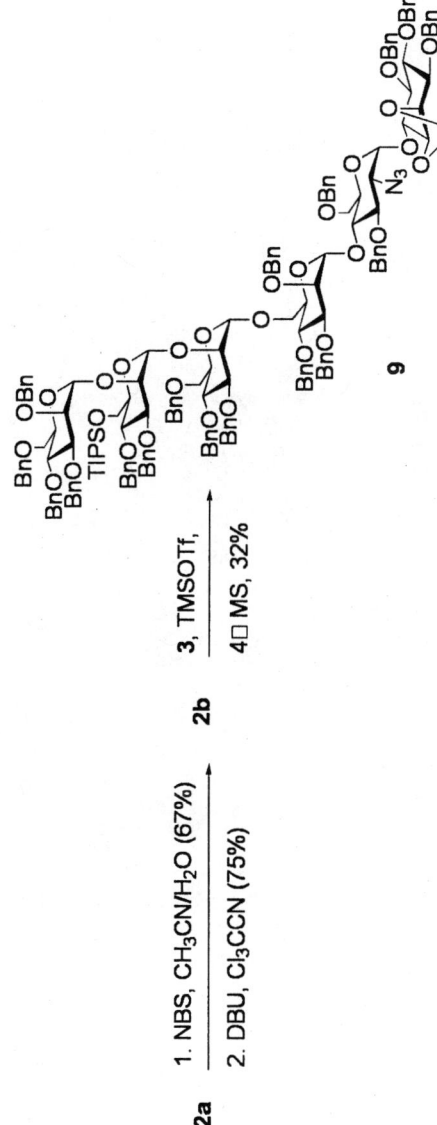

Scheme 3. Synthesis of the malaria hexasaccharide

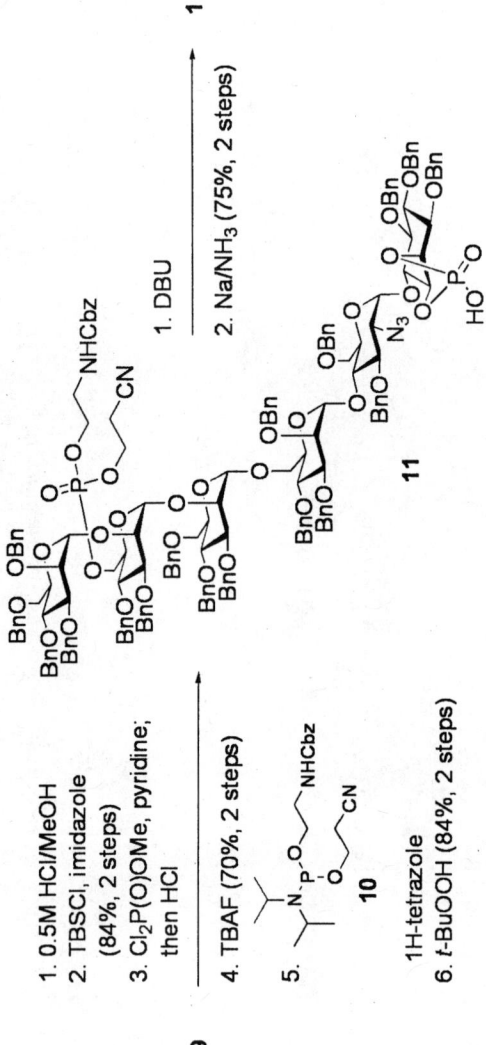

Scheme 4. Synthesis of malarial toxin 1

147

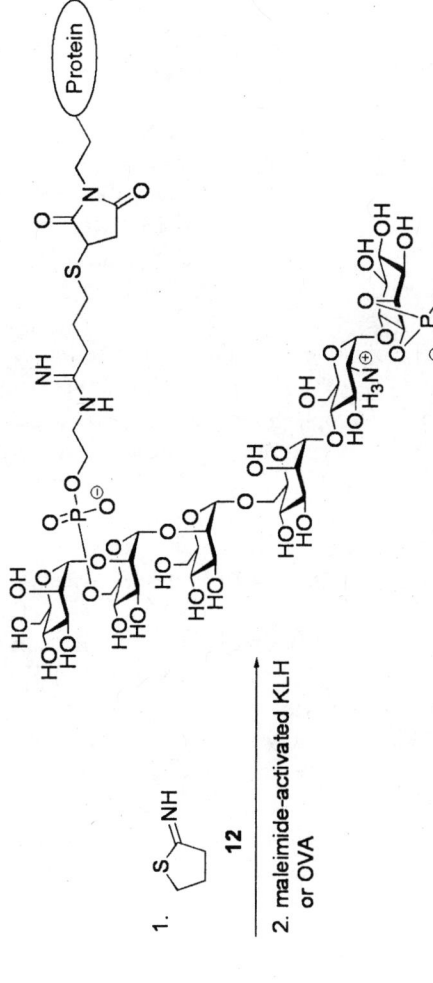

Scheme 5. Synthesis of the malarial vaccine candidates

Immunogenic Testing

Gratifyingly, the synthetic malarial GPI glycan was immunogenic in mice *(45)*. Antibodies from animals immunized with the KLH-glycan gave positive immunoglobulin-γ (IgG) titres against the OVA-glycan, yet no reactivity was observed in pre-immune sera or in animals receiving sham-conjugated KLH or OVA. Notably, anti-glycan IgG bound to native *P. falciparum* GPI. Anti-GPI however, failed to bind to uninfected erythrocytes despite the expression of endogenous GPIs by mammalian cells. This lack of cross-reactivity may arise as a result of differences between the malarial GPI and the mammalian GPI core glycan *(48)*. Although encouraging, it must however be kept in mind that non-binding to endogenous GPI does not exclude the possibility of serological cross-reactions with other tissues. In erythrocytes infected with *P. falciparum*, anti-GPI glycan IgG detected multiple molecular species, consistent with the presence in mature parasites of multiple GPI-modified proteins and their processing products. Thus protein-specific features do not greatly influence the binding of anti-glycan IgG to native GPI anchors. Furthermore, antibodies from mice immunized with the KLH-glycan were able to neutralize the tumor-nercosis factor (TNF)-α level from macrophages induced by crude total extracts of *P. falciparum*. The GPI therefore appeared sufficient, and necessary, for the induction of the malarial parasites hosts' pro-inflammatory response.

Additional immunogenic studies were performed using murine *P. berghei* ANKA malaria - the best available small animal model of clinically sereve malaria *(49-50)*. C57BL6/J mice primed and boosted twice with 6.5 μg of the KLH-glycan, or the KLH-cysteine, in Freud's adjuvant were challenged with *P. berghei* ANKA. At day 12, a 0–8.7% survival rate for all sham-immunized and naïve control mice was observed while those mice immunized with the synthetic KLH-glycan had a 58 to 75% survival rate ($n = 50$, 5 separate experiments). All dead mice showed severe neurological damage – observations consistent with the malarial toxin being the causative agent of death. There were no differences between the naïve and sham-induced mice indicating that exposure to KLH in Freud's adjuvant did not influence the rate of the disease. Parasitaemia levels were not significantly different between the immunized and control mice, demonstrating that the prevention of fatality by anti-GPI vaccination does not operate through effects on parasite replication.

Thus, synthetic GPIs conjugated to a protein were shown to act as an anti-toxin vaccine candidate against malaria by providing significant protection against malaria fatalities and pathogenesis. Further immunological and clinical studies using the conjugated GPI adduct **1** are currently being performed.

The Synthesis of Second Generation Malaria Vaccine Candidates

In view of the aforementioned immunological studies, a series of second generation vaccine candidates were developed. As the native malaria toxin is linked to the cell membrane via an inositol phosphate diester, it was envisioned that presenting the antigen in the proper orientation, such as in **13**, could result in better vaccination (Fig. 2). Again, a convergent and modular solution-phase approach, utilizing a [4+2] glycosylation strategy, similar to that described for the synthesis of toxin **1**, was used *(51)*. The direct incorporation of a thiol group onto the inositol moiety allows for the rapid conjugation of the glycans to carrier proteins and for convenient screening on carbohydrate chips *(52)*. The fully lipidated GPI oligosaccharide **14** was also constructed *(53, 54)*.

Malarial GPI adducts containing differing numbers of mannose residues have also been synthesized (Fig. 3) *(55)*. These compounds are currently being used as molecular tools for the examination of the biosynthesis, antigenicity, and serology of GPIs, and will provide the basis for the first detailed structure/activity relationship studies of GPI toxins *(56)*. A better understanding of malarial pathogenesis will undoubtedly aid in the discovery of an even more effective anti-toxin vaccine.

In summary, a variety of efficient synthetic methods for the preparation of malaria GPI glycans have been developed. Using these strategies sufficient quantities of pure oligosaccharides can be produced for biochemical, biological, immunological, and medicinal investigations. In all of the aforementioned strategies, synthesis of the mannan caps is amenable to automated solid-phase synthesis.

Development of a Carbohydrate-based Vaccine against Leishmaniasis

Vaccine Design Considerations

Lipophosphoglycans (LPGs) are the ubiquitous epitopes found on the surface of Leishmania parasites (Fig. 4) *(57)*. The LPG contains three constituents: a glycosylphosphatidylinositol (GPI) anchor, a repeating phosphorylated disaccharide, and a tetrasaccharide cap. Different leishmania species are distinguished by minor modifications in their backbone and by the number of repeating units. The phosphoglycan portion of the LPG has been shown to be a disease-promoting antigen *(58-63)* sparking much interest in the LPGs. However the size (MW *ca.* 1200) and the heterogeneity of the LPG precludes its use as a vaccine candidate. Consequently, several fragments of the Leishmania LPG have been investigated for their antigenicity including

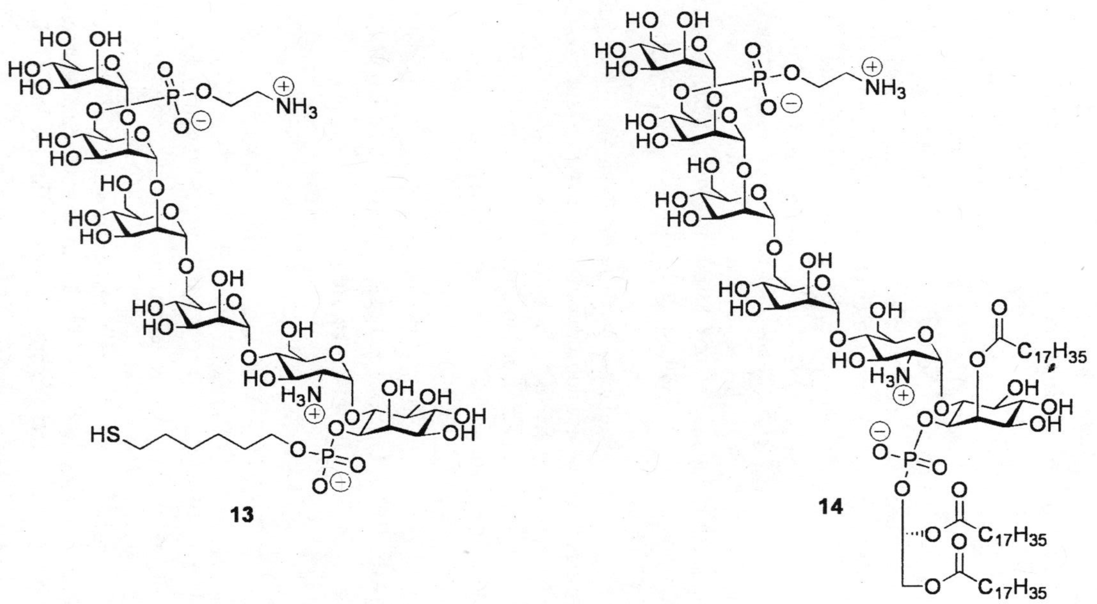

150

Figure 2. Second generation malaria GPIs

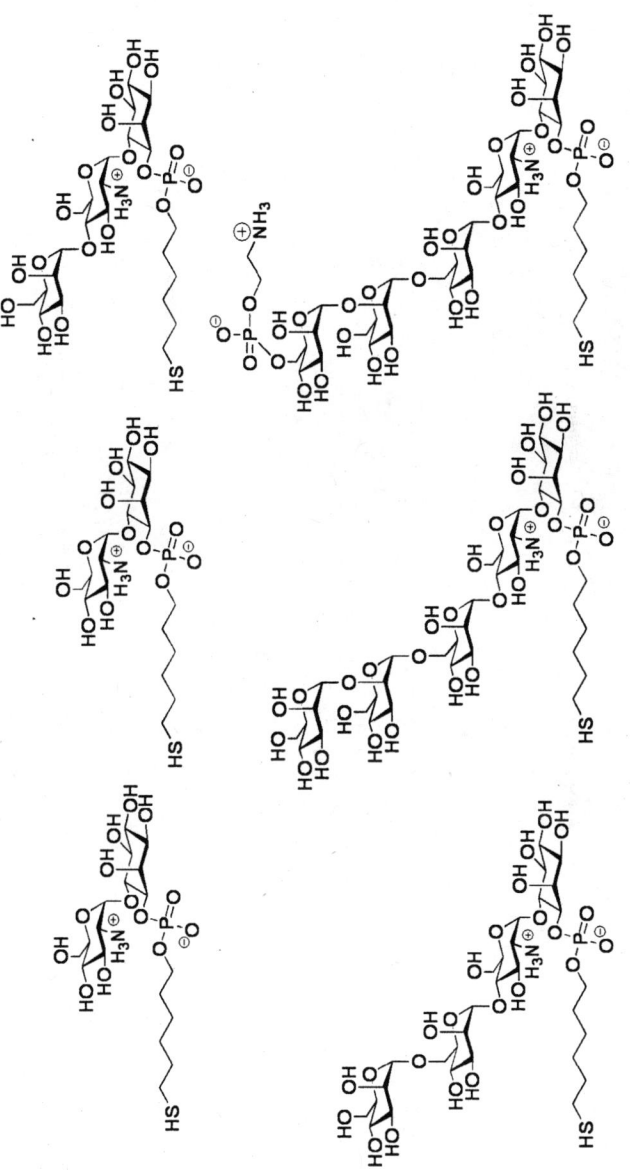

Figure 3. Target malaria GPI glycans of differing lengths

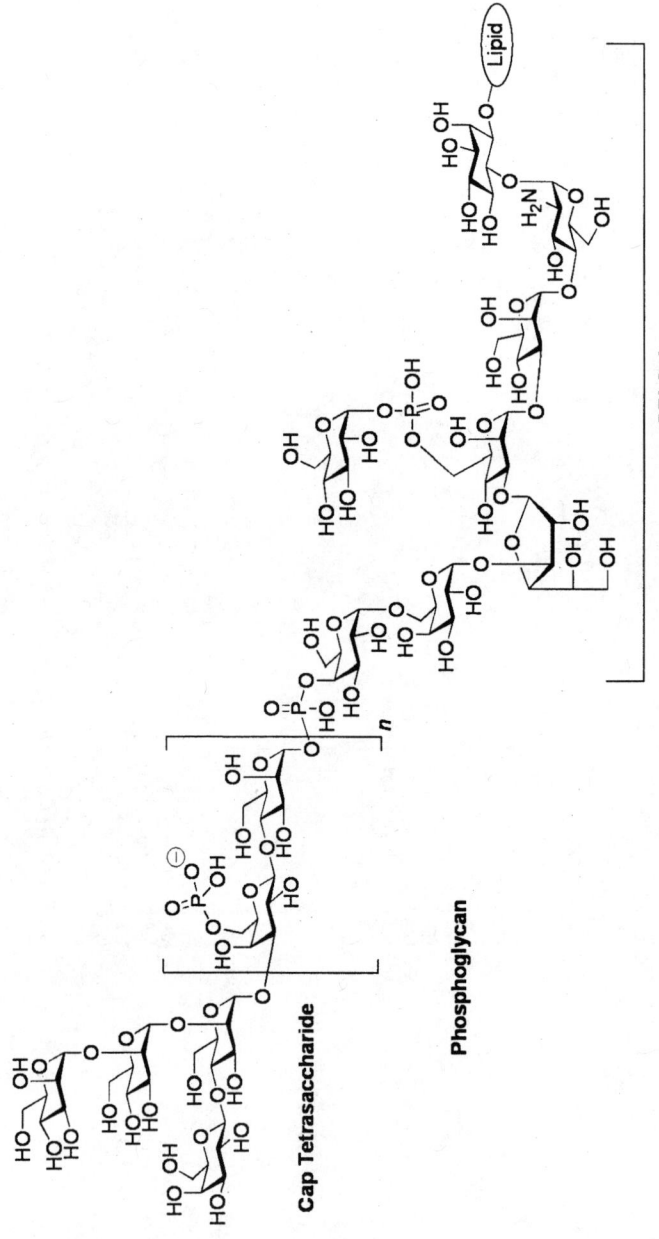

Figure 4. Leishmania lipophosphoglycan

phosphoglycan fragments *(64, 65)*, a GPI heptasaccharyl *myo*-inositol adduct *(66)*, and the tetrasaccharide cap containing the unusual galactose β-(1→4) mannosidic linkage *(67-70)*.

Carbohydrate-Based Vaccine Constructs against Leishmaniasis

The first chemical synthesis of the protected LPG tetrasaccharide cap was accomplished using the *n*-pentenyl glycoside (NPG) protocol, in both a linear and convergent approach *(69)*. With respect to yield, the convergent strategy, whereby a pentenyl α-(1→2)-mannosyl-α-mannoside was coupled to the appropriately protected 1,2-dibromopentyl β-(1→4)-galactosyl-α-mannoside, was the method of choice. A similar disconnection pattern, incorporating a dimannan trichloroacetimidate building block and the Gal-Man methyl glycoside (synthesized from the corresponding Gal-Glu disaccharide), yielded the fully deprotected methyl tetrasaccharide cap *(70)*. When converting the glucose sugar to the required galactose configuration radiolabelled NaB^3H_4 was used allowing for the isotopic labelling of the glycan.

A linear, efficient solution phase synthesis of the tetrasaccharide has been achieved using thioglycoside building blocks. Key in this synthesis was the conversion of hexabenzyl lactal **15** *(71)* to the ethyl thiolactoside **16** following treatment with 2,2'-dimethyldioxirane then ethanethiol (Scheme 6) *(67)*. A two-step oxidation-reduction process, to invert the stereochemistry at the C2 position, followed by pivaloylation gave the desired thioethyl galactose-β-(1→4)-mannose **17**. Coupling to the *N*-protected aminohexanol **18**, removal of the pivaloyl group, mannosylation with **19**, pivaloyl cleavage, a second mannosylation with **19** and global deprotection gave the desired tetrasaccharide **20**.

Two Leishmania immunogens were then prepared by joining the tetrasaccharide antigen **20** with the immmunostimulator tripalmitoyl-*S*-glycerylcysteine (Pam₃Cys) and with the keyhole limpet haemocyanin (KLH) (Scheme 7). The fully synthetic immunogen **21** was synthesized in 60% yield using HATU/HOAt activation *(72, 73)*, and purified by normal phase silica gel chromatography. Semisynthetic vaccine **22** was prepared by condensation of **20** with *S*-acetylmercaptoacetate pentafluorophenyl ester (SAMA-OPfp) and subsequent conjugation to the bromoacetate-modified KLH. Both constructs are currently undergoing initial immunological evaluation in Balb-C mice and have demonstrated an ability to elicit strong B- and T-cell responses *(74)*.

The first automated synthesis of the LPG tetrasaccharide cap employed a glycosyl phosphate and several glycosyl trichloroacetimidate building blocks containing different ester protecting groups to achieve branching (Scheme 8) *(68)*. This methodology built on work conducted in a previous solid-phase study *(67)*. Each coupling cycle relied on double glycosylation, to ensure high

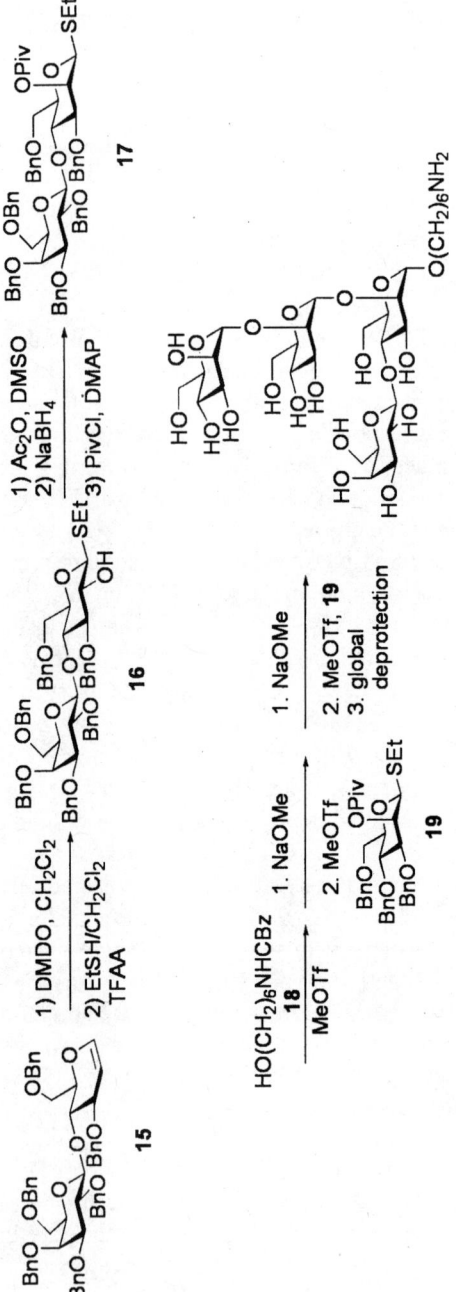

Scheme 6. Solution phase synthesis of the fully deprotected leishmania cap

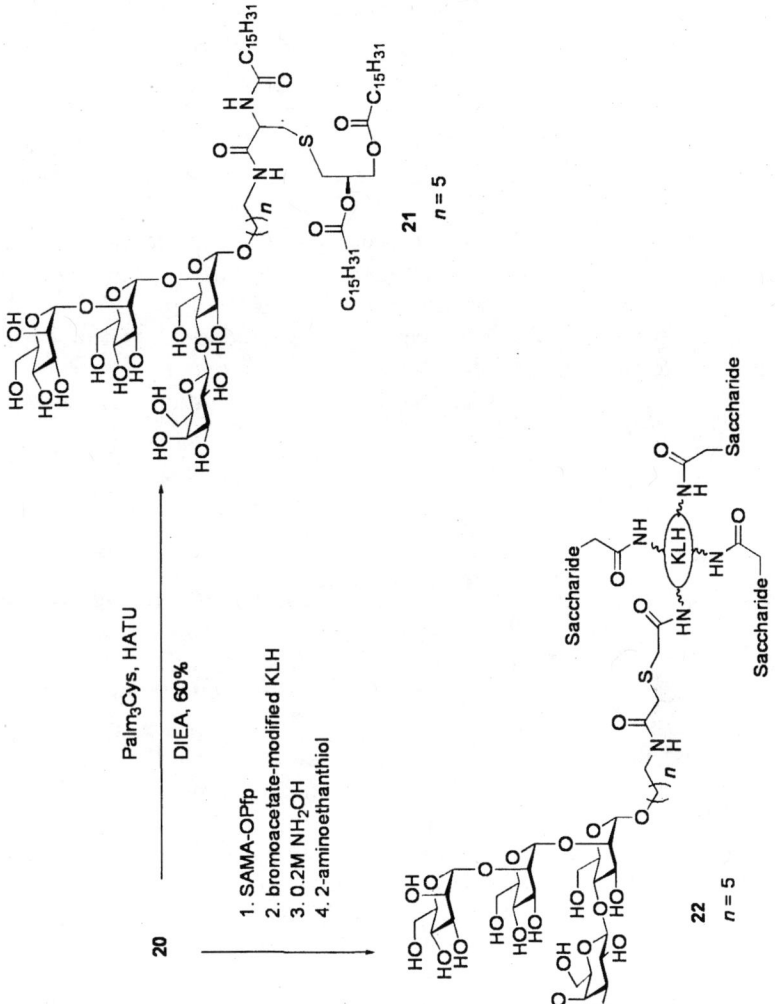

Scheme 7. Formation of synthetic and semisynthetic leishmaniasis vaccine constructs

coupling efficiencies, and a single deprotection event. Imidate **23** was first coupled to the octenediol-functionalized Merrifield's resin using catalytic TMSOTf and the levulinate ester was removed upon exposure to hydrazine. Formation of the central β-(1→4) linkage was achieved by reaction of the resin-bound alcohol with phosphate **24** and stoichiometric amounts of TMSOTf. Selective removal of the acetate was achieved using NaOMe and the resulting alcohol mannosylated with imidate **25**. Repetition of the acetate cleavage and mannosylation with **26** completed the assembly of the resin-bound tetrasaccharide in a total assembly time of approximately 9 hours. The tetrasaccharide was then cleaved from the resin, following exposure to Grubbs' catalyst and ethylene, to provide the crude *n*-pentenyl glycosidic tetrasaccharide **27**.

HPLC analysis of the crude reaction products revealed the presence of three major peaks: the styrene, the trisaccharide (*n-1*) deletion sequence (19% relative peak area), and the desired tetrasaccharide **27** (50% relative peak area). Purification by preparative HPLC yielded the pure tetrasaccharide. The terminal *n*-pentenyl glycoside may then be further elaborated into an aldehyde or carboxylic acid that would allow for attachment to immunostimulators such as proteins or lipopeptides.

Conclusion and Outlook

Parasitic infections, such as malaria and leishmaniasis, are major public health concerns particularly for developing countries. Vaccines are the most effective means by which to control these diseases. Immunization of mice with a GPI malarial-KLH vaccine construct has shown promising results in initial studies – leading to an increased survival rate of vaccinated mice compared to sham-immunized mice – and has prompted the search for new and better second generation vaccine candidates. These second generation vaccine candidates are also progressing through preclinical studies towards clinical evaluation. Leishmaniasis immunogens have also been prepared by coupling the tetrasacharide cap of the leishmaniasis LPG to a protein (KLH) or a semi-synthetic lipid carrier (Pam$_3$Cys).

The advent of automated solid-phase oligosaccharide synthesis allows for the rapid preparation of many vaccine adducts, including the malarial and leishmaniasis oligosaccharides discussed herein. Although much work remains, the ultimate goal is to develop automated solid-phase oligosaccharide synthesis to such a level that it allows for the rapid preparation of most biologically important oligosaccharides and is suitable for use by non-specialist laboratories.

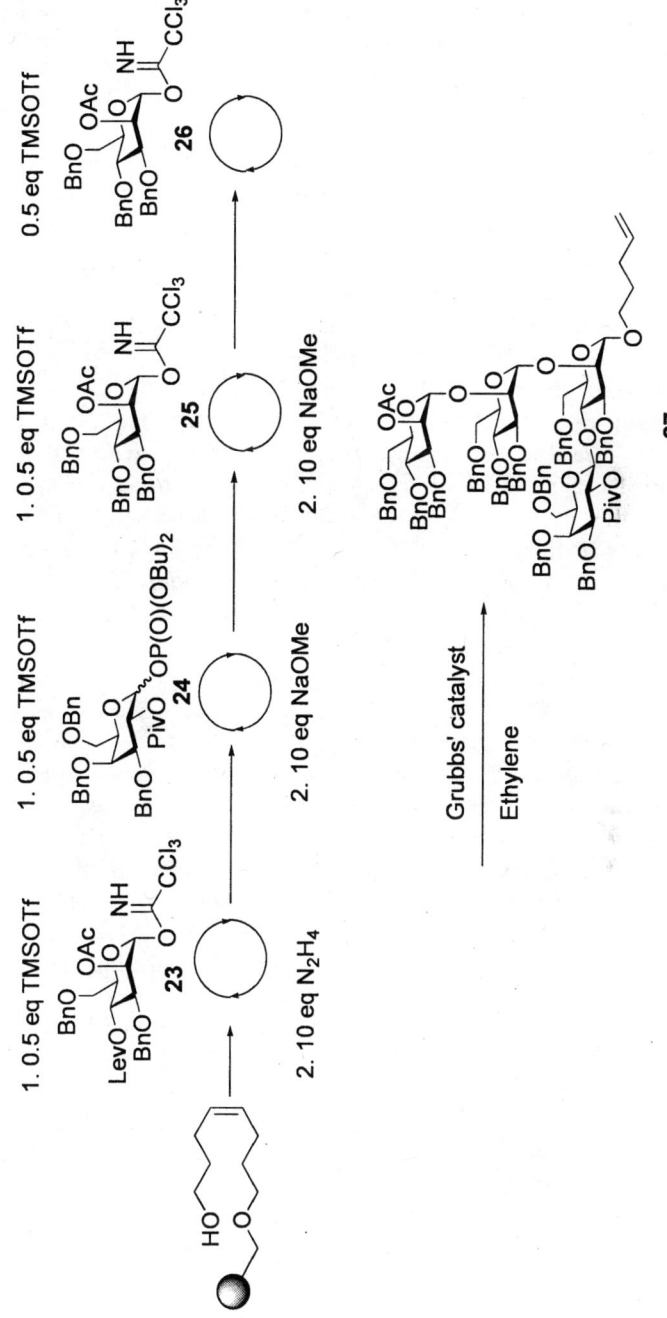

Scheme 8. *Automated synthesis of leishmania cap tetrasaccharide 27*

Experimental

***n*-Pentenyl 2,3,4,6-tetra-*O*-benzyl-α-D-mannopyranosyl-(1→2)-3,4-di-*O*-benzyl-6-*O*-triisopropylsilyl-α-D-mannopyranosyl-(1→6)-2,3,4-tri-*O*-benzyl-α-D-mannopyranoside (2a).** Octenediol-functionalized resin 4 (50 μmol, 50 mg, 1.00 mmol/g loading) was loaded into a reaction vessel and inserted into a modified ABI-433A peptide synthesizer. The resin was glycosylated using donor 5 (5 equiv., 0.25 mmol, 160 mg loaded into cartridges) delivered in CH_2Cl_2 (3 mL) and TMSOTf (0.5 equiv., 2.0 mL, 0.0125 M TMSOTf in CH_2Cl_2) at room temperature. Mixing of the suspension was performed (10 s vortex, 50 s rest) for 15 min. The resin was then washed with CH_2Cl_2 (6 x 4 mL) and the glycosylation repeated (double glycosylation). Deprotection of the acetyl ester was carried out by treating the glycosylated resin with sodium methoxide (8 equiv., 0.5 mL, 0.75 M NaOMe in MeOH) in CH_2Cl_2 (5 mL) for 30 minutes at room temperature. The resin was then washed with CH_2Cl_2 (4 mL) and subjected to the deprotection conditions a second time for 30 min. The deprotected polymer-bound C6-OH monosaccharide was then glycosylated using building block 6 (5 equiv., 0.25 mmol, 160 mg loaded into cartridges) delivered in CH_2Cl_2 (3 mL) and TMSOTf (0.5 equiv., 2.0 mL, 0.0125 M TMSOTf in CH_2Cl_2) at room temperature. The resin was then washed with CH_2Cl_2 (6 x 4 mL) and the glycosylation repeated (double glycosylation). Deprotection of the acetyl ester was carried out by treating the glycosylated resin with sodium methoxide (8 equiv., 0.5 mL, 0.75 M NaOMe in MeOH) in CH_2Cl_2 (5 mL) for 30 minutes at room temperature. The resin was then washed with CH_2Cl_2 (4 mL) and subjected to the deprotection a second time for 30 min. The deprotected polymer-bound disaccharide was then glycosylated using building block 7 (5 equiv., 0.25 mmol, 175 mg loaded into cartridges) delivered in CH_2Cl_2 (3 mL) and TMSOTf (0.5 equiv., 2.0 mL, 0.0125 M TMSOTf in CH_2Cl_2) at room temperature. Mixing of the suspension was performed (10 s vortex, 50 s rest) for 15 min. The resin was then washed with CH_2Cl_2 (6 x 4 mL) and the glycosylation repeated (double glycosylation). Deprotection of the acetyl ester was carried out by treating the glycosylated resin with sodium methoxide (8 equiv., 0.5 mL, 0.75 M NaOMe in MeOH) in CH_2Cl_2 (5 mL) for 30 minutes at room temperature. The resin was then washed with CH_2Cl_2 (4 mL) and subjected to the deprotection a second time for 30 min. The deprotected polymer-bound triisaccharide was then glycosylated using building block 8 (5 equiv., 0.25 mmol, 171 mg loaded into cartridges) delivered in CH_2Cl_2 (3 mL) and TMSOTf (0.5 equiv., 2.0 mL, 0.0125 M TMSOTf in CH_2Cl_2) at room temperature. Mixing of the suspension was performed (10 s vortex, 50 s rest) for 15 min. The resin was then washed with CH_2Cl_2 (6 x 4 mL) and the vessel removed from the synthesizer.

The glycosylated resin (50 µmol was dried *in vaccuo* over phosphorous pentoxide for 12 h and transferred to a round bottom flask. The flask was purged with ethylene and Grubbs' catalyst (bis(tricyclohexylphosphine)benzylidine ruthenium (IV) dichloride, 12 mg, 30 mol%) was added. The reaction mixture was diluted with CH_2Cl_2 (3 mL) and stirred under 1 atm ethylene for 36 h. Triethylamine (100 µL, 160 equiv.) and tris(hydroxymethyl)phosphine (50 mg, 80 equiv.) were added and the resulting solution stirred at room temperature for 1 h. The pale yellow reaction mixture was diluted with CH_2Cl_2 (5 mL) and washed with water (3 x 5 mL). The aqueous phase was extracted with CH_2Cl_2 (3 x 5 mL) and the combined organic phase dried over Na_2SO_4, filtered and concentrated. The crude product was analyzed by HPLC using a Waters model 600 pump and controller coupled to a Waters model 2487 dual λ absorbance detector.

Analytical HPLC was performed on a Waters Nova-Pak© silica column (3.9 x 150 nm) using a gradient of 5→20% EtOAc/Hexanes (20 min) and a flow rate of 2.5 mL/min, monitoring at 260 nm. Fractions collected during semi-preparative HPLC were checked by analytical HPLC for purity. Clean fractions were concentrated to give **2a** as a clear oil. $[\alpha]^{24}_D$ +21.7° (*c* 0.63, CH_2Cl_2); IR (thin film) 2361, 2338, 1095, 668 cm^{-1}; ^1H NMR ($CDCl_3$, 500 MHz): δ 7.35-7.03 (m, 60H), 5.77-5.71 (m, 1H), 5.31 (s, 1H), 5.21(s, 1H), 4.98-4.86 (m, 7H), 4.82 (d, J = 10.7 Hz, 1H), 4.77 (s, 1H), 4.72-4.61 (m, 5H), 4.60-4.53 (m, 8H), 4.50-4.46 (m, 6H), 4.44 (d, J = 12.2 Hz, 1H), 4.36 (d, J = 11.6 Hz, 1H), 4.28 (d, J = 12.2 Hz, 1H), 4.05 (t, J = 9.5 Hz, 1H), 4.01-3.98 (m, 1H), 3.96-3.84 (m, 11 H), 3.82-3.65 (m, 7H), 3.64-3.49 (m, 8H), 3.32-3.27 (m, 1H), 2.08-2.00 (m, 4H), 1.57-1.55 (m, 2H), 1.10-1.02 (m, 24 H); HSQC 13C (125 MHz)/^1H (500 MHz): 100.5/5.31, 73.1/4.15, 80.1/3.94, 99.3/4.90, 75.1/4.16, 80.7/3.89, 98.0/4.77, 74.8/3.75, 80.7/3.88, 99.8/5.21, 74.9/3.86, 71.2/3.69, 66.8/3.59; ESI MS *m/z* ($M^+ + Na^+$) calcd: 1993.971, found 1993.975.

Acknowledgements

We are grateful to all present and past members of the Seeberger group as well as to our collaborators contributing to the results reported herein. We also thank the ETH, the Swiss National Science Foundation (Grant Nr. 200021-101593, "Automated Solid-Phase Synthesis of Oligosaccharides"), the Foundation of Research, Science and Technology, New Zealand (Postdoctoral Fellowship to B. L. S) and the Deutsche Forschungsgemeinschaft (DFG, Emmy Noether Postdoctoral Fellowship to A. H.).

References

1. Merrifield, R. B. *Angew. Chem. Int. Ed. Engl.* **1985**, *97*, 801-812.
2. Caruthers, M. H. *Science* **1985**, *230*, 281-285.
3. Plante, O. J.; Palmacci, E. R.; Seeberger, P. H. *Science* **2001**, *291*, 1523-1527.
4. Seeberger, P. H. *Chem. Comm.*, **2003**, 1115-1121.
5. Palmacci, E. R.; Plante, O. J.; Hewitt, M. C.; Seeberger, P. H. *Helv. Chim. Acta* **2003**, *86*, 3975-3990.
6. Palmacci, E. R.; Plante, O. J.; Seeberger, P. H. *Eur. J. Org. Chem.* **2002**, 595-606.
7. Plante, O. J.; Andrade, R. B.; Seeberger, P. H. *Org. Lett.* **1999**, *1*, 211-214.
8. Schmidt, R. R.; Kinzy,W. *Adv. Carbohydr. Chem. Biochem.* **1994**, *50*, 21-123.
9. Plante, O. J.; Palmacci, E. R.; Seeberger, P. H. *Science* **2001**, *291*, 1523-1527.
10. Palmacci, E. R.; Plante, O. J.; Hewitt, M. C.; Seeberger, P. H. *Helv. Chim. Acta* **2003**, *86*, 3975-3990.
11. Ratner, D. M.; Swanson, E. R.; Seeberger, P. H. *Org. Lett.* **2003**, *5*, 4717-4720.
12. Ratner, D. M.; Plante, O. J.; Seeberger, P. H. *Eur. J. Org. Chem.* **2002**, 826-833.
13. Routtenberg-Love, K.; Seeberger, P. H. *Agnew. Chem. Int. Ed. Engl.* **2004**, *116*, 612-615.
14. The World Heath Organisation, *World Health Report* **2004**. http://www.who.int/whr/2004/en/report04_en.pdf
15. World Health Organization, Geneva, *State of the Art of New Vaccines Research and Development, Initiative for Vaccine Research*, **2003**.
16. Pukrittaayakamee, S.; Supanaranond, W.; Looareesuwan, S.; Vanijanonta, S.; White, N. J. *Trans. R. Soc. Trop. Med. Hyg.* **1994**, *88*, 324-327.
17. Zalis, M. G.; Pang, L.; Silveria, M. S.; Milhous, W. K.; Wirth, D. F. *Am. J. Trop. Med. Hyg.* **1998**, *58*, 630-637.
18. Klymann, D. L. *Science* 1985, **228**, 1049-1055.
19. Magill, A. J.; Grogl, M.; Gasser, R. A.; Sun, W.; Oster, C. N. *N. Engl. J. Med.* **1993**, *328*, 1383-1387.
20. Alvar, J.; Canavate, C.; Gutierrez-Solar, B.; *et. al. Clin. Microbiol. Rev.* **1997**, *10*, 298-319.
21. Enserink, M. *Science* **2000**, *290*, 1881-1883.
22. Golgi, C. *Arch. Sci. Med. (Torino)*, **1886**, *10*, 109-135.
23. Golgi, C. *Arch. Sci. Med. (Torino)*, **1889**, *13*, 173-196.
24. Marchiafava, E.; Bignami, A. *On Summer-Autumn Malaria Fevers.* London, New Sydenham Society, **1894**.

25. Maegraith, B. G. *Pathological Processes in Malaria and Blackwater Fever.* Oxford, Blackwell, **1948**.
26. Clark, I. A. *Lancet* **1978**, *ii*, 75-77.
27. Clark, I. A.; J. Virelizier, J. L.; Carswell, E. A.; Wood, P. R. *Infect. Immun.* **1981**, *32*, 1058-1066.
28. Bate, C.A.; Taverne, J.; Playfair, J. H. *Immunology* **1988**, *64*, 227-231.
29. Bate, C. A.; Taverne, J.; Playfair, J. H. *Immunology*, **1989**, *66*, 600-605.
30. See: L. Schofield, *Chem. Immunol. (Malarial Immunology)*, **2002**, *80*, 322-342; and references therein.
31. Gupta, S.; R. W. Snow, R. W.; Donnelly, C. A.; Marsh, K.; Newbold, C. *Nat. Med.* **1999**, *5*, 340-343.
32. Good, M .F. *Nat. Rev. Immunol.* **2001**, *1*, 117-125.
33. Bate, C. A. W.; Taverne, J.; Playfair, J. H. L. *Infect. Immun.* **1992**, *60*, 1894-1901.
34. Schofield, L.; Hackett, F. *J. Exp. Med.* **1993**, *177*, 145-153.
35. Tachado, S. D.; Gerold, P.; McConville, M. J.; Baldwin, T.; Quilici, D.; Schwarz, R. T.; Schofield, L. *J. Immunol.* **1996**, *156*, 1897-1907.
36. Tachado, S. D.; Gerold, P.; Schwarz, R.; Novakovic, S.; McConville, M.; Schofield, L. *Proc. Natl. Acad. Sci. USA* **1997**, *94*, 4022-4027.
37. Schofield, L.; Vivas, L.; Hackett, F.; Gerold, P.; Schwarz, R. T.; Tachado, S. *Ann. Trop. Med. Parasitol.* **1993**, *87*, 617-626.
38. Tachado, S.; Schofield, L. *Biochem. Biophys. Res. Commun.* **1994**, *205*, 984-991.
39. Schofield, L.; Tachado, S. D. *Immunol. Cell Biol.* **1996**, *74*, 555-563.
40. Schofield, L.; Novakovic, S.; Gerold, P.; Schwarz, R. T.; McConville, M. J.; Tachado, S. D. *J. Immunol.* **1996**, *156*, 1886-1896.
41. Gerold, P.; Dieckmann-Schuppert, A.; Schwarz, R. T. *J. Biol. Chem,* **1994**, *269*, 2597-2606.
42. Gerold, P.; Schofield, L.; Blackman, M, Holder, A. A.; Schwartz, R. T. *Mol. Biochem. Parasitol.*, **1996**, *75*, 131-143.
43. Gerold, P.; Vivas, L.; Ogun, S. A.; Azzouz, N.; Brown, K. N.; Holder, A. A.; Schwarz, R. T. *Biochem. J.* **1997**, *328*, 905-911.
44. Gowda, D. C. *Microbes. Infect.* **2002**, *4*, 983-990.
45. Schofield, L.; Hewitt, M. C.; Evans, K.; Slomos, M.-A.; Seeberger, P. H. *Nature* **2002**, *418*, 785-789.
46. Hewitt, M. C.; Snyder, D. A.; Seeberger, P. H. *J. Am. Chem. Soc.* **2002**, *124*, 13434–13436.
47. Madsen, R.; Fraser-Reid, B. in *Modern Methods in Carbohydrate Synthesis*, Khan, S. H., O'Neill, R. A. Eds.; Harwood Academic Publishers; Switzerland, **1995**, Chapter 4.
48. Gowda, D. C.; Gupta, P.; Davidson, E. A. *J. Biol. Chem.* **1997**, *272*, 6428-6439.

49. Miller, L. H.; Barunch, D. I.; Marsh, K.; Doumbo, O. K. *Nature* **2002**, *415*, 673-679.
50. de Souza, B. J.; Riley, E. M. *Microbes. Infect.* **2002**, *4*, 291-300.
51. Seeberger, P. H.; Soucy, R. L.; Kwon, Y.-U.; Snyder, D. A.; Kanemitsu, T. *Chem. Commun.* **2004**, 1706-1707.
52. Disney, M. D.; Seeberger, P. H. *Drug. Discov. Today: Targets* **2004**, *3*, 151-158.
53. Liu, X.; Kwon, Y.-U.; Seeberger, P. H. *J. Am. Chem. Soc.* **2005**, *127*, 5004-5005.
54. Liu, X.; Seeberger, P. H. *Chem. Comm.* **2004**, 1708-1709.
55. Kwon, Y.-U.; Soucy, R. L.; Snyder, D. A.; Seeberger, P. H. *Chem. Eur. J.* **2005**, *11*, 2493-2504.
56. Kamena, F. J.; Childs-Disney, J. L.; Seeberger, P. A. Unpublished results.
57. For a review see: Turco, S. J.; Descoteaux, A. *Annu. Rev. Microbiol.* **1992**, *46*, 65-94.
58. Mitchell, G. F.; Handman, E. *Parasite Immun.* **1986**, *8*, 255-263.
59. Moll, H.; Mitchell, G. F.; McConville, M. J.; Handmann, E. *Infect. Immun.* **1989**, *57*, 3349-3356.
60. McConville, M. J.; Bacic, A.; Mitchell, G. F.; Handman, E. *Proc. Nat. Acad. Sci. USA* **1987**, *84*, 8941-8954.
61. Russell, D. G.; Alexander, J. *J. Immunol.* **1988**, *140*, 1274-1279.
62. Tonui, W.K.; Mbati, P. A.; Anjili, C. O.; Orago, A. S.; Turco, S. J.; Githure, J. I.; Koech, D. K. *East Afr. Med. J.* **2001**, *78*, 84-89.
63. Tonui, W. K.; Mbati, P. A.; Anjili, C. O.; Orango, S.; Turco, S. J.; Githure, J. I.; Koech, D. K. *East Afr. Med. J.* **2001**, *78*, 90-92.
64. Nikolaev, A. V.; Chudek, J. A.; Ferguson, M. A. J. *Carbohydr. Res.* **1995**, *272*, 179-189.
65. Nikolaev, A. V.; Rutherford, T. J.; Ferguson, M. A. J.; Brimacombe, J. S. *Bioorg. Med. Chem. Lett.* **1994**, *4*, 785-788.
66. Ruda, K.; Lindberg, J.; Garegg, P. J.; Oscarson, S.; Konradsson, P. J. *J. Am. Chem. Soc.* **2000**, *122*, 11067-11072.
67. Hewitt, M. C.; Seeberger, P. H. *J. Org. Chem.* **2001**, *66*, 4233-4243.
68. Hewitt, M. C.; Seeberger, P. H. *Org. Lett.* **2001**, *3*, 3699-3702.
69. Arasappan, A.; Fraser-Reid, B. *J. Org. Chem.* **1996**, *61*, 2401-2406.
70. Upreti, M.; Ruhela, D.; Vishwakarma, R. A. *Tetrahedron*, **2000**, *56*, 6577-6584.
71. Kinzy, W.; Schmidtt, R. R. *Carbohyrate Res.* **1987**, *164*, 265-276.
72. Carpino, L. A. *J .Am. Chem. Soc.* **1993**, *115*, 4397-4398.
73. Carpino, L. A.; El-Faham, A. *J. Org. Chem.* **1996**, *61*, 2460-2465.
74. Schofield, L.; Seeberger, P. H. Unpublished results.

Chapter 8

A Uniquely Small, Protective Carbohydrate Epitope May Yield a Conjugate Vaccine for *Candida albicans*

David R. Bundle[1,*], M. Nitz[1,2], Xiangyang Wu[1], and Joanna M. Sadowska[1]

[1]Department of Chemistry, University of Alberta, Edmonton, Alberta T6G 2G2, Canada
[2]Current address: Department of Chemistry, University of Toronto, 80 St. George Street, Toronto, Ontario M5S 3H6, Canada

The cell wall phosphomannan of *C. albicans* is a promising target for the induction of immunity by development of a conjugate vaccine. It contains a unique antigen, a β1,2-mannan that affords active protection to mice following immunization and subsequent challenge with live organisms. Disaccharide and trisaccharide fragments of the β1,2-mannan antigen optimally inhibit two murine monoclonal antibodies that confer protection in a mouse model of candidiasis implying that epitopes of this size might constitute viable vaccine components. Short oligosaccharides conjugated to suitable immunogenic proteins have been synthesized by two distinct approaches one of which is well suited to multigram synthesis. Tetanus toxoid was chosen as a carrier protein for its ability to induce a vigorous hapten specific IgG response and for compatibility with human vaccine applications. Preliminary data show that rabbits immunized three times with a trisaccharide glycoconjugate produce sera with ELISA titers of 1:500,000 for the *Candida albicans* cell wall mannan. These rabbit antibodies also bind the antigen when it is present on the fungal cell wall. Only slightly lower antibody responses to the trisaccharide epitope were observed when 2 injections of these tetanus toxoid conjugates was performed with alum, an adjuvant acceptable for use in humans.

© 2008 American Chemical Society

Introduction

Carbohydrate antigens occupy an enigmatic role with respect to the mammalian immune system but have nevertheless played a significant role in the development of the disciplines of immunochemistry and practical aspects of immunology (*1-3*). The identification of the carbohydrate epitopes of the human ABO blood groups established that the specificity of carbohydrate-antibody interactions could be, in the case of a mismatched blood transfusion, a life threatening situation. On the other hand, immunization with the capsular polysaccharides antigens of pneumococcal or meningococcal bacteria induced antibody that at least for adults conferred life saving immunity against these potentially deadly bacterial infections (*4,5*). Subsequent work on the cells and molecules of the mammalian immune system that are responsible for processing antigens has shown that the limited immunogenicity of carbohydrate antigens is directly related to the manner in which antigens are presented by specialized immunocompetent cells (*6*). The immune system is highly evolved and tailored to process protein antigens into peptide fragments which are then bound to MHC molecules that present the peptide to T-cell receptors (*7,8*), thereby initiating a cascade of signals that instruct B-cells to secrete antibody of increasing affinity (*6*). The T-cell independent nature of polysaccharide antigens is explained by this finding, which also logically leads to the concept of carbohydrate conjugate vaccines. When polysaccharides and oligosaccharides are colaventry linked to immunogenic carrier proteins such as *Diptheria* toxin or tetanus toxoid (*5,9,10*) antigen processing creates T-cell peptides from the carrier protein. These peptides when presented to T-cells by MHC molecules recruit T-cell help that results in an antibody response to the carbohydrate epitope of the conjugate. Furthermore this immune response can be boosted by a second injection, in which case it is said to exhibit secondary response characteristics such as antibody class switch, including affinity maturation (*6*). Commercial conjugate vaccines of this type have been widely adopted in the last two decades and are highly effective in reducing the incidence of diseases caused by *Haemophilus influenzae*, *Neisseria meningitis* and *Streptococcus pneumoniae* (*11*). Recently a significant break through was reported for *Haemophilus influenzae* vaccine. The carbohydrate capsular antigen component of this conjugate vaccine that originated from bacterial fermentation was substituted by a totally synthetic oligosaccharide (*12*). In this way a fully semi-synthetic conjugate vaccine was prepared from an oligomeric *H. influenzae* heptameric construct conjugated to tetanus toxoid. The resultant vaccine was demonstrated to be as effective as the polysaccharide-protein conjugate (*12*).

We report here preliminary findings that suggest a uniquely small, synthetic trisaccharide epitope conjugated to tetanus toxoid will be sufficient to secure recognition of the β-mannan of the *Candida albicans* cell wall phosphomannan complex and eventually a protective immune response against *Candida* infections. A conjugate vaccine composed of a carbohydrate epitope consisting of only 3 hexose residues would be a radical departure from previously held views regarding the minimum size of protective epitopes (*13*). Several lines of evidence have suggested that the saccharide component of conjugate vaccines should consist of multiple repeating units ranging in size from 8 up to 200 saccharide residues (*13-16*). If it proved immunologically active, a trisaccharide epitope for a *Candida* conjugate vaccine would not only be a significant break with conventional wisdom but would also be highly desirable from the perspective of the economics of chemical synthesis.

In this paper we also report the potential utility of oligosaccharides in which the glycosidic oxygen atom bridging adjacent hexose residues may be exchanged for sulphur without significance loss of the fidelity of binding to native *O*-linked antigens. It could be envisaged that terminally linked sugar residues attached by such metabolically stable linkages would, in special circumstances, provide glycosidase resistant epitopes that might otherwise be susceptible to endogenous glycosidases (*17,18*).

Candidate vaccines to treat fungal infections are the focus of growing intetest for a variety of reasons. *Candida albicans* is the most common etiologic agent in candidiasis, a serious infection with high morbidity rates especially for immunocompromised patients (*19-21*). Immunotherapeutic strategies to increase host resistance are now attracting attention since antifungal agents with excellent *in vitro* activity have significant toxicity issues (*22*).

A unique β1,2-mannan antigen present in the cell wall phosphomannan of *C. albicans* is a promising target for the induction of protective immunity (Figure 1). Cutler's group showed that immunization with a conjugate vaccine prepared from the β1,2-mannan antigen conferred active and passive protection in a mouse model of disseminated candidiasis (*23,24*) and also that passive protection could be achieved with two monoclonal antibodies raised to similar antigen preparations (*25,26*).

Employing a series of oligosaccharides from di- up to hexasaccharide we investigated the binding specificity of the two monoclonal antibodies from Cutler's group by inhibition of their binding to immobilized native mannan (Figure 2). The surprising data showed that synthetic disaccharide and trisaccharide fragments of the β1,2-mannan antigen were optimal inhibitors for these monoclonal antibodies, while larger structures were significantly less active in the order tetrasaccharide> pentasaccharide>hexasaccharide (*27*). We also noted that a tetrasaccharide derivative in which the terminal hexose was attached via a thio linkage gave better inhibition than the corresponding *O*-linked tetrasaccharide. These observations are a dramatic departure from a paradigm that has held sway for over 30 years. Kabat had demonstrated that human sera raised against dextran, the α1,6 linked polymer of glucose, exhibited specificity for oligosaccharides of glucose that increased with the size of the oligo-

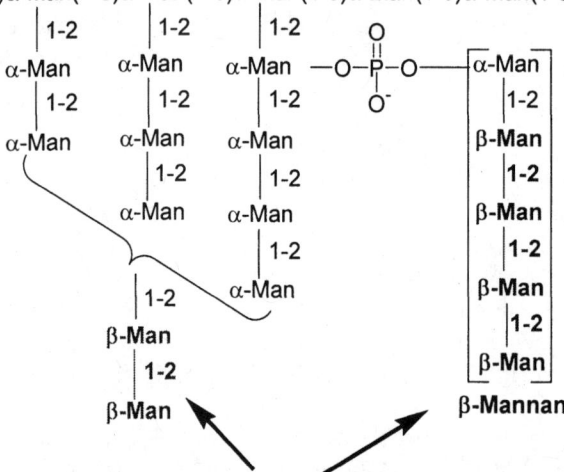

Figure 1. Gross structure of the C. albicans phosphomannan illustrating the two types of β-mannan, one attached to the α-mannan via phosphodiester and via a glycosidic linkage of α-mannan side chains.

saccharide (2, 28). Although tri- and tetrasaccharide were frequently very effective inhibitors there was a universal trend in which inhibitor power increased with the size of the oligosaccharide generally reaching a maximum when the size of the oligosaccharide approached hexa- to octasaccharide. Although many different oligosaccharide and antibody systems have been studied since Kabat's initial work, it has always been observed that inhibitory power either increased with hapten molecular weight or remained constant on a molar basis, and to the best of our knowledge we are unaware of any study that showed a significant reduction in binding as inhibitor length increased between 3-6 hexose residues. Since a trisaccharide appears to fill the binding site of the protective antibodies (27) we reasoned that a synthetic conjugate vaccine composed of short oligosaccharides ranging in size from hexasaccharide to perhaps as small as a trisaccharide epitope should be capable of raising antibodies that bind native phosphomannan and possibly offer a viable, chemically defined, cost effective, conjugate vaccine. We now report the efficient synthesis of these oligosaccharides epitopes and immunochemical data for rabbits that have been immunized with such oligosaccharide conjugate vaccines.

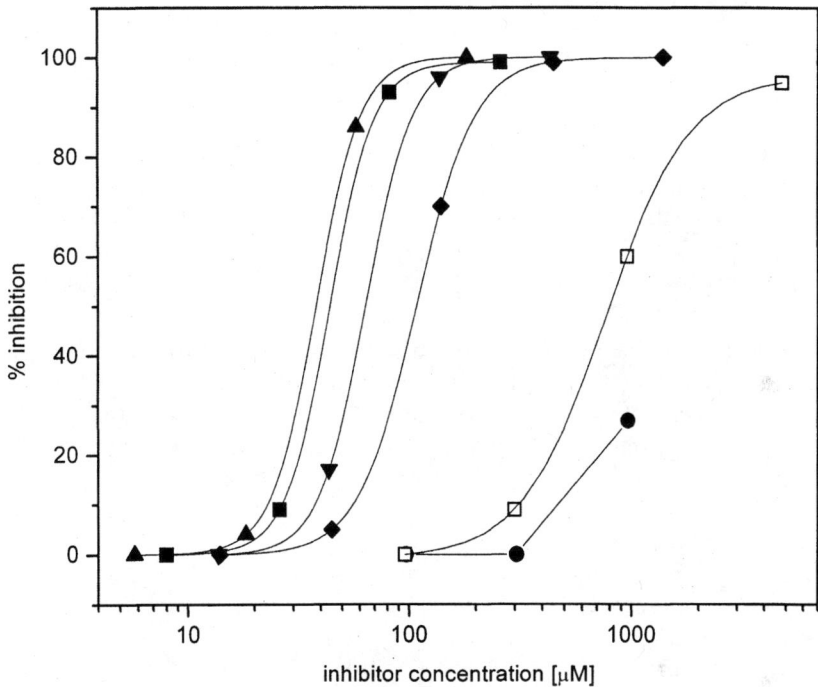

Figure 2. ELISA ihibition by synthetic oligosaccharides for the mouse IgG monoclonal antibody C3.1.

- ■ propyl (1→2)-β-D-mannopyranobioside,
- ▲ propyl (1→2)-β-D-mannopyraontrioside,
- ▼ propyl (1-thio-β-D-mannopyranosyl)-(1→2)-β-D-mannopyranotrioside,
- ◆ propyl (1→2)-β-D-mannopyranotetroside,
- □ propyl (1→2)-β-D-mannopyranopentoside,
- ● propyl (1→2)-β-D-mannopyranohexoside.

Synthesis of Oligosaccharide Epitopes

The ulosyl bromide **1** introduced by Lichtenthaler's group for β-mannopyranoside synthesis was investigated for its utility in the synthesis of β1,2 mannan oligomers (*29*). This glycosyl donor reacted stereospecifically with the acceptor **2** to afford a β-linked disaccharide **3**, which was subsequently reduced stereoselectively by L-selectride to give the *manno* configuration and hence the required disaccharide **4** (Scheme 1).

The conditions employed for disaccharide synthesis failed to yield significant amounts of trisaccharide when donor **1** was reacted with acceptor **4**. Exploration of different activation protocols resulted in the use of the soluble promoter, silver triflate, with 2,6-di-*tert*-butyl-4-methylpyridine (DtBMP) as an acid scavenger, and acetonitrile as a participating solvent. These conditions followed by similar L-Selectride mediated reduction gave a 40-45% yield of trisaccharide **5**, and 10% yield of the α-gluco epimer **6** together with a significant portion of 3,4-di-*O*-benzyl-1,6-anhydro-β-D-mannopyranose **7**. It was hypothesized that the glycosyl donor reacts with acetonitile to yield an α-nitrilium intermediate and that intramolecular cyclizaton gives rise to the anhydrosugar as well as the intended glycosylation product **5** (Scheme 2). Increasing the stability of the ether protecting groups would disfavor intramolecular reaction and the formation of 1,6-anhydro sugar side product and in turn increase the yield of the desired oligosaccharide. The *p*-chlorobenzyl protecting group was explored for this purpose since it has been shown to be more acid stable than the parent benzyl group but to have otherwise similar properties.

The more reactive glycosyl donor **8** afforded trisaccharide intermediates in acceptable yield that were also easier to deprotect (Scheme 3). However, with acetonitile as solvent attempted chain extension of a trisaccharide to give a tetrasaccharide was met by the observation of an interesting and unexpected product. Using the same conditions as those employed to make the trisaccharide, up to 20% of the unexpected 2-*O*-acetyl trisaccharide acceptor was isolated. We hypothesized that this product resulted from attack of the acceptor on the nitrile carbon of the proposed α-nitrilium intermediate (e.g. pathway b), instead of reaction at the anomeric center of the donor (pathway a). This postulate was supported by the isolation of the chloroacetylated acceptor (**11**) when chloroacetonitrile was employed as the solvent (Scheme 3). Initial formation of an imidate intermediate (**10b**) that is hydrolyzed during aqueous work-up accounts for the observed acylated acceptor (**11**). Literature precedence for this type of intermediate exists (*30*). In this case the resulting side product must be favored due to a sterically hindered acceptor (**9**) as well as the electron deficient nitrilium uloside intermediate. When the sterically more hindered pivaloyl nitrile was chosen as the solvent, the glycosidic linkage was synthesized in 48% yield to give the desired tetrasaccharide **13**, 10% of the α-gluco epimer along with

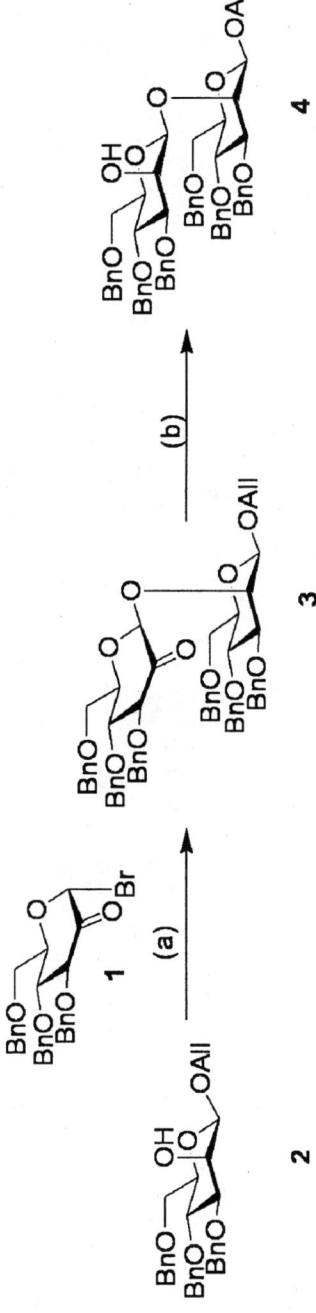

Scheme 1. a) Ag(zeolite), CH$_2$Cl$_2$, (78%), b) L-Selectride, THF

170

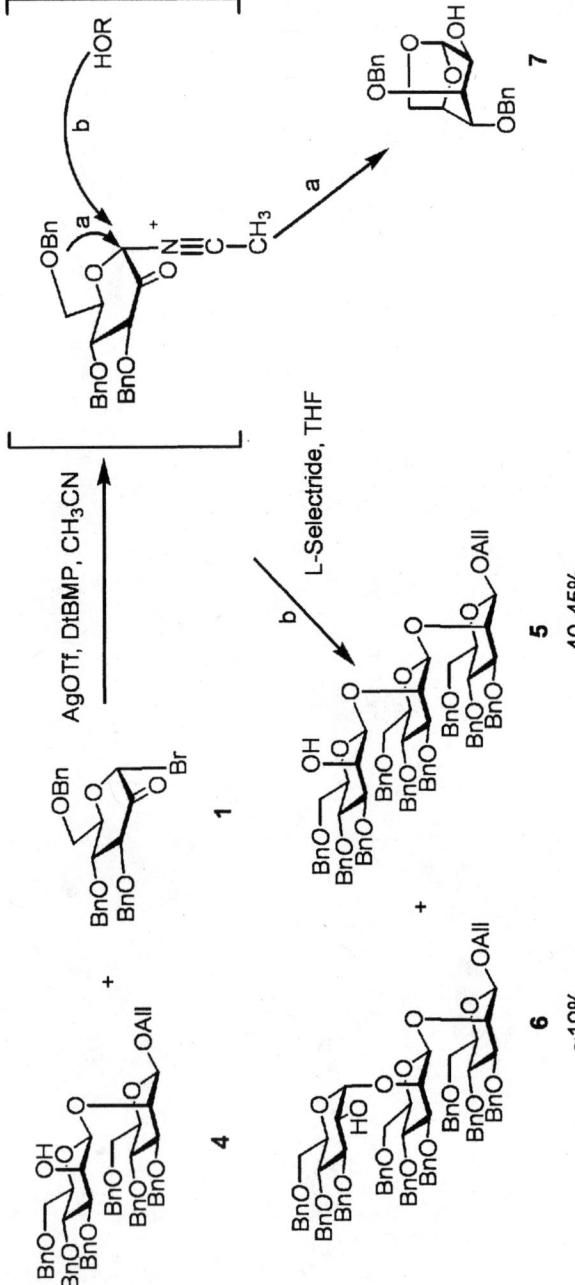

Scheme 2. Initial glycosylations indicating a 1,6-anhydrohexose side product indicative of intramolecular cyclization of the activated glycosyl donor intermediate.

small amounts of the trisaccharide pivaloyl ester also being formed (Scheme 4). The glycosyl donor **8** demonstrates high diastereoselectivity over both the glycosylation and subsequent reduction step and minimizes the number of protecting group manipulations necessary for the synthesis. Thus *p*-chlorobenzyl protected ulosyl bromide (**8**) in combination with the sterically hindered, participating solvent, pivaloyl nitrile were considered to be the optimized conditions for this new approach to the synthesis of these unique homo-oligomers ranging from disaccharide up to hexasaccharide (*31*) (Scheme 4).

Congeners of the (1→2)-β-D-mannotetraose were also synthesized containing a terminal *S*-linked (1→2)-β-D-mannopyranosyl residue (*31*) (Scheme 5). The 4,6-*O*-benzylidene-glucopyranosyl imidate **15** was employed to create a trisaccharide bearing a thiol group at C-2. This was achieved by removal of acetate **16** and conversion of the alcohol **17** to the triflate **18**. Displacement with thioacetate gave **19**. After conversion to the thiol **20** reaction with **8** followed by reduction of the uloside gave the protected tetrasaccharide **21**.

An alternate synthetic approach more suited to multi-gram synthesis of oligo-mannan epitopes was developed with the expectation that *C. albicans* conjugate vaccines of this type could find commercial application (Scheme 6). The synthetic strategy is related to the ulosyl donor method but employs a glucosyl imidate donor **22** with a participating but temporary protecting group at *O*-2 (*32*). The glycosyl imidate **22** gives excellent yields of β-glucopyranosides **23** and **24** under the stereoselective control of the acetyl group. After removal of the acetate group, the 2-hydroxy derivative is oxidized to an uloside which is stereoselectively reduced to the *manno* configuration. Although the procedure introduces an extra oxidation step, the yields of product are superior and the glucosyl donor **22** is easier to prepare and more stable than either of the ulosyl bromides **1** or **8**.

In order to functionalize oligosaccharides **9, 12, 13, 14, 21** and **25** to permit conjugation to protein the allyl group of the protected oligosaccharides was subjected to photochemical addition of cysteineamine prior to removal of the benzyl ether protecting groups (Scheme 7). This created a seven atom tether terminated by an amino group which was reacted either with diethyl squarate (*31*) or the di-*p*-nitrophenyl ester of adipic acid (*33*). These homo-bifunctional coupling reagents each gave a half ester type intermediate that could be worked up and added directly to the protein to which the oligosaccharide is to be conjugated. Whereas, the squarate activation and coupling appeared to be the most efficient giving rise to conjugates with degree of hapten incorporation of ~20-30 with tetanus toxoid (*31*), the adipic ester approach gave hapten incorporation of ~10 haptens per molecule of tetanus toxoid (*33*). These conjugates were used to immunize rabbits.

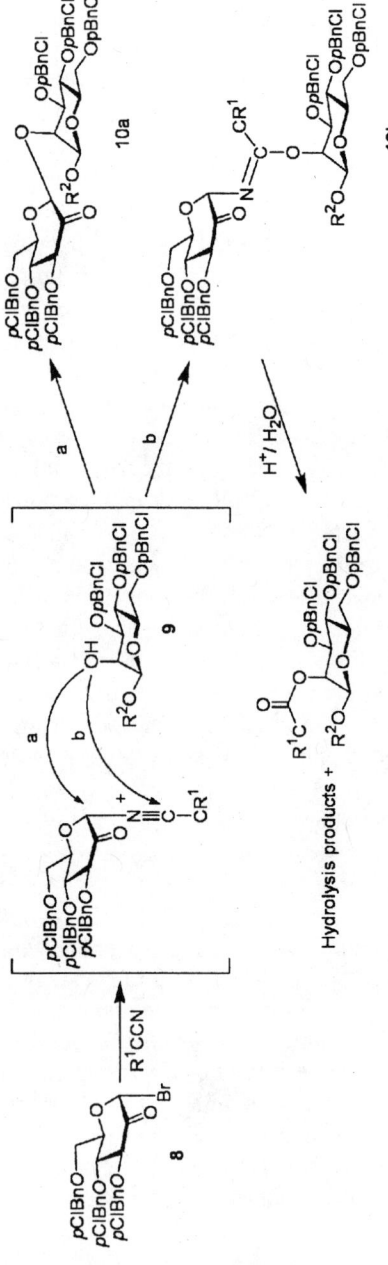

Scheme 3. *Acylated glycosyl acceptor side products suggest the need for a sterically demanding nitrile solvent.*

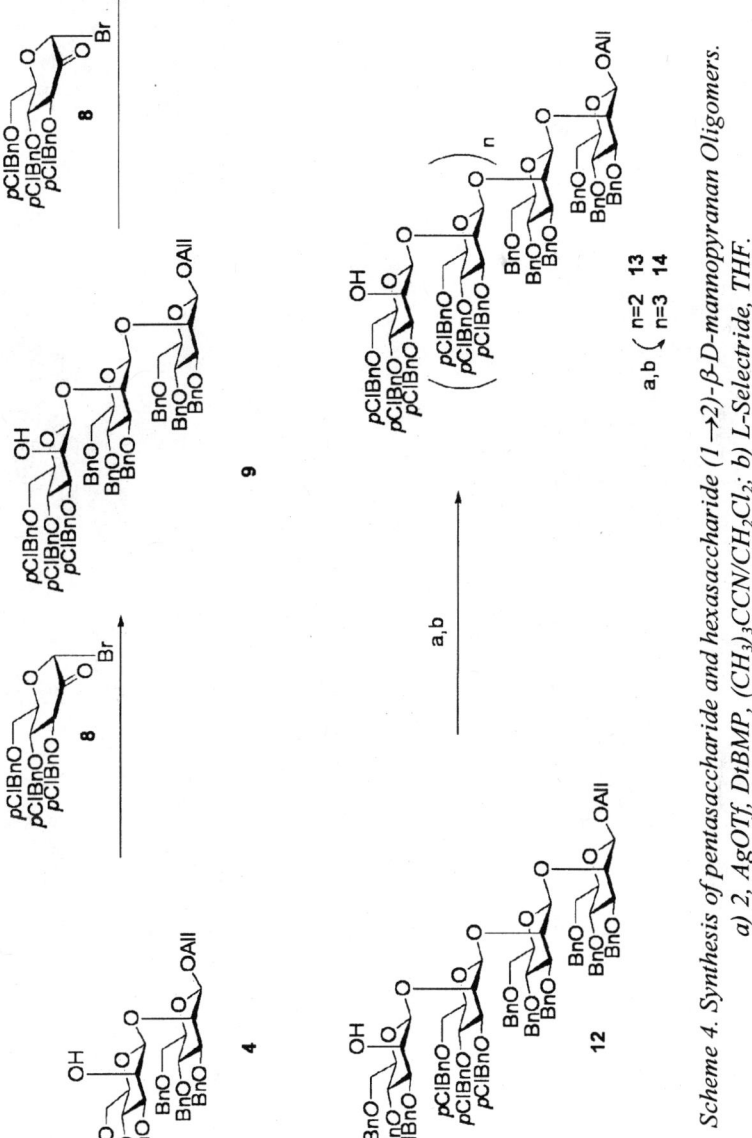

Scheme 4. *Synthesis of pentasaccharide and hexasaccharide (1→2)-β-D-mannopyranan Oligomers.*
a) **2**, *AgOTf, DtBMP, (CH₃)₃CCN/CH₂Cl₂*; b) *L-Selectride, THF.*

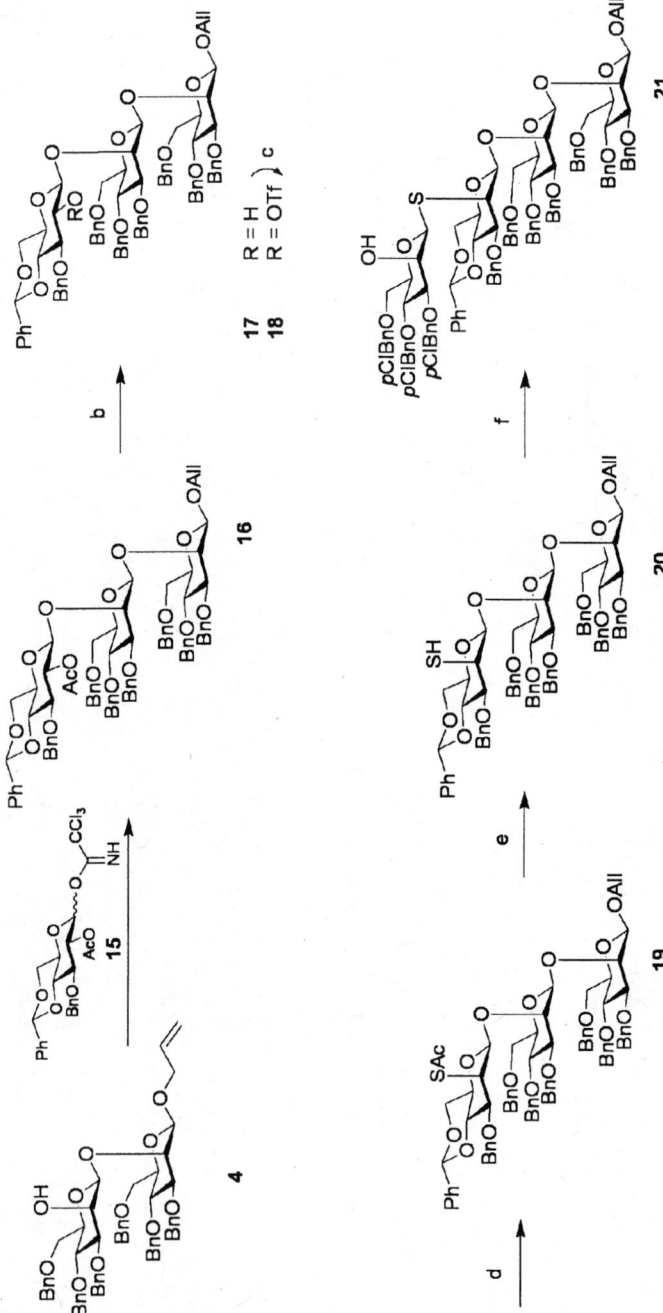

Scheme 5. *Synthesis of the thioglyoside mimetic of (1→2)-β-D-mannopyranotetraose. a) 23, TMSOTf, CH₂Cl₂, 76%; b) MeONa, MeOH/THF, 94%; c)Tf₂O, pyridine, 89%; d) KSAc, DMF, 63%; e) Hydrazine hydrate, cyclohexene, EtOH/THF, 89%; f) i. 2, lutidine, CH₂Cl₂, ii. L-Selectride, THF, 49%.*

Conjugation to the heterologous protein, bovine serum albumin (BSA) provided conjugates for monitoring the level of anti-hapten antibody response in immunized animals by ELISA.

Rabbits were routinely immunized 2 or 3 times with the tetanus toxoid conjugates and the titration of the immune sera is shown in Figures 3 and 4 against both native hapten extracted form *C. albicans* or oligosaccharide-BSA conjugates. It can be seen that while immunization of rabbits with tetanus toxoid conjugates delivered with the powerful, Freund's adjuvant gave antibody titres in excess of 1,000,000 (Figure 3), just 2 immunizations with a related trisaccharide-tetanus toxoid conjugate delivered with alum still gave remarkably high titres in the range 200,000 (Figure 4). This is significant when considering the use of a conjugate vaccine since alum is approved for use in humans while Freund's adjuvant is not suitable for such applications.

Tetanus toxoid conjugates of the deprotected tetrasaccharide hapten **21**, where sulphur replaces oxygen in the terminal non-reducing glycosidic linkage were also used as an immunogen. The immune response to this modified antigen raised high titre sera that reacted with the corresponding *O*-linked tetrasaccharide-BSA conjugate and with the *C. albicans* native antigen (Figure 3). Although the fidelity of the recognition for the native *O*-linked antigen was surprisingly high by sera raised to the *S*-linked antigen, the titres of this sera were nevertheless slightly lower than titres for sera raised against the homologous *O*-linked epitope.

As a prelude to active challenge experiments in an animal model of *C. albicans* infections, we sought to establish whether the antibodies raised to synthetic vaccine constructs were able to recognize the β-mannan antigen displayed on the cell wall of *C. albicans*. Anti-sera to conjugate vaccines constructed from trisaccharide **5** and both *O*-linked and S-linked tetrasaccharides at dilutions ranging between 1:1,000-10,000 were all able to bind the native β-mannan when present on the cell wall of *C. albicans*. The experiment involved detecting antibody labeled cell wall with a fluorescein labeled goat anti-rabbit antibody. The labeling experiments showed that the β-mannan could be detected on both *Candida* hypia and budding cells.

These labeling experiments provide promising evidence that a vaccine designed to generate antibodies to the β-mannan should have potential as a therapeutic vaccine in agreement with data generated in mice to a related vaccine created from isolated cell wall components (23,24). Preliminary data from our ongoing experiments suggest that rabbits immunized with a trisaccharide-tetanus toxoid conjugate show an increased ability to clear *Candida albicans*. The results of these ongoing studies will be the subject of forthcoming manuscripts.

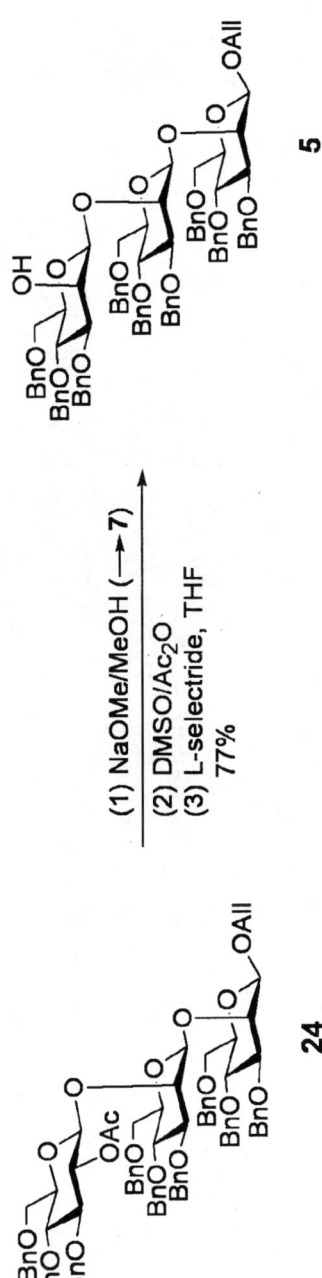

Scheme 6. Efficient process for large scale synthesis of β1,2-linked mannans

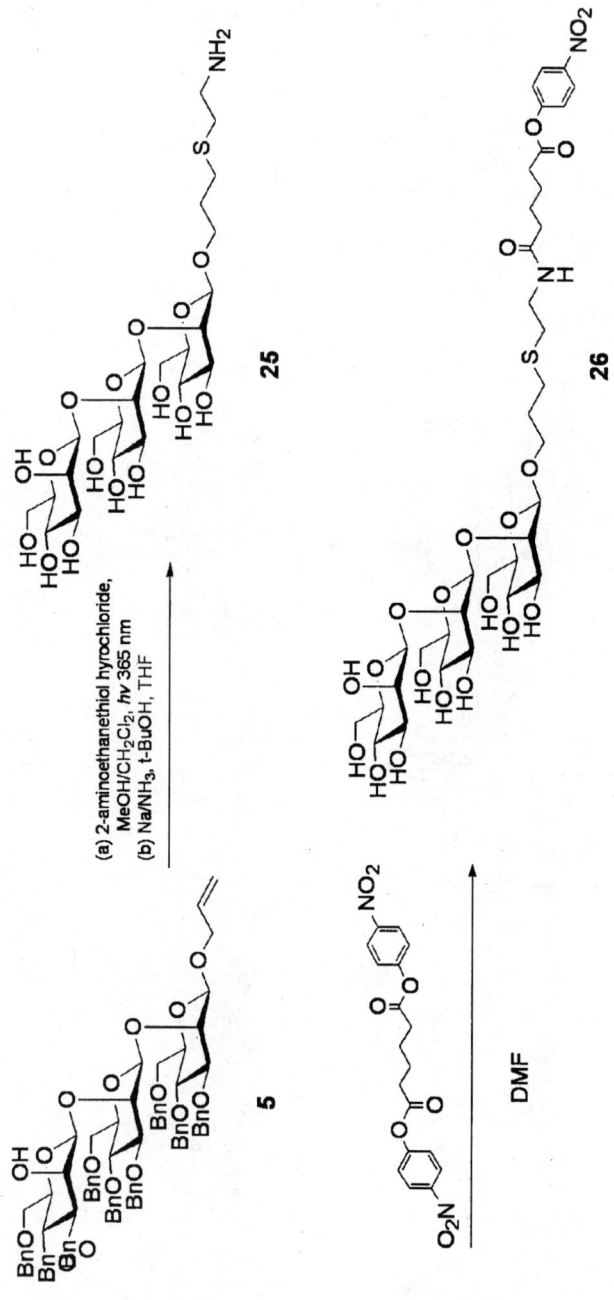

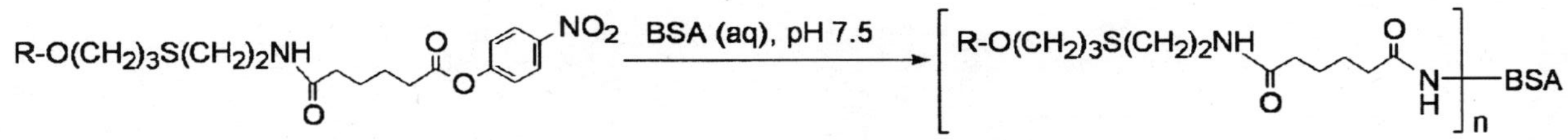

R-O(CH$_2$)$_3$S(CH$_2$)$_2$NH-CO-(CH$_2$)$_4$-CO-O-C$_6$H$_4$-NO$_2$ →[BSA (aq), pH 7.5] [R-O(CH$_2$)$_3$S(CH$_2$)$_2$NH-CO-(CH$_2$)$_4$-CO-NH-BSA]$_n$

R = trisaccharides **5 and 9**, tetrasaccharides **12 and 21**, pentasaccharide **13** and hexasaccharide **14**.

R-O(CH$_2$)$_3$S(CH$_2$)$_2$NH-CO-(CH$_2$)$_4$-CO-O-C$_6$H$_4$-NO$_2$ →[TT (aq), pH 7.2] [R-O(CH$_2$)$_3$S(CH$_2$)$_2$NH-CO-(CH$_2$)$_4$-CO-NH-TT]$_n$

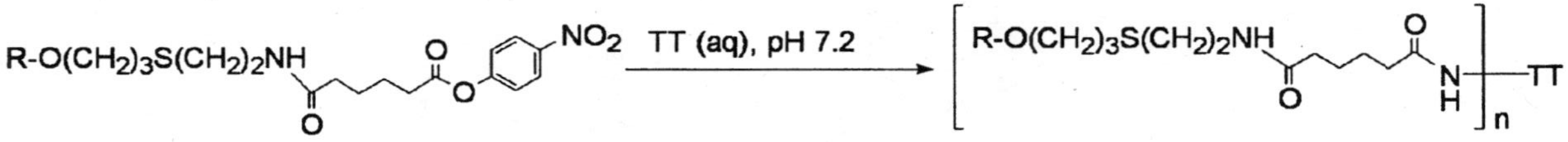

R = trisaccharide **26**

Scheme 7. Preferred strategy for conjugation of oligosaccharide to protein carrier.

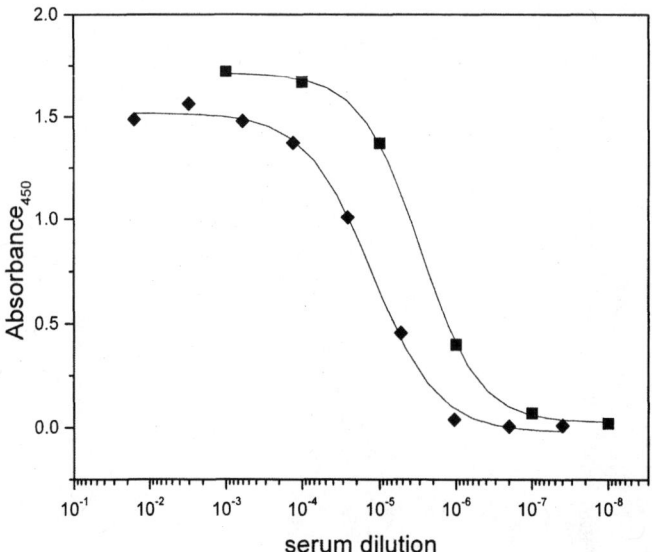

Figure 3. ELISA titration of rabbit serum against an extract of native β-mannan; ■ *rabbit immunized 3 times with a tetrasaccharide-tetanus toxoid conjugate, and* ◆ *rabbit immunized 3 times with a thio-tetrasaccharide-tetanus toxoid conjugate. Both antigens were given with Freund's adjuvant.*

Experimental

Antigens

Glycoconjugates were synthesized as previously described (31-33). The tetanus toxoid conjugates were dissolved in phosphate buffered saline (PBS). BSA conjugates were dissolved in PBS and used to coat ELISA plates.

Immunization

Protocol A: tetanus toxoid conjugate (50 μg) was diluted in 500 μL of PBS and mixed with 500 μL of Freund's complete adjuvant or with 500 μL of Freund's incomplete adjuvant. Each rabbit was injected with 1ml of vaccine: 0.5 mL intramusculary in one rear thigh and 2 x 0.25 mL at subcutaneous injections. Three injections were given at monthly intervals. For the second and third injections, the antigen was given in Freunds incomplete adjuvant. Rabbits were bled 9 days after the last injection.

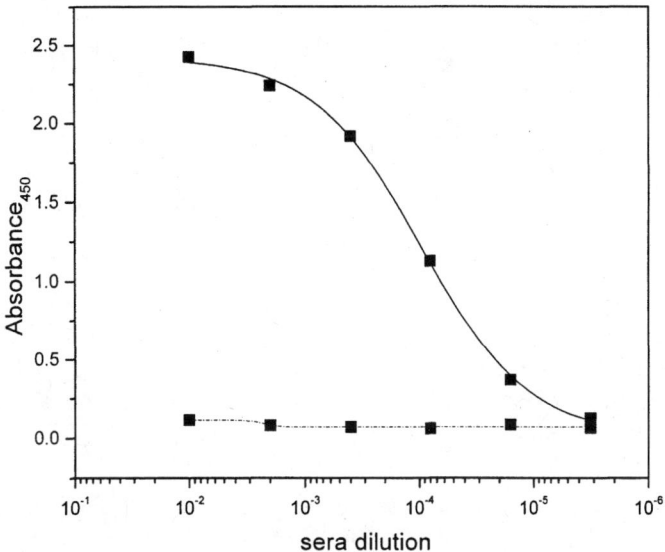

Figure 4. ELISA titration of a rabbit serum on trisaccharide-BSA coated plates: antibody titers achieved by two injections of the trisaccharide-tetanus toxoid conjugate are representative of rabbit responses. Dashed lines are titration curves of pre-imune sera

Protocol B: The trisaccharide-tetanus toxoid conjugate (0.5 mg) was dissolved in 0.5 mL of PBS, 2 mL of alum suspension was added and the sample was rotated for 1 hour at ambient temperature. To maintain sterility 250µg of thimerosal at 10mg/ml was added. Rabbits were immunized with the antigen absorbed alum suspension (0.5 mL /rabbit) distributed over 5 sites (0.1 mL/ site). Two injections were given intramuscularly in the rear thigh and 3 subcutaneously along the back.

Enzyme Immunoassays

Antibody levels in sera were established by ELISA experiments utilizing trisaccharide BSA conjugates coated on 96 well microtitration plates. Carbohydrate-protein conjugates (10 µg/mL; 100 µL; 4 °C, overnight) were used to coat 96-well ELISA plates (MaxiSorp, Nunc). The plate was washed (Molecular Devices Skan Washer 400) 5 times with PBST (PBS containing Tween 20, 0.05% v/v). Sera were diluted with PBST containing 0.15% BSA and the solutions were added to the plate and incubated at room temperature for 2 hours. The plate was washed with PBST (5 times) and goat-anti-rabbit IgG antibody conjugated to horseradish peroxidase (Kirkegaard & Perry Laboratories; 1:2000 dilution in PBST) was added (100 μL) and incubated for a period of 1 hour. The plate was washed 5 times with PBST before addition of a 1:1 mixture of 3,3',5,5'-tetramethylbenzidine (0.4 g/L) and 0.02% H2O2 solution (Kirkegaard & Perry Laboratories; 100 μL). After 2 minutes the reaction was stopped by addition of 1 M phosphoric acid (100 μL). Absorbance was read at 450 nm (Molecular Devices Spectra Max 190 plate reader).

C. albicans mannan was obtained by 2-mercaptoethanol extraction of whole cells without subsequent affinity fractionation (23) and was dissolved in PBS (10 μg/ml), and the solution was used to coat 96-well ELISA plates (100 μl/well, 18 h at 4 °C). Plates were washed five times with PBST and blocked for 1 h at room temperature (2% bovine serum albumin/PBS, 100 μL/well).

The dilution of sera at which a significant absorbance reading above background (~OD = 0.2) is recorded at the antibody titer.

Acknowledgements

We thank the Natural Sciences and Engineering Research Council of Canada (NSERC), the Canadian Institutes of Health Research (CHIR) and Alberta Ingenuity for support of this work.

References

1. Kabat, E.A. *Federation Proc.* **1962**, *21*, 694-701.
2. Kabat, E.A. *J. Immunol.* **1966**, *97*, 1-11.
3. Kabat, E.A. *Structural concepts in immunology & immunochemistry;* 2nd edition, Holt, Rinehart & Winston, New York, NY **1976**.
4. Gotschlich E.C.: Goldschneider, I.; Artenstein, M.S. *J. Exp. Med.* **1969**, *129*, 1367-1384.
5. Jennings, H.J. *Adv. Carbohydr. Chem. Biochem.* **1983**, *41*, 155-208.
6. *Immunobiology*. 6th ed. The Immune System in Health and Disease Janeway, C.A.; Travers, P.; Walport, M.; Shlomchik, M. Chapters 3 and 5, Garland Publishing · 2004.
7. Bjorkman, P.J.; Saper, M.A.; Samraoui, B.; Bennett, W.S.; Strominger, J.L.; Wiley, D.C. *Nature*, **1987**, *329*, 512-518.
8. Brown, J.H.; Jardetzky, T.S.; Gorga, J.C.; Stern, L.J.; Urban R.G.; Strominger, J.L. Wiley, D.C. *Nature*, **1993**, *364*, 33-39.
9. Schneerson, R.; Barrera, O.; Sutton, A.; Robbins, J.B. *J. Exp. Med.* **1980**, *152*, 361-376.
10. Schneerson, R.; Robbins, J.B. *J. Infect. Dis.*, **1990**, *161*, 821-832.

11. Lucas, A.H.; Apicella, M.A.; Taylor, C.E. *Clinical Infect. Dis.*, **2005**, *41*, 705-712.
12. Verez-Bencomo, V.; Fernandez-Santana, V.; Hardy, E.; Toledo, M.E.; Rodriguez, M.C.; Heynngnezz, L.; Rodriguez, A.; Baly, A.; Herrera, L.; Izquierdo, M.; Villar, A.; Valdes, Y.; Cosme, K.; Deler, M.L.; Montane, M.; Garcia, E.; Ramos, A.; Aguilar, A.; Medina, E.; Torano, G.; Sosa, I.; Hernandez, I.; Martinez, R.; Muzachio, A.; Carmenates, A.; Costa, L.; Cardoso, F.; Campa, C.; Diaz, M.; Roy, R. *Science.* **2004**, *305*, 522-525.
13. Svenson, S.B.; Lindberg, A.A. *Infect. Immun.*, **1981**, *32*, 490-496.
14. Mäkelä, O.; Peterfy, F.; Outschoorn, I.G.; Richter, A.W.; Seppälä, I. *Scand. J. Immunol.*, **1984**, *19*, 541-550.
15. Anderson, P.W.; Pichichero, M.E.; Insel, R.A.; Betts, R.; Eby, R. *J. Immunol.*, **1986**, *137*, 1181-1186.
16. Pozsgay, V.; Chu, C.; Pannell, L.; Wolfe, J.; Robbins, J. B.; ;Schneerson, R.; *Proc. Natl. Acad. Sci. USA* **1999**, *96*, 5194-5197.
17. Bundle, D.R.; Rich, J.R.; Jacques, S.; Yu, H.N.; Nitz, M.; Ling, C.-C. *Angew. Chem. Int. Ed.,* **2005**, *44*, 7725-7729.
18. Rich, J.R.; Wakarchuk, W.W.; Bundle, D.R. *Chem. Eur. J.* **2006**, *12*, 845-858.
19. Maksymiuk, A.W.; Thong Prasert, S.; Hopfer, R.; Luna, M.; Fainstein, V.; Bodey, G.P. *Am. J. Med.*, **1984**, *77*, 20-27.
20. Crawford, S.W. *Semin. Respir. Infect.* **1993**, *8*, 183-190.
21. Thaler, M.; Pastakia, B.; Shawker, T.H.; O'Leary, T.; Pizzo, P.A.. *Ann. Intern. Med.*, **1988**, *108,* 88-100.
22. Espinel-Ingroff, A.; Pfaller, M.A., *In* Murray, P.R., Baron, E.J., Pfaller, M.A., Tenover, F.C., Yolken, R.H. Eds.; *Manual of clinical microbiology*, 6th ed. Amer. Soc. Microbiol., Washington, D.C. 1995; pp1405-1414.
23. Han, Y.; Cutler, J.E. *Infect. Immun.*, **1995**, *63*, 2714-2719.
24. Han, Y.; Ulrich, M.A.; Cutler, J.E. *J. Infect. Dis.*, **1999**, *179*, 1477-1484.
25. Han, Y.; Rieselman, M.H.; Cutler, J.E. *Infect. Immun.*, **200**, *68*, 1649-1654.
26. Han, Y.; Morrison, R.P.; Cutler, J.E. *Infect. Immun.*, **1998**, *66*, 5771-5776.
27. Nitz, M.; Ling, C.-C.; Otter, A.; Cutler, J.E.; Bundle, D.R. *J. Biol. Chem.*, **2002**, *277*, 3440-3446.
28. Kabat, E. A. *J. Immunol.*, **1960**, *84,* 82-85.
29. Nitz, M.; Purse B.W.; Bundle, D.R. *Org. Let.*, **2000**, *2*, 2939-2942.
30. Briner, K.; Vasella, A. *Helv. Chim. Acta,* **1992**, *75*, 621-637.
31. Nitz, M.; Bundle, D.R. *J. Org. Chem.*, **2001**, *66*, 8411- 8423.
32. Wu, X., and D.R. Bundle. *J. Org. Chem.,* **2005**, *70*, 7381-7388.
33. Wu, X., Ling, C.-C., and D.R. Bundle. *Org. Lett.*, **2004**, *6*, 4407-4410.

Chapter 9

Studies toward the Development of Anti-Tuberculosis Vaccines Based on Mycobacterial Lipoarabinomannan

Hyo-Sun Kim[1,2], Ella S. M. Ng[1,3], Ruixiang Blake Zheng[1,2], Randy M. Whittal[2], David C. Schriemer[1,3], and Todd L. Lowary[1,2,*]

[1]Alberta Ingenuity Centre for Carbohydrate Science and [2]Department of Chemistry, The University of Alberta, Gunning-Lemieux Chemistry Centre, Edmonton, Alberta T6G 2G2, Canada
[3]Department of Biochemistry and Molecular Biology, The University of Calgary, 3330 Hospital Drive NW, Calgary, Alberta T2N 4N1, Canada

Recent investigations suggest that oligosaccharide-based conjugate vaccines containing fragments of a mycobacterial polysaccharide, lipoarabinomannan (LAM), have potential utility in the prevention of tuberculosis. We report here work towards the synthesis of oligosaccharide fragments of LAM and studies directed towards identifying the minimum epitopes required for vaccine generation.

Introduction

Tuberculosis is the world's most lethal bacterial disease, killing more than 2 million people each year (*1*). The impact of tuberculosis on world health has been the source of increased concern due to the emergence of multi-drug resistant strains of the organism that causes the disease, *Mycobacterium tuberculosis*, and problems in treating tuberculosis in HIV-positive patients. (*2,3*) As a consequence, there has been renewed interest both in the identification of new antibiotics for the treatment of tuberculosis (*4,5*), as well as in the development of novel vaccines that protect against the disease (*6*).

The only vaccine currently used to protect against tuberculosis is the Bacille Calmette-Guerin (BCG) vaccine, a live-attenuated strain of *Mycobacterium bovis*, which has been in use for more than 80 years. More than 100 million newborn children receive this vaccine each year (*7*). However, despite this widespread administration, its efficacy has been shown to vary greatly. In addition, it is also now generally accepted that although the BCG vaccine protects children against tuberculosis, it offers little protection against the disease in adults (*8*). A number of reasons for the variability of the BCG vaccine have been suggested, including variations in the strains used to produce the vaccine, differences (environmental or genetic) in the populations involved in the protection studies, previous sensitization by exposure to saprophytic mycobacteria present in the environment and less than optimal delivery of the vaccine (*9*). Regardless of the reasons, the poor efficacy of the BCG vaccine has led to international efforts to identify new vaccines for preventing this disease, supported by both governmental and private concerns (*6–10*).

Historically, antibody-mediated immunity in protection against tuberculosis has been thought to be of only limited importance (*11,12*). This view, which became dogma in the field, was based on the lack of studies demonstrating a natural protective antibody against *M. tuberculosis* in addition to the belief that an immune response to a particular pathogen was either primarily cell-mediated or antibody-mediated. Further undermining the prospect of antibody-mediated protection against tuberculosis was the assumption that serum antibodies could not access intracellular pathogens. Therefore, most efforts in the design of anti-tuberculosis vaccines have focused on those that stimulate cell-mediated immunity. However, an increasing number of studies have suggested that antibodies recognizing mycobacterial antigens do have a role in protection against tuberculosis (*11,12*). Furthermore, recent successes (*13*) in the use of a cell wall polysaccharide-based conjugate vaccine against *Cryptococcus neoformans*, an intracellular fungal pathogen, also bolster the case for antibodies as players in protection against tuberculosis.

A predominant antigen of the mycobacterial cell wall is lipoarabinomannan (LAM), a glycoconjugate consisting of phosphadtidylinositol (PI) functionalized with an arabinomannan polysaccharide (*14*). A schematic depiction of the structural domains in LAM is provided in Figure 1. The mannan core of LAM, which is attached to the PI moiety, consists of a chain of α-(1→6)-linked mannopyranose residues, approximately half of which are further elaborated with α-(1→2)-linked mannopyranose caps. Attached to this mannan is the arabinan domain, which contains α-(1→3), α-(1→5) or β-(1→2)-linked arabinofuranose residues. The arabinan portion of the polysaccharide is terminated with either a branched hexasaccharide (**1**, Figure 2) or a related linear tetrasaccharide (**2**). These motifs are usually substituted at the 5-hydroxyl group of the β-arabinofuranosyl residues with capping motifs that are species specific (Figure 3). One example is ManLAM, which has mannopyranose-containing oligosaccharide caps, and which is present in *M. tuberculosis*, *M. avium*, *M. leprae* and *M. bovis* (*15,16*). LAM capped with inositol phosphate moieties (PILAM) is found in *M. smegmatis* (*17*). Recently, LAM capped with 5-methylthio-α-D-xylofuranose residues has been identified in *M. tuberculosis* and *M. kansasii* (*18,19,20*), and LAM from *M. chelonae* has been shown to have no capping motifs (*21*). These capping motifs are believed to be key players in the interaction of the organism with the immune system, however a detailed understanding of their role is not currently available (*22*).

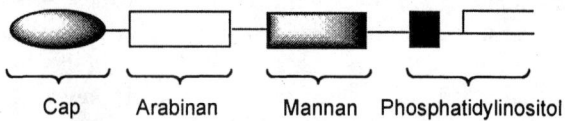

Figure 1. *Schematic depiction of structural domains in mycobacterial LAM.*

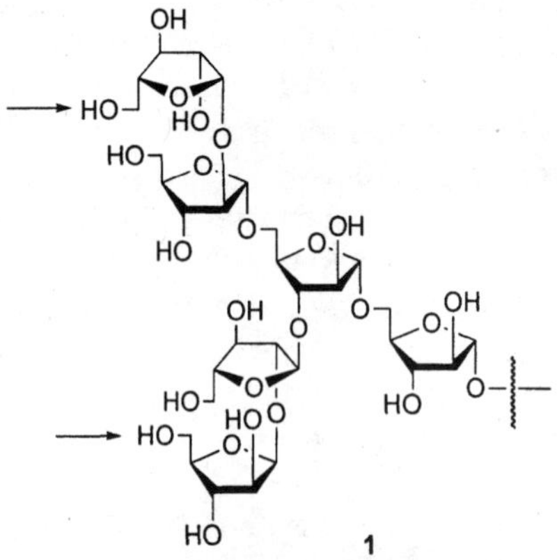

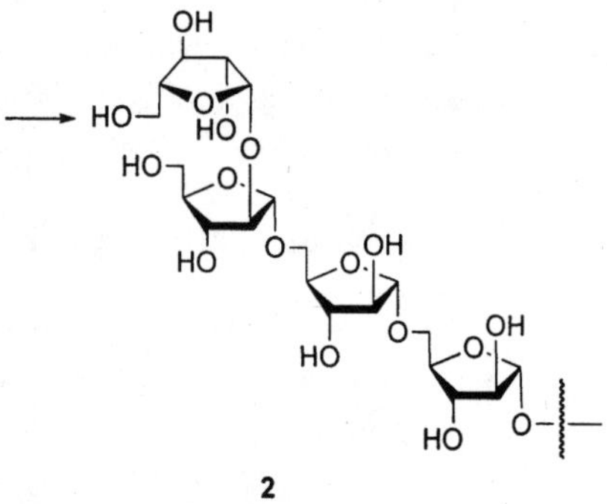

Figure 2. *Terminating structures in the arabinan domain of mycobacterial LAM. The attachment sites of the capping motifs are indicated with arrows.*

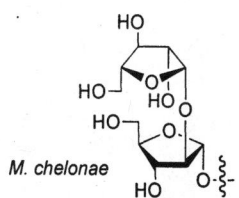

Figure 3. Capping motifs in mycobacterial LAM identified to date.

Recent studies point to the potential of LAM for the development of anti-tuberculosis vaccines. Svenson and coworkers have shown that partial degradation of LAM from *M. tuberculosis* (Harlingen strain), followed by coupling of the resulting oligosaccharide fragments to proteins, provides conjugates that protect mice against tuberculosis at levels comparable to those observed with the BCG vaccine (*23*). Further studies demonstrated that passive immunization of mice with antibodies generated against LAM also protected against tuberculosis challenge (*24*). Related to these investigations is work by Hoft and coworkers, who have demonstrated that vaccination with BCG led to large increases in LAM-specific IgG and that these antibodies can enhance both cell-mediated and innate immunity to tuberculosis (*25,26*). Together, this body of work suggests that neoglycoproteins containing synthetic fragments of LAM have the potential to be useful anti-tuberculosis vaccines.

Over the past ten years, a number of syntheses of oligosaccharide fragments of LAM have been reported. Among these are the impressive synthesis of a pseudo-dodecasaccharide related to the PI core (**3**, Figure 4) by Fraser-Reid and coworkers (*27*). This synthesis follows earlier work in this area by the van Boom group, which was the first to synthesize phosphoinositide fragments of LAM (*28,29*). In addition, a variety of arabinofuranose-containing oligosaccharides have been synthesized, representing structures that are found both internally in the arabinan domain (*30*) as well as at the termini. With regard to the synthesis of terminal fragments, hexasaccharide **1** has been synthesized as its methyl (*31,32,33*) and 8-aminooctyl glycoside (*34*). Related pentasaccharide (*35,36*) and octasaccharide (*37*) derivatives, all containing the key β-arabinofuranosyl residues, have also been synthesized. Finally, oligosaccharides containing the capping units have been prepared. Among them are those with the mannopyranose oligosaccharide capping structures (*34, 38,39,40*) as well as those containing the inositol phosphate (*41*) and 5-methylthio-xylofuranose (*20*) motifs.

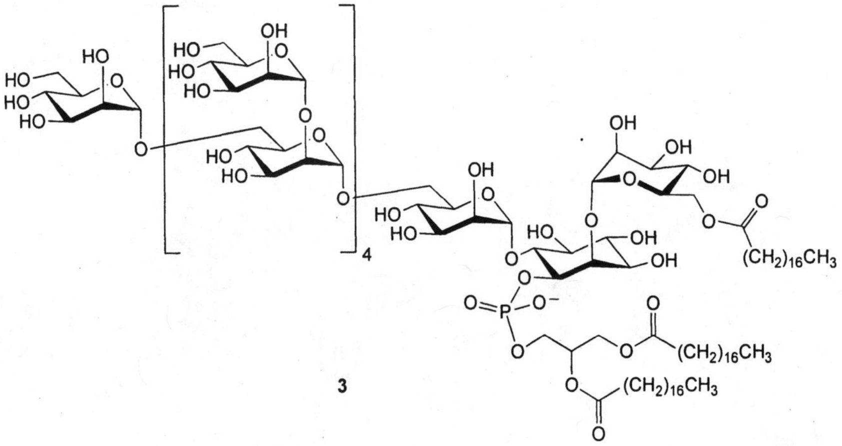

Figure 4. Structure related to the PI core of mycobacterial LAM synthesized by Fraser-Reid and coworkers (27).

Discussion

In choosing oligosaccharide fragments of LAM that could be used in the generation of neoglycoconjugates for vaccine development, we reasoned that a logical approach would be to select motifs present in the terminus of the polysaccharide as these are expected to be displayed at the periphery of the cell wall complex where they would interact with the immune system of the host. Thus, we developed methods for the synthesis of LAM fragments containing the mannopyranose capping motifs (*34, 38*), as well as the terminal hexa- and tetraarabinofuranoside structures (*32–34*), and those containing the 5-methylthio-xylofuranose residues (*20*).

Our first syntheses (*38*) of the mannopyranose-capped fragments were the preparation of a homologous a series of octyl glycosides containing a single β-arabinofuranosyl residue (**4–6**, Figure 5). Subsequent work (*34*) in this area involved the preparation of the corresponding 8-aminooctyl glycosides (7–9), which are appropriately functionalized for the synthesis of neoglycoconjugates. The route to these targets involved the sequential addition of mannopyranose residues to the growing glycan chain, making use of a mannopyranosyl thioglycoside donor. As an example of the approach, the synthesis of **7** and **8** is outlined in Figure 6. In addition to their potential use in the generation of glycoconjugate vaccines, these oligosaccharides found application in studies on elucidating the interaction of mycobacterial polysaccharides with the C-type lectins DC-SIGN, L-SIGN and SIGNR1 (*42*).

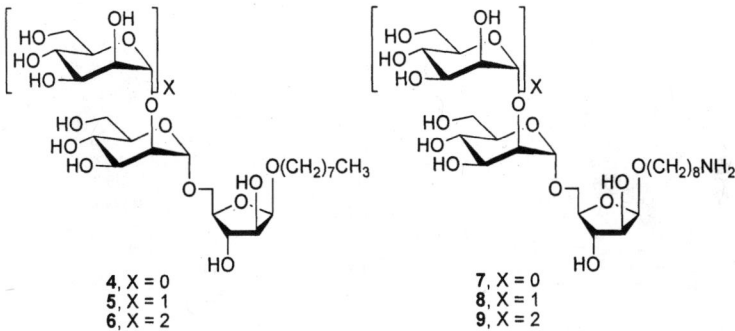

Figure 5. Synthetic fragments of ManLAM (34,38).

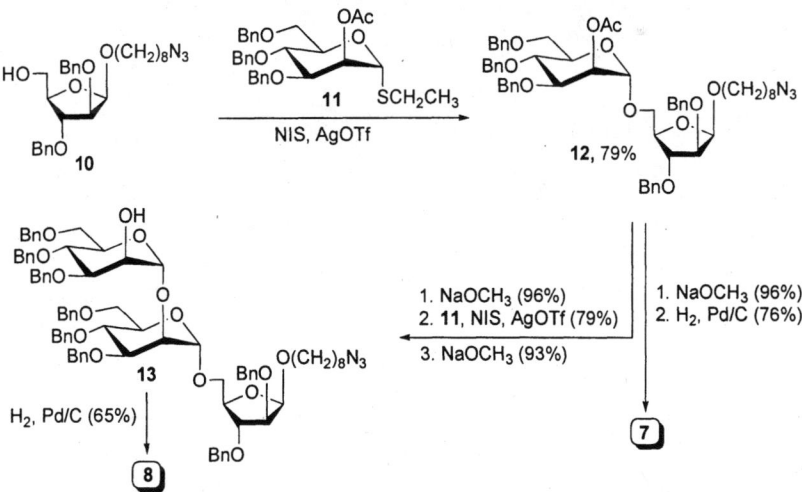

Figure 6. Synthesis of 7 and 8 (34). NIS = N-iodosuccinimide; AgOTf = silver trifluoromethanesulfonate.

Our work (*31,32*) on the synthesis of the branched hexasaccharide motif **1** was initiated with the preparation of its methyl glycoside (**14**, Figure 7). The key step in this synthesis was the simultaneous introduction of the two β-arabinofuranoside residues by the low-temperature reaction of tetrasaccharide diol **15** with an excess of thioglycoside **16** (Figure 7). This same approach was subsequently used to synthesize all di-, tri-, tetra-, and pentasaccharide fragments of **1**, as methyl glycosides (*32*). A later extension of this work was the synthesis of an 8-aminooctyl glycoside analog of **14** (*34*), which has recently been used to develop a highly selective diagnostic method for detecting tuberculosis in human sera (*43*).

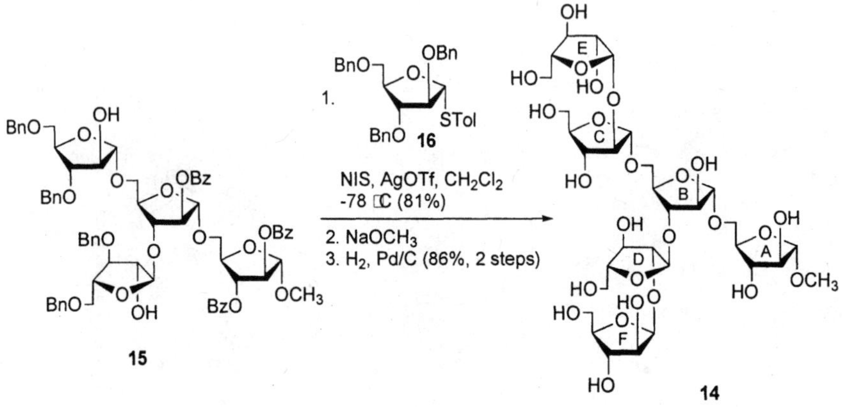

*Figure 7. Synthesis of hexasaccharide **14** (31,32). NIS = N-iodosuccinimide; Tol = p-tolyl; AgOTf = silver trifluoromethanesulfonate.*

The low-temperature glycosylation approach illustrated in Figure 7 is effective for the preparation of hexasaccharide **14** and its 8-aminooctyl counterpart. However, this method was less effective for the synthesis of some of the fragments (*32*), which underscores the challenge in the stereoselective synthesis of β-arabinofuranosides. Often, even seemingly modest changes in the structures of either of the coupling partners lead to dramatic differences in stereoselectivity (*37,44*). We have therefore developed an alternate approach, in which the two β-arabinofuranosyl residues are introduced in a two-step approach as illustrated in Figure 8 (*33*). Thus, reaction of tetrasaccharide diol **17** with an excess of the 2,3-anhydrosugar glycosyl sulfoxide **18** yielded hexasaccharide **19** in excellent yield and stereoselectivity. Mechanistic investigations have confirmed that the stereoselectivity of these glycosylations arises from the in situ generation of a glycosyl triflate intermediate (*45*). The epoxide rings were then opened upon reaction with lithium benzylate in the presence of (−)-spartiene and the product cleanly deprotected affording **14** (*33*).

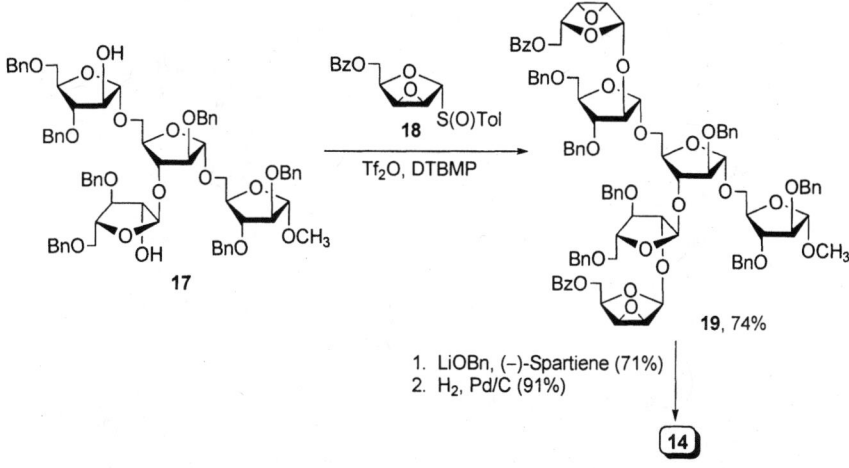

*Figure 8. Synthesis of hexasaccharide **14** via 2,3-anhydrosugar glycosyl sulfoxide donors (33). Tf$_2$O = trifluoromethanesulfonic acid anhydride; Tol = p-tolyl; DTBMP = 2,6-di-t-butyl-4-methylpyridine.*

With synthetic routes to oligosaccharide fragments of LAM in place, we began to explore the possibility of using these compounds in the development of anti-tuberculosis vaccines. To determine which structures would be the most suitable epitopes, we chose to probe the specificity of an antibody that is known to recognize mycobacterial LAM. Therefore, using ELISA techniques, we collaborated with scientists at Colorado State University to demonstrate that the CS-35 antibody (*46*), which was generated against LAM from *M. leprae*, recognizes hexasaccharide **14** (*47*). In addition, we also showed that tetrasaccharide **20** (Figure 9), a fragment of **14** is bound by the antibody, but that another tetrasaccharide fragment, **21** is not, therefore suggesting that the protein recognizes an extended portion of the glycan involving residues A, B, C, and E.

Recently, we have further probed the interaction between CS-35 and its epitope, through the evaluation of a larger panel of oligosaccharides against the protein. Thus, a panel of 14 di-, tri-, tetra-, and pentasaccharide fragments of **14**, (**20–33**, Figures 9 and 10) were evaluated as ligands for the protein using Frontal-Affinity Chromatography Mass Spectrometry (FAC-MS) as the screening method. FAC-MS is a rapid method for determining the relative affinity of ligands for a given receptor and can also be used to determine binding constants (*48*). Screening of this panel of oligosaccharides against the antibody

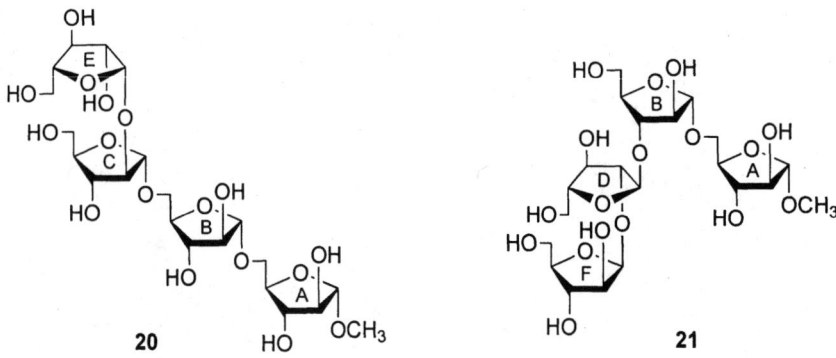

*Figure 9. Tetrasaccharides evaluated as ligands for the CS-35 antibody by ELISA (46). Rings are lettered to facilitate comparison with **14** (see Figure 7).*

revealed that only two of them, pentasaccharide **22** and tetrasaccharide **20** bound, which is in line with previous observations (47). Interestingly, an isomeric pentasaccharide (**23**) did not bind to CS-35 thus demonstrating the critical importance of residue E in **14**, **20** and **22** for recognition by the protein. Furthermore, tetrasaccharide **24**, which contains residues B, C and E, also did not bind to the antibody, thus indicating the residue A in **14**, **20** and **22** is essential for recognition by CS-35. Dissociation constants (K_d's) were measured for **14**, **20** and **22**, and were shown to be 4.9 µM, 32.6 µM and 6.2 µM, respectively.

On the basis of these data, we propose that the protein binds to hexasaccharide **14** in an extended binding pocket, making significant contact with residues A, B, C and E and that additional affinity to the protein is provided by residue D. In contrast, given the nearly identical K_d values of pentasaccharide **22** and hexasaccharide **14** (4.9 µM for **14** vs. 6.2 µM for **22**) it appears that residue F does not interact significantly with the protein. These results therefore suggest that conjugation of the glycan portion of pentasaccharide **20** to protein will provide a suitable hapten for the generation of vaccines and that the entire hexasaccharide motif may not be required. Efforts to generate the appropriate neoglycoconjugates for vaccine trials are currently on-going, as are studies aimed at further elucidating the binding of these oligosaccharides by the protein. These structural studies involve X-ray crystallographic investigations of the Fab fragment of CS-35 bound to hexasaccharide **14**, as well as STD-NMR (49) and FT-ICR (50) studies of the interaction between the antibody and these oligosaccharides.

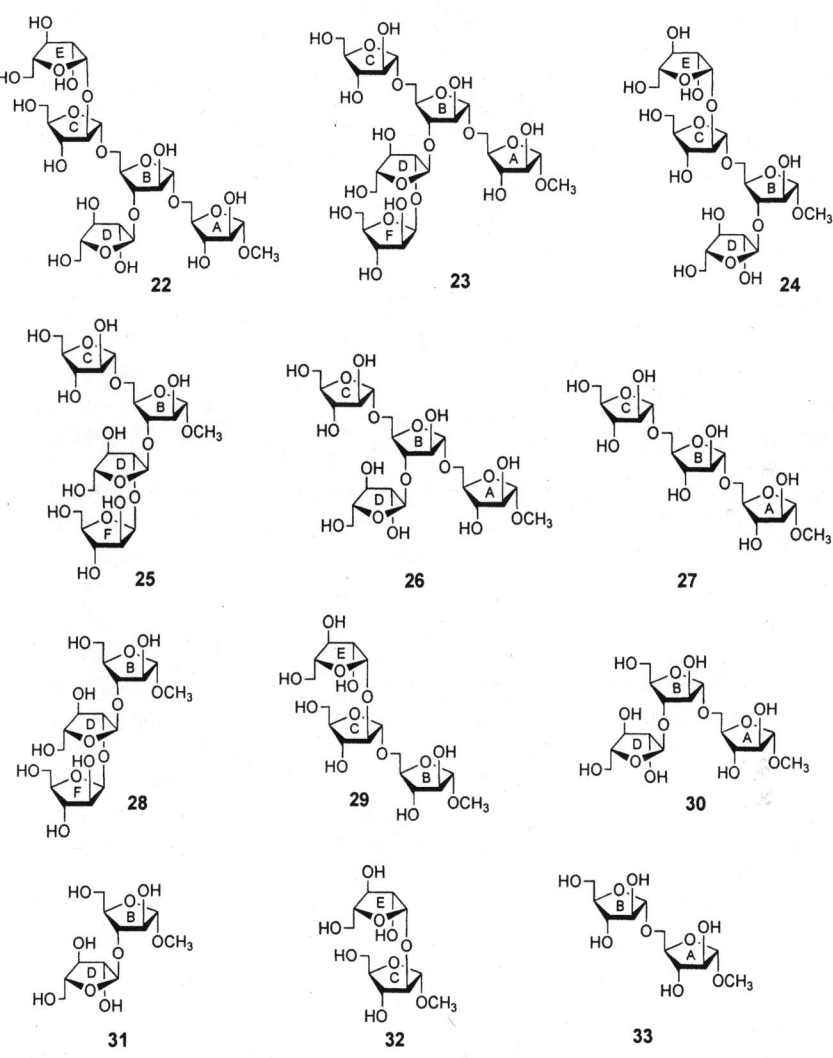

*Figure 10. Oligosaccharide fragments of LAM screened as ligands for the CS-35 antibody using FAC-MS. Rings are lettered to facilitate comparison with **14** (see Figure 7).*

Conclusions

In summary, recent investigations suggest that oligosaccharide-based conjugate vaccines employing fragments of mycobacterial LAM have potential utility in the prevention of tuberculosis. Methods are now in place in our and other laboratories for the synthesis of the required oligosaccharides and preliminary studies have been undertaken to identify the minimum epitopes required for protection.

Experimental

Production of mAb CS-35. A stock vial of mAb CS-35 was removed from storage in liquid nitrogen, and quickly thawed in a water bath. The cells were transferred to a 15 mL conical tube, and 10 mL of BD media (BD Bioscience) was added slowly. The number of viable cells was determined using 0.1% trypan blue stain in a 1:1 ratio and a hemocytometer. The cells were resuspended in the appropriate volume of media to obtain the desired cell concentration (500,000–1,000,000 cells/mL) after centrifugation. The cells (10 mL) were transferred to a 25 mL flask (T-25) and placed in a humidified CO_2 (8%) incubator for 2 days. Cells were split in a 1:10 ratio (established culture:fresh media) until cell viability was greater than 95%. The cells (3×10^7) were resuspended in 15 mL BD media with 20% heat inactivated fetal bovine serum (HI-FBS) and transferred to the small compartment of an artificial mouse (a CELLLine device, BD Bioscience). BD media (1 L) was added to the large compartment, and the artificial mouse was placed in the incubator. After one week, the cells were removed and centrifuged. The supernatant containing the Ab was collected and frozen. The cell pellets were resuspended in BD media with 20% HI-FBS (15 mL) and returned to the the small compartment of the artificial mouse. These processes were repeated until cell viability dropped below 20%.

Purification of mAb CS-35. The collected supernatant was adjusted to pH 7–9 with 1M Tris (pH 8.0), and ammonium sulfate was added to a final concentration of 50%. The solution was stirred at 4 °C for 2 h and centrifuged. The pellets were resuspended in 0.75 M Tris buffer, 0.15 M NaCl, pH 8.0, and ammonium sulfate was added again to a final concentration of 50%. The solution was stirred at 4 °C for 2 h. After centrifugation, the pellets were resuspended in 0.05 M Tris buffer, 0.15 M NaCl, pH 8.0 and dialyzed overnight at 4 °C against 2 L of this buffer (pH 8.0). The dialyzed antibody was loaded on to a protein A column with dialysis buffer, and the antibody was eluted with 0.1 M citric acid buffer (pH 4.0). The eluate was neutralized and dialyzed to obtain the purified Ab as verified by SDS–PAGE (12.5%). The amount of protein was determined by UV spectrometry (A_{280}). The OD value was divided by the extinction coefficient for a monoclonal antibody (1.3) to determine the amount of protein.

Biotinylation of mAb CS-35. Biotinylated mAb CS-35 was prepared using either a commercially available biotinylation kit (Amersham) or a biotinylating reagent (Pierce). When using the biotinylation kit, 2.5 mg of mAb CS-35 was diluted with 0.04 M sodium bicarbonate (1 mg/mL). The biotinylation reagent was added to the solution and the mixture was agitated for 1 h at rt. The biotinylated antibody was purified on a Sephadex G25 column. To biotinylate mAb CS-35 using the Biotin-LC-Hydrazide reagent, cold 0.02 M sodium metaperiodate solution (1 mL) was added to a 2 mg/mL solution of mAb CS-35 in 0.1 M NaOAc buffer, pH 5.5, and the oxidation reaction was allowed to proceed for 30 min at 0 °C in the dark. Glycerol was added to the reaction mixture up to a final concentration of 15 mM, and the solution was incubated for 5 min at 0 °C. The reaction mixture was dialyzed overnight against 0.1 M NaOAc buffer, pH 5.5. After dialysis, 0.05 M biotin hydrazide in DMSO was added to the solution up to a final concentration of 0.005 M, and then the reaction was left to proceed for 2 h at rt with continuous agitation. The reaction solution was centrifuged to remove unreacted reagent using a Centricon filter (molecular weight cut off: 10,000) and the resulting solution was stored in Phosphate Buffered Saline (PBS) buffer. Confirmation of biotinylation with mAb CS-35 was performed by a Western Blot.

Preparation of affinity column for FAC-MS analysis. An affinity column was prepared with silica capillary tubing. Either 2.5 or 5 cm capillary tubing (250 µM ID, 360 µM OD) was packed with Streptavidin-CPG (controlled-pore glass) beads (Purebiotech, 37–74 µm), and mAb CS-35 was immobilized by infusion in PBS buffer (pH 7.2, 0.5 mL of ~1000 pM) onto the column. To prevent nonspecific binding by substrates any unoccupied sites on the streptavidin were blocked with *d*-biotin.

FAC-MS assay. The FAC-MS system used was an integrated module composed of two high flow rate precision nanofluidic delivery systems and a six-port valve (Scivex Confluent Nano Fluidic Module, Upchurch). The eluent buffer, 0.01 M NH$_4$OAc buffer (pH 7.2), was delivered by pump A at a flow rate of 1 µL/min through the six port injection valve, where library compounds were injected. The FAC column was connected from the outlet of the injection valve to a micro tee. A 90% methanol solution was introduced through pump B of the FAC system at a flow rate of 4 µL/min.[51] In online mode, the combined flow was directed into an electrospray mass spectrometer (Agilent 1100 MSD, Model A). For characterization of the compound set (quasimolecular ion determination), the spectrometer scanned from 150–1000 *m/z* in positive-ion mode. The elution of the compounds from the FAC column was continuously monitored by selected ion monitoring (SIM) of the (M+Na)$^+$ ion for each compound in the mixture. Breakthrough volumes were measured as midpoints in the extracted ion chromatograms. All data were processed with Microsoft Excel software.

Acknowledgements

We thank the Alberta Ingenuity Centre for Carbohydrate Science, NSERC and the NIH (USA) for partial support of this work. The cell line producing mAb CS-35 was obtained from Dr. John Spencer (Colorado State University) under the terms of the NIH-sponsored contract Tuberculosis Research Materials and Vaccine Testing Contract (NO1, AI-75320).

References

1. Paolo, W. F. Jr.; Nosanchuk, J. D. *Lancet Infect. Dis.* **2004**, *4*, 287–293.
2. Davies, P. D. O. *Ann. Med.* **2003**, *35*, 235–243.
3. Coker, R. J. *Trop. Med. Int. Health* **2004**, *9*, 25–40.
4. Sharma, K.; Chopra, P.; Singh, Y. *Expert Opin. Ther. Targets* **2004**, *8*, 79–93.
5. Kremer, L.; Besra, G. S. *Expert Opin. Inv. Drugs* **2002**, *11*, 153–157.
6. Haile, M.; Källenius, G. *Curr. Opin. Infect. Dis.* **2005**, *18*, 211–215.
7. Fine, P. E. M.; Carneiro, I. A. M.; Milstien, J. B.; Clements, J. C. *WHO/V&B/99.23* **1999**, *23*, 1–42.
8. Fine, P. E. M. *Lancet* **1995**, *346*, 1339–1345.
9. Fine, P. E. M. *Scand. J. Infect. Dis.* **2001**, *33*, 243–245.
10. Hewinson, R. G. *Tuberculosis* **2005**, *85*, 1–6
11. Glatman-Freedman, A.; Casadevall, A. *Clin. Microbiol. Rev.* **1998**, *11*, 514–532.
12. Glatman-Freedman, A. *FEMS Immunol. Med. Microbiol.* **2003**, *39*, 9–16.
13. Casadevall, A.; Pirofski, L. *Curr. Molec. Med.* **2005**, *5*, 421–433.
14. Nigou, J.; Gilleron, M.; Puzo, G. *Biochemie* **2003**, *85*, 153–166.
15. Nigou, J.; Gilleron, M.; Cahuzac, B.; Bournery, J. D.; Herold, M.; Thurnher, M.; Puzo, G. *J. Biol. Chem.* **1997**, *272*, 23094–23103.
16. Khoo, K.-H.; Tang, J. B.; Chatterjee, D. *J. Biol. Chem.* **2001**, *276*, 3863–3871.
17. Khoo, K.-H.; Dell, A.; Morris, H. R.; Brennan, P. J.; Chatterjee, D. *J. Biol. Chem.* **1995**, *270*, 12380–12389.
18. Turnbull W. B.; Shimizu, K. H.; Chatterjee, D.; Homans, S. W.; Treumann, A. *Angew. Chem. Intl. Ed.* **2004**, *43*, 3918–3922.
19. Treumann, A.; Feng, X. D.; McDonnell, L.; Derrick, P. J.; Ashcroft, A. E.; Chatterjee, D.; Homans, S.W. *J. Mol. Biol.* 2002, 316, 89–100.

20. Joe, M.; Sun, D.; Taha, H.; Completo, G. C.; Croudace, J. E.; Lammas, D. A.; Besra, G. S.; Lowary, T. L. *J. Am. Chem. Soc.* Submitted.
21. Guérardel, Y.; Maes, E.; Elass, E.; Leroy, Y.; Timmerman, P.; Besra, G. S.; Locht, C.; Strecker, G.; Kremer, L. *J. Biol. Chem.* **2002**, *277*, 30635–30648.
22. Briken, V.; Porcelli, S. A.; Besra, G. S.; Kremer, L. *Mol. Microbiol.* **2004**, *53*, 391–403.
23. Hamasur, B.; Haile, M.; Pawlowski, A.; Schröder, U.; Williams, A.; Hatch, G.; Hall, G.; Marsh, P.; Källenius, G.; Svenson, S. B. *Vaccine* **2003**, *21*, 4081–4093.
24. Hamasur, B.; Haile, M.; Pawlowski, A.; Schröder, U.; Källenius, G.; Svenson, S. B. *Clin. Exp. Immunol.* **2004**, *138*, 30–38.
25. De Vallière, S.; Abate, G.; Blazevic, A.; Heurtz, R. M.; Hoft, D. F. *Infect. Immun.* **2005**, *73*, 6711–6720.
26. Brown, R. M.; Cruz, O.; Brennan, M.; Gennaro, M. L.; Schlesinger, L.; Skeiky, Y. A.; Hoft, D. F. *J. Infect. Dis.* **2003**, *187*, 513-517.
27. Jayaprakash, K. N.; Lu, J.; Fraser-Reid, B. *Angew. Chem. Intl. Ed.* **2005**, *44*, 5894–5898.
28. Elie, C. J. J.; Verduyn, R.; Dreef, C. E.; Brounts, D. M.; van der Marel, G. A.; van Boom, J. H. **1990**, *Tetrahedron 46*, 8243–8254.
29. Elie, C. J. J.; Verduyn, R.; Dreef, C. E.; van der Marel, G. A.; van Boom, J. H. **1992**, *J. Carbohydr. Chem. 11*, 715–739.
30. Lu, J.; Fraser-Reid, B. *Chem. Commun.* **2005**, 862–864.
31. D'Souza, F. W.; Lowary, T. L. *Org. Lett.* **2000**, *2*, 1493–1495.
32. Yin, H.; D'Souza, F. W.; Lowary, T. L. *J. Org. Chem.* **2002**, *67*, 892–903.
33. Gadikota, R. R.; Callam, C. S.; Wagner, T.; Del Fraino, B.; Lowary, T. L. *J. Am. Chem. Soc.* **2003**, *125*, 4155-4165
34. Gadikota, R. R.; Callam, C. S.; Appelmelk, B. J.; Lowary, T. L. *J. Carbohydr. Chem.* **2003**, *22*, 459-480.
35. Mereyala, H. B.; Hotha, S.; Gurjar, M. K. *J. Chem. Soc. Chem. Commun.* **1998**, 685-687.
36. Sanchez, S.; Bamhaoud, T.; Prandi, J. *Tetrahedron Lett.* **2000**, 41, 7447-7452.
37. Lee, Y. J.; Lee, K.; Jung, E. H.; Jeon, H. B.; Kim, K. S. *Org. Lett.* **2005**, *7*, 3263–3266.
38. Subramaniam, V.; Lowary, T. L. *Tetrahedron,* **1999**, *55*, 5965–5976
39. Bamhaoud, T.; Sanchez, S.; Prandi, J. *Chem. Commun.* **2000**, 659–660.
40. Marotte, K.; Sanchez, S.; Bamhaoud, T.; Prandi, J. *Eur. J. Org. Chem.* **2003**, 3587–3598.
41. Désiré, J.; Prandi, J. *Carbohydr. Res.* **1999**, *317*, 110–118.
42. Koppel. E. A.; Ludwig, I. S.; Sanchez-Hernandez, M.; Lowary, T. L.; Gadikota, R. R.; Bovin, N.; Vandenbroucke, C. M. J. E, van Looyk, Y.; Appelmelk, B. J.; Geijtenbeek, T. B. H. *Immunobiol.* **2004**, *209*, 117–127.

43. Tong, M.; Jacobi, C. E.; van de Rijke, F. M.; Kuijper, S.; van de Werken, S.; Lowary, T. L.; Hokke, C. H.; Appelmelk, B. J.; Nagelkerke, N. J. D.; Hans J. Tanke, van Gijlswijk, R. P. M.; Kolk, A. H. J.; Raap, A. K. *J. Immunol. Meth.* **2005**, *301*, 154–163.
44. Yin, H.; Lowary, T. L. *Tetrahedron Lett.* **2001**, *42*, 5829–5832.
45. Callam, C. S.; Gadikota, R. R.; Krein, D. M.; Lowary, T. L. *J. Am. Chem. Soc.* **2003**, *125*, 13112–13119.
46. Hunter, S. W.; Gaylord, H.; Brennan, P. J. *J. Biol. Chem.* **1986**, *261*, 12345–12351.
47. Kaur, D.; Lowary, T. L.; Vissa, V. D.; Crick, D. C.; Brennan, P. J. *Microbiology UK*, **2002**, *148*, 3059–3067.
48. Palcic, M. M.; Zhang, B.; Qian, X. P.; Rempel, B.; Hindsgaul, O. *Methods Enzymol.* **2003**, *362*, 369–376.
49. Mayer, M.; Meyer, B. *Angew. Chem. Int. Ed.* **1999**, *38*, 1784–1788.
50. Kitova, E. N.; Bundle, D. R.; Klassen, J. S. *J. Am. Chem. Soc.* **2002**, *124*, 5902–5913.
51. Ng, E. S. M.; Yang, F.; Kameyama, A.; Palcic, M. M.; Hindsgaul, O.; Schriemer, D. C. *Anal. Chem.* **2005**, *77*, 6125–6133.

Chapter 10

The Lipoarabinomannan Glycolipid of *Mycobacterium tuberculosis*: Progress in Total Synthesis via n-Pentenyl Orthoesters

Bert Fraser-Reid, Jun Lu, K. N. Jayaprakash, and Siddhartha Ray Chaudhuri

Natural Products and Glycotechnology Research Institute Inc. (NPG), 595F Weathersfield Road, Pittsboro, NC 27312

The lipoarabinomannan (LAM) cell surface glycolipid of *Mycobacterium tuberculosis* anchors a complex, highly branched glycan comprising arabinofuranan, mannan, and mannose-capped arabinan domains. This multifaceted architecture mirrors a multifaceted biological profile that implicates LAM in a wide range of health disorders, tubercular and non-tubercular, in addition to serving its principal role of protecting the bacterium against assault by diagnostic and therapeutic agents. Syntheses of the various domains of LAM have been undertaken to assist in structure-activity analysis. The approach revives Paulsen's concept of donor/acceptor MATCH as a prime requirement in oligosaccharide assembly. A major advantage of the strategy is that if MATCH can be applied to effect monoglycosylation of a polyol, the most labor-intensive synthetic tasks can be avoided, since misMATCHed hydroxyls need not be protected - and subsequently deprotected. *n*-Pentenyl orthoesters (NPOEs) are pivotal for the MATCH approach, and extending the method to arabinofuranosyl donors presents a new tool for furanoside syntheses. Notably, the reducing end of the arabinofuranan oligomers can be liberated by oxidative hydrolysis, thereby avoiding the perils of the acid-catalysed alternative.

Introduction

Tuberculosis is a leading cause of morbidity and mortality worldwide with more than 8 million new cases per year, and causing approximately 2 million deaths (*1*). Approximately 1.86 billion people, a third of the world's population, are infected (*2*). In the "third world", tuberculosis causes an estimated annual death toll of 400,000 children (*3*), and those who survive face an uncertain future, since infection acquired during childhood can serve as a reservoir for future illness, resulting in active disease during adolescence and adulthood.

The treatment of tuberculosis is a complex process, requiring the use of multiple medications for prolonged periods of time. However, the demand of such regimens often results in poor compliance, and this has led to the emergence of multiple drug resistant (MDR) (*4a*) and extremely drug resistant (XDR) (*4b*) strains.

The depredations of tuberculosis are exponentially increased by its synergy with AIDS, for tuberculosis is one of the main causes of death for the afflicted patients (*5*). The convergence of the AIDS epidemic with burgeoning MDR tuberculosis, poses a formidable threat to overcrowded populations centered in prisons, homeless shelters and such at-risk communities (*6*). However, since tuberculosis is air-borne, no segment of the world's population can be thoroughly insulated from it.

Directly observed therapy programs, established to address many of the difficulties associated with the care of infection and disease caused by *Mycobacterium tuberculosis*, have been shown to be effective in certain areas of the world (*7*). However, such programs are costly and logistically complex, and may not be feasible in many parts of the world, particularly for underserved populations. BCG, the only available vaccine against tuberculosis (*8*), was shown to prevent dissemination of the disease in young children, but has not been effective in preventing pulmonary tuberculosis (*9*), the main form of the disease. Furthermore, BCG is losing its efficacy against MDR strains (*6*).

The Complex Lipoarabinomannan Cell Surface Glycolpid

Brennan has described the cell envelope of *Mycobacterium tuberculosis* as a "marvel of chemistry, an absolute treasure-house of unusual compounds" (*10*). Historically, this "treasure house" inhibited diagnosis of the disease until, too late, there was blood in the sputum. That envelope was breached in 1882 by Koch (*11*) which enabled the first systematic assault on the bacillus.

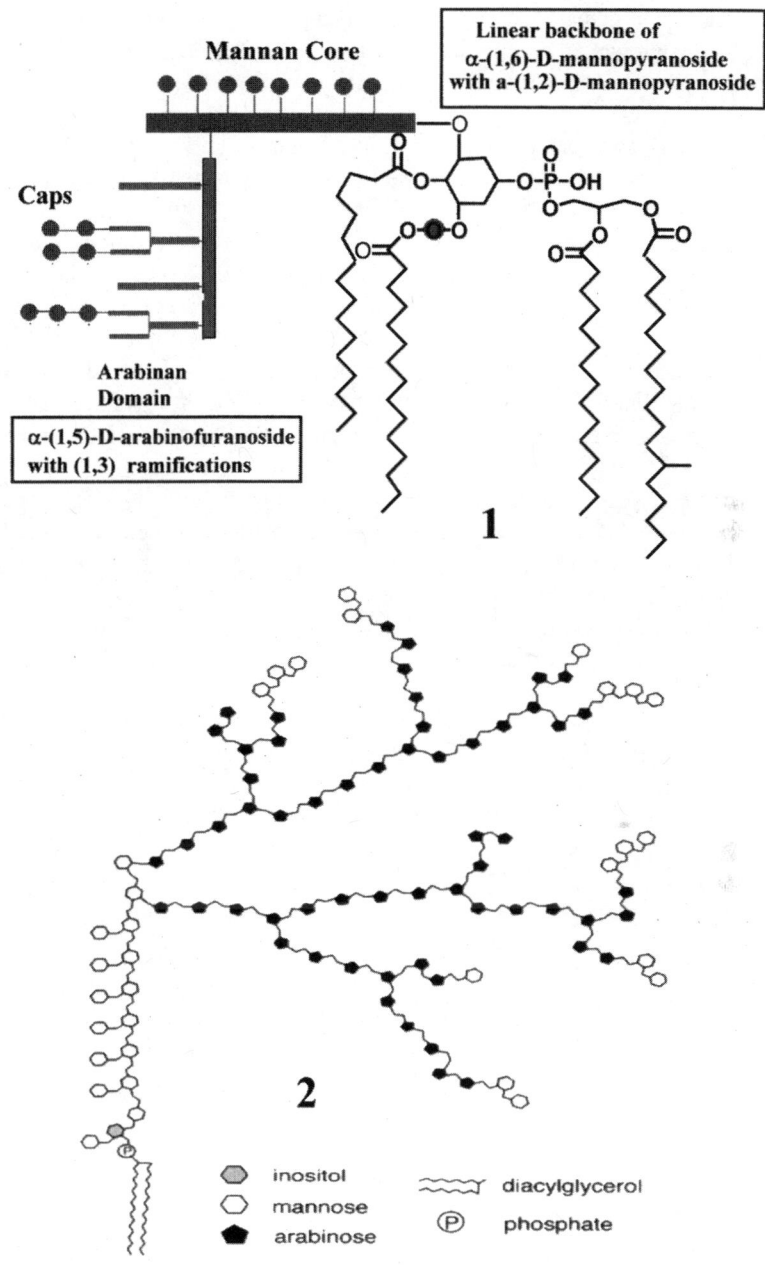

Scheme 1. Cartoons of the lipoarabinomannan from Mycobacterium tuberculosis

The major structural and physiological component of the coat turns out to be a glycolipid known as lipoarabinomannan (LAM). Recent studies reveal that LAM exhibits profound immunomodulatory propensities (*12*), exhibits widespread physiological activity, being able to enhance resistance to cancers (*13*) and herpes (*14*), and also to potentiate HIV retrovirals (*15*).

This multifaceted biological profile of LAM is matched by its multifaceted architecture captured by the cartoons **1** (*16*) and **2** (*17*) (Scheme 1), devised at two major TB research centers. Subunits of LAM, representing biosynthetic stages in its elaboration, are dispersed throughout the cell wall and are of independent interest. For example, the mannose capped arabinans that festoon the distal non-reducing end of LAM (see Scheme 1) are major virulence factors that are suspected of engaging in intracellular signaling (*18*, *19*), and may be the host's earliest line of defense against pulmonary infections. Additionally, the lipomannan domain exhibits strong proinflammatory and apoptosis inducing activity (*20*).

In view of the overwhelming problem of tuberculosis, its health-related tentacles, the difficulties associated with its treatment, and the new potentially harmful prospects, there is an urgent need to develop more effective strategies for the prevention of tuberculosis. The Global Alliance on TB, The Bill and Melinda Gates Foundation, and AERAS the Global TB Vaccine Foundation are collaborating in the effort to develop urgently needed therapeutic agents and vaccines.

Syntheses of LAM, in part and in whole, are a pressing need. A chemist's rendition of cartoons **1** and **2** leads to compound **3** shown in Scheme 2, and articulates, in structural terms, the challenges to synthesis.

Selectivities and Concepts of "MATCH"

With any synthesis of a complex organic target, the most daunting challenges center around selectivities, four modes of which, as identified by Trost, are chemo, regio, (dia)stereo and enantio (*21*). The last is usually not of concern with carbohydrates since the D or L genus of most targets of interest is designated by nature. However, it is interesting to note a study by Spijker and van Boeckel which showed that when D/L donor/acceptor partners "match" anomeric stereoselectivity is high, but low when there is a "mismatch" (*22*). Thus the reaction of **6** and **L5** constitutes a "match" because the expected product, **4β**, predominates (Scheme 3b). However **6** and **D5** are a "mismatch" because product **7β** is minor.

The forgoing usage of "match" is Masamune's concept of double sterodifferentiation (*23*). However, the term "match" had been introduced earlier by Paulsen (*24*) in connection with the coupling of glycosyl donors and

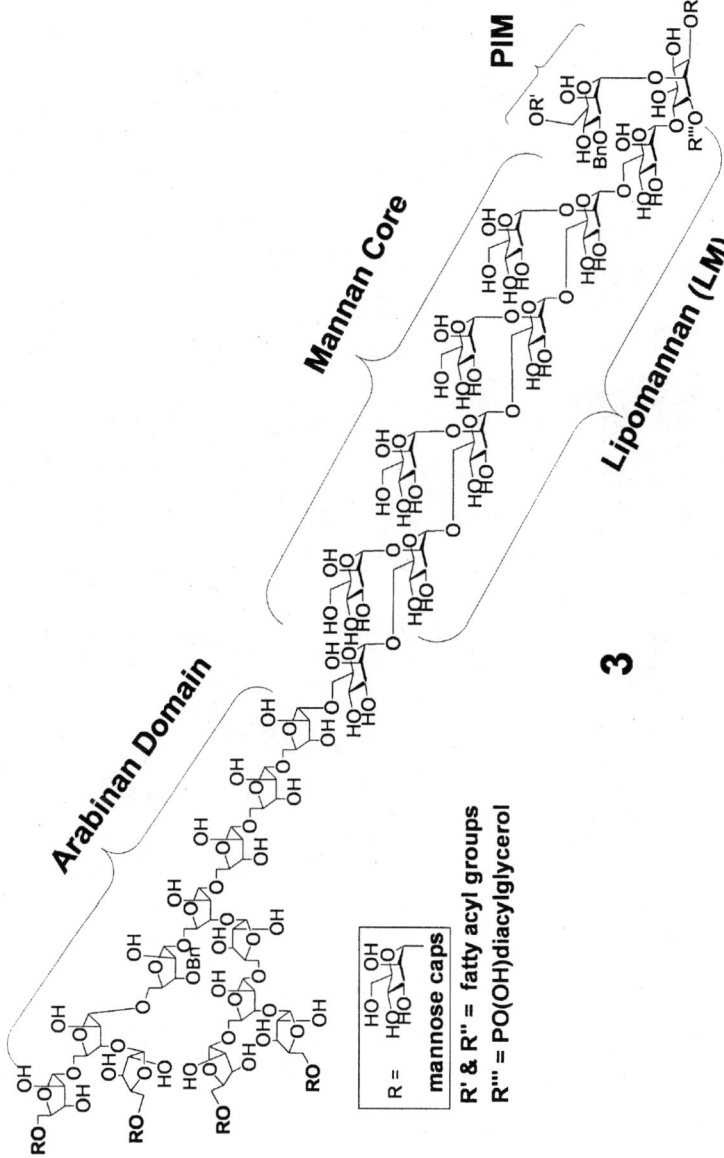

Scheme 2. A Chemists rendition of Cartoons 1 and 2

204

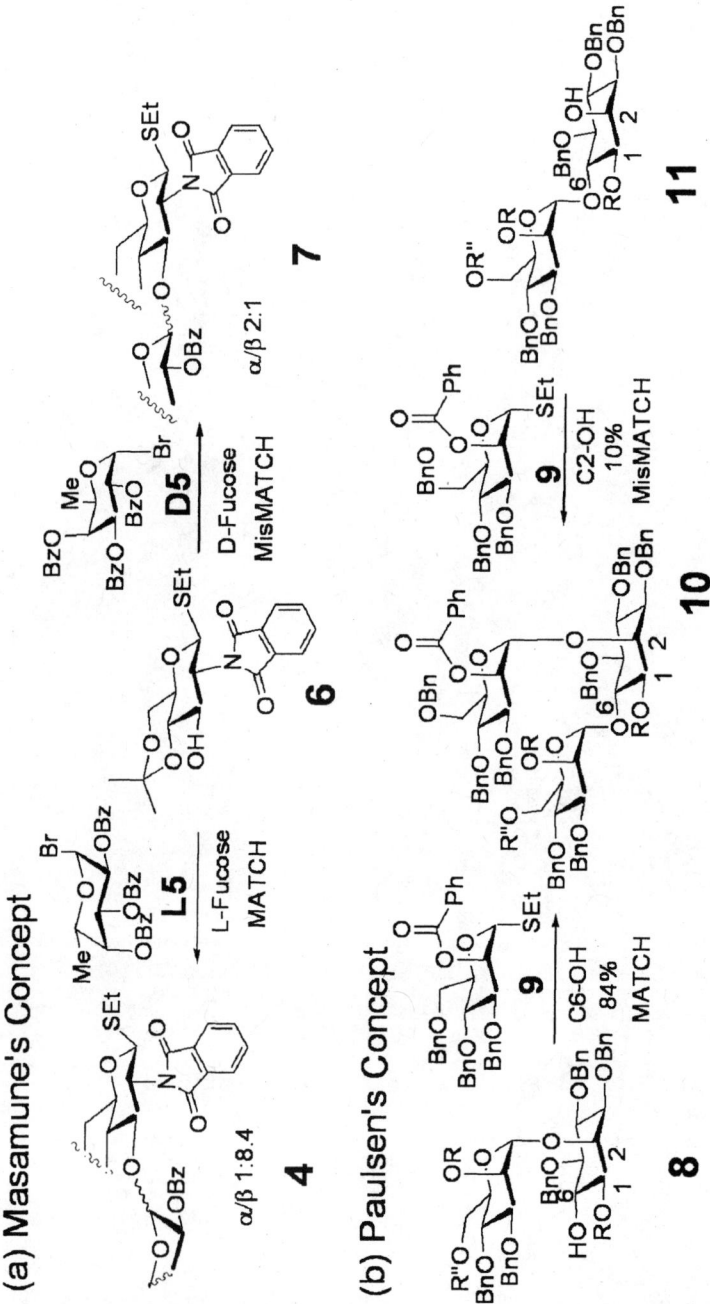

Scheme 3. Concepts of MATCH

acceptors, some of which were more successful than others, as exemplified in Scheme 3b.

Thus, for the synthesis of *pseudo*trisaccharide **10**, there is a "match" between donor **9** and the 6-OH of the acceptor *pseudo*disaccharide **8**, but not between **9** and the 2-OH of the analogous *pseudo*trisaccharide **11**. Explanations for such a match was not pursued by Paulsen because as he noted recently, 25 years ago "we did not have all the new coupling reagents" that are available today (*25*). Consequently discovery of the best match required trial and error.

However such a strategy is in keeping with the best traditions of organic synthesis, because "trial and error" usually serves until a body of critical knowledge is acquired, or careful investigations provide mechanistic insight.

Similarly, rationalization of Paulsen's "match" will emerge, and we hope that this article will contribute to that eventuality.

The Three "Pertinent" Selectivities in Glycoside Coupling

Since enatioselectivity, such as illustrated in Scheme 3a, is usually irrelevant, we are concerned with the other three of Trost's selectivities. The hypothetical example in Scheme 4a shows that all three are encountered in the fundamental process of glycoside syntheses. That stereochemical outcomes can be influenced by the donor's O2 functionality, as indicated for **15** (Scheme 4a), is familiar to carbohydrate chemists ever since Isbell's postulate of neighboring group participation in 1940 (*26*). This monumental observation was early evidence of the fact that with carbohydrate structures, protecting groups do more than protect, foreshadowing our current awareness that all three selectivities shown in Scheme 4a can be influenced by protecting groups (*27*).

Chemoselectivity requires that both donor and acceptor be susceptible to activation by the same reagent(s), but that one of them reacts preferentially. As an example, the term applies in the coupling of two thioglycosides (*28*), two n-pentenyl glycosides (NPGs) (*29*) or one of each (*30*) the reagent in question being halonium. Orthogonal strategies, by contrast, involve, for example, a thioglycoside and a glycosyl fluoride reacting with different reagents (*31*).

Some Problems of Regioselectivity in "Protection" and Glycosidation

Regioselectivity remains the most challenging of the three most pertinent selectivities shown in Scheme 4a. For acceptor polyols such as **19a→22a** (Scheme 5), the primary–OHs can be regioselectively protected by tritylation and silylation (*32*). Some regioselectivity can also be achieved in benzoylation (*33*), as shown for **19** and **20**. However, the well-known vagaries of benzylation are demonstrated with diols **21** and **22** (*34*).

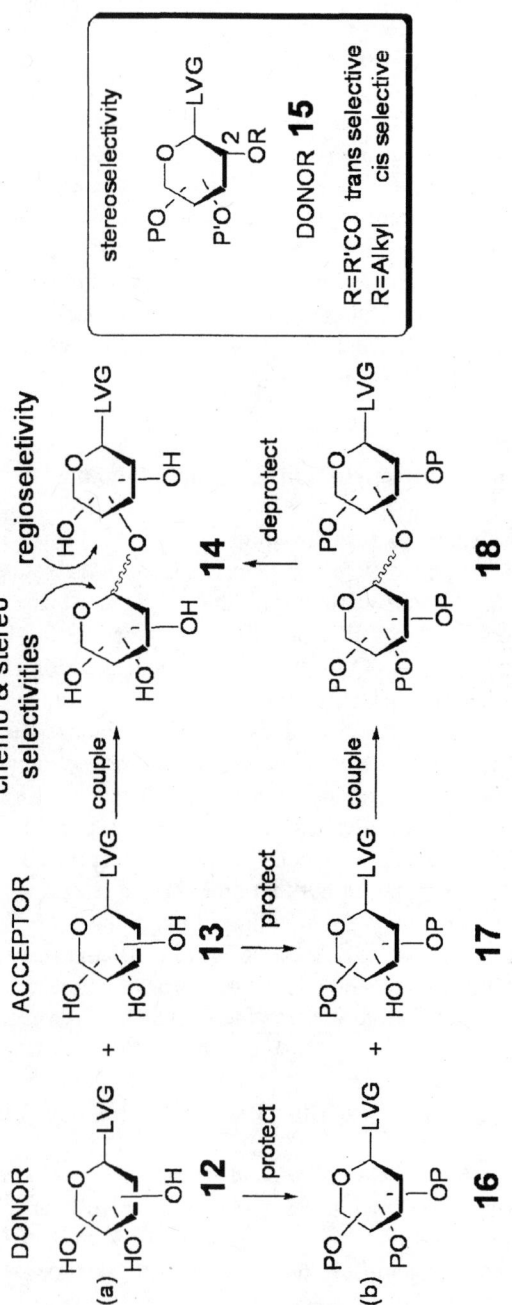

Scheme 4. *The deceptively simple basis of oligosaccharide synthesis*

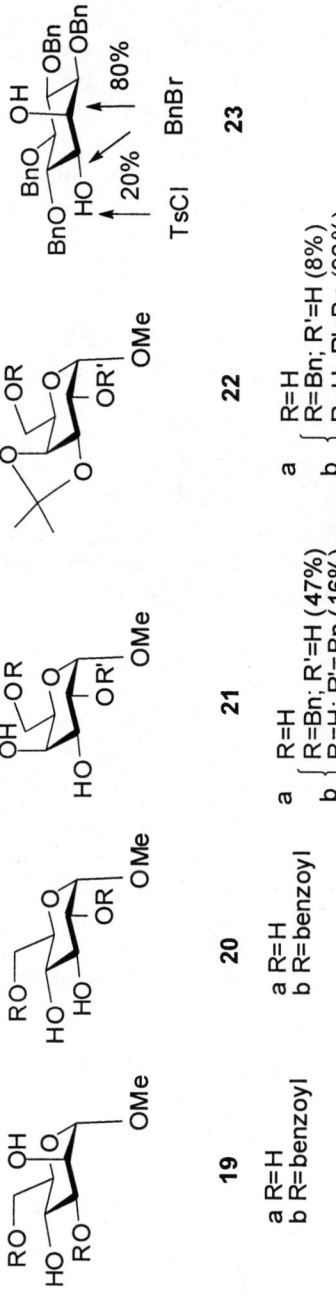

Scheme 5. Problems of regioselective acylation/akylation

With regard to secondary/secondary competitions, Angyal and coworkers found that with diol **23**, the classical equatorial preference applies to tosylation, but not to benzylation, where the axial-OH is preferred (*35*).

Not surprisingly, the problems illustrated in Scheme 5 for simple reagents are exacerbated in glycosidation reactions. Thus, in their "random glycosidation" studies, Hindsgaul and co-workers exposed several hexopyranoside acceptors to perbenzylated trichloroacetimidate or phosphite donors and found that the expectation of α–O6 regiopreference was not fulfilled (*36*). Thus in the example involving **24** and **25** in Scheme 6, taken from their work, there was no substantial difference in the amounts of α-glycosidation at O6, O4 and O3 of the product **26**.

In view of such uncertainties, the accepted practice is to protect all-OHs on the donor and acceptor, except the one that is targeted for coupling.

The *n*-Pentenyl Family of Donors

Reports of regioselective glycosidations are scattered throughout the literature; but in view of the varied experimental conditions that were used, meaningful comparisons cannot be drawn. To provide a level playing field, we confined our early attention to the *n*-pentenyl family of glycosyl donors. Their ready preparation and (some) of their fundamental transformations are illustrated in Scheme 7 (*37*).

For hexoses, e.g **27**, perbenzoylation is followed by conversion to the glycosyl bromide which, upon treatment with pent-4-enol in the presence of lutidine, gives the orthoester **28** (*38*). The three-OH groups of the latter once freed by saponification, can be selectivity differentiated as required. Lewis acid treatment (see also Scheme 7 below) leads to the disarmed n-pentenyl glycoside (NPG$_{AC}$), **29**, and thence the armed counterpart (NPG$_{ALK}$), **30**. Donors **28→30** upon activation with I$^+$ releases iodomethylfuran **31** and generates cationic intermediates **32→34**.

However, a particularly attractive feature as illustrated in Scheme 7c, is that NPOEs can provide ready access to thioglycosides (*39*) and trichloroacetimidates (*40*), the most widely used glycosyl donors. Thus, the anomeric center can be liberated in **35** by mild acid treatment of NPOE **28**, or by prior conversion of **28** into NPG$_{AC}$ (**29**) or NPG$_{ALK}$ (**30**) followed by oxidative hydrolysis (*41*) to **35** (Ar-Bz or Bn). Conversion into **36a** or **b** then follows established procedures (*42*).

Discovery of MATCH

Our interest in regioselective glycosidations emanated from the observation that benzoylation of diol **38** (Scheme 8) occurred exclusively at O6 to give **37**,

Scheme 6. An example of random glycosidation

210

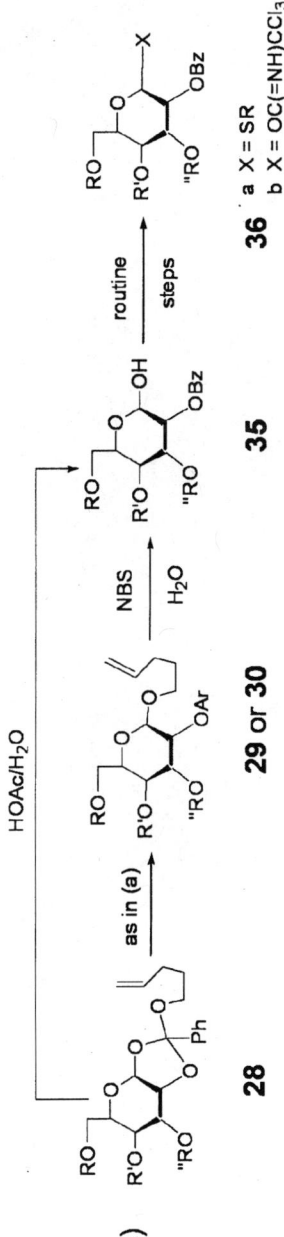

Scheme 7. *Family of n-pentenyl donors*

36 a X = SR
b X = OC(=NH)CCl$_3$

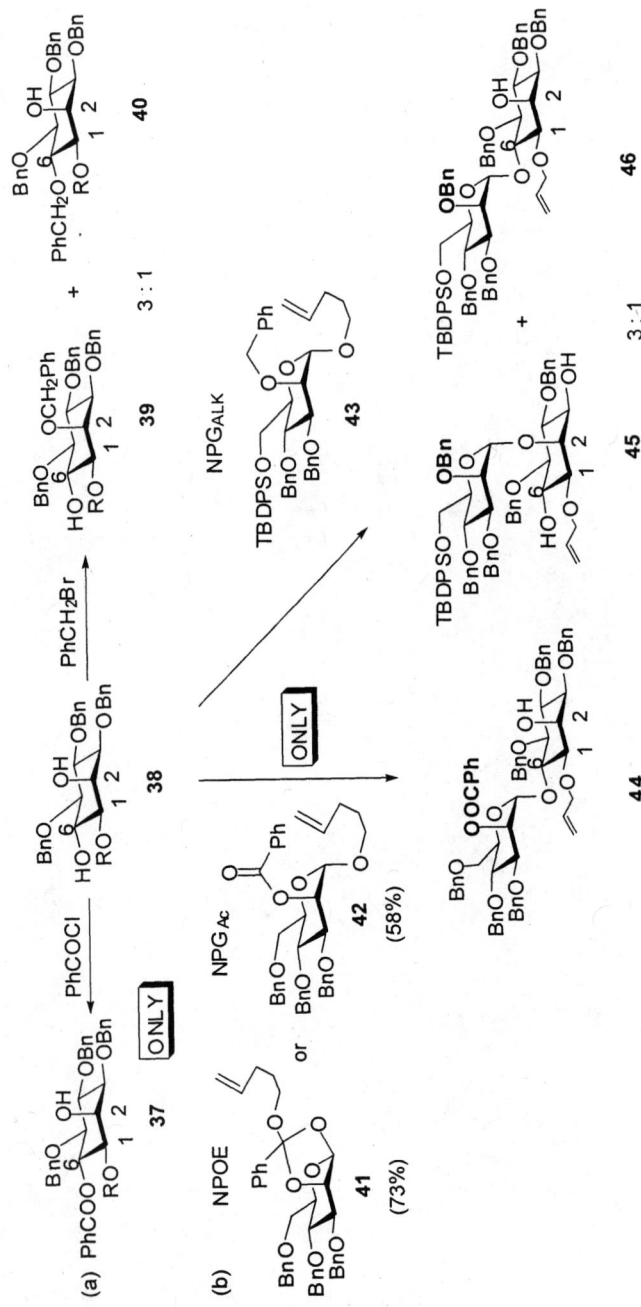

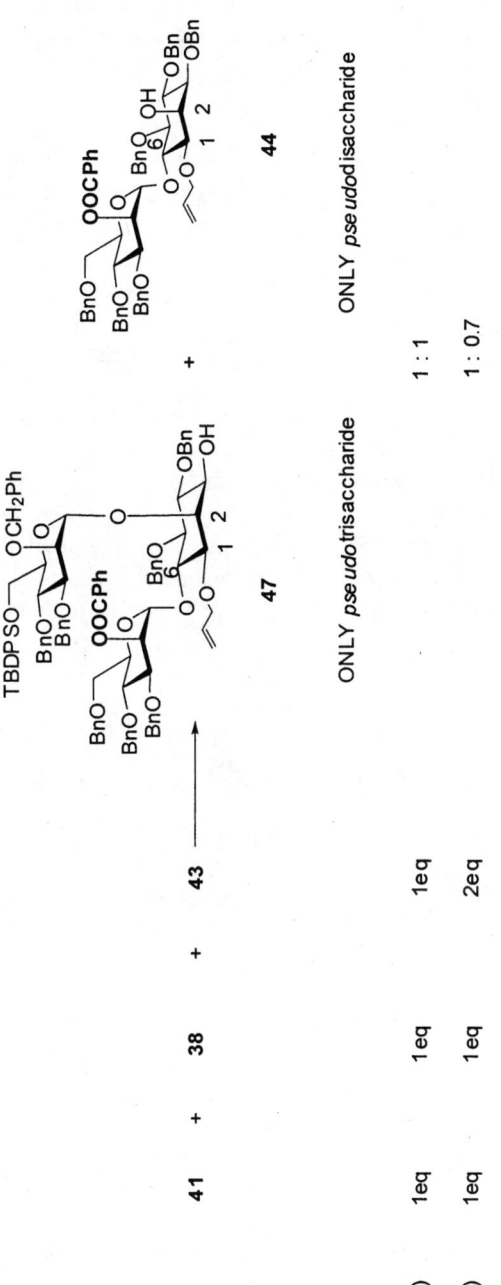

Scheme 8. An example of reciprocal donor acceptor selectivity (RDAS)

whereas benzylation was 3:1 in favor of O2 (*43*), with **39** being the major product. These results were reminiscent of those reported by Angyal and coworkers for the diastereomer **24** (*35*) shown in Scheme 5.

Could these selectivities be extended to glycosylations? If so, the MATCH that had been postulated by Paulsen (*24*), as referenced in Scheme 3b, could be extended to diol acceptors. A clear advantage would be that the diol-OH that does not MATCH would not have to be protected, thereby abandoning the general practice, in oligosaccharide synthesis, of protecting all acceptor-OHs, except for the targeted one, (as in **17** Scheme 4b). The result would be a reduction in the elaborate procedures for orchestrating protecting group deployments which are the most time consuming aspects of oligosaccharide synthesis of any modality, solution (*44*), solid-phase (*45*), programmed (*46*) or automated (*47*).

With diol **38**, we found that disarmed n-pentenyl donors (NPG_{AC}) **42** reacted exclusively at O6 to give **44** only, in parallel with acylating agent in Scheme 8a. By contrast, the armed donor NPG_{ALK}, **43**, was parallel with the alkylating agent, giving a 3:1 ratio of **45** and **46** (*48*).

Since, from Scheme 7b, NPOEs and $NPGs_{AC}$ share the same intermediates **32** and **33**, it came as no surprise that exclusive glycosidation at O6 was also observed in the reaction of NPOE **41** with **38** (Scheme 8b). However, the much higher yield with the NPOE attracted our attention, and we will return to this below.

Evidence for Reciprocal Donor Acceptor Selectivity (RDAS)

The O6-OH of **38 selects** NPOE and NPG_{AC} donors, while the O2-OH does not. On the other hand NPOE and NPG_{AC} donors **select** the O6-OH, and **not** the O2-OH, of **38**. Therefore, the O6-OH of **38** and NPOE/NPG_{AC} mutually select one another; i.e. they exhibit **Reciprocal Donor Acceptor Selectivity (RDAS)** (*49*).

The same RDAS holds true, albeit with less exclusivity, for O2-OH of **38** and $NPG_{ALK.}$

A simple test of the RDAS preferences would be to present BOTH donors simultaneously to diol **38**. All reacting entities would then be able to express their preferences **competitively**. In the event (Scheme 8c), when equimolar amounts of the three reactants were mixed, both donors went to their RDAS preferred-OHs, with the result that ONLY ONE of the four possible *pseudo*trisaccharides, **47**, was formed (*50*).

An even more stringent test came from attempts to enhance the yield of **47**. The *pseudo*disaccharide **44** had been formed in equal amount; but applying standard optimization practice, we reasoned that increasing the ratio of the armed donor **43** would enhance formation of **47** at the expense of **44**. Indeed,

Scheme 8d shows that doubling the ratio of **43** led to an increase in the yield of **47** with concomitant diminution in the amount of **44** (*51*).

These results are notable because in the **non-competitive** situation of Scheme 8b, NPG$_{ALK}$ **43** reacted at O6 to give a substantial amount of **46**. However, when put in competition against NPOE **41** in Scheme 8c the NPG$_{ALK}$ was completely shut out---**even when used in excess as in Scheme 8d.**

Primary versus Secondary Selectivity

That the observations in Scheme 8 were not isolated instances, has been established with several diols (*52*); but of particular relevance to primary/secondary hydroxyl selectivity, is a recent study by Lopez and co-workers (*53*) on diols **48** and **52** using armed and disarmed thioglycoside donors, **49** and **50** (Scheme 9). **The results in entries (i) and (iv) of Scheme 9 show that armed donor 49 preferred the secondary-OH groups of both diols, whereas entries (ii) and (v) show that disarmed donor 50 preferred the primary-OHs.**

Lopez' study indicates further, that NPOE **51**, entries (iii) and (vi), exaggerates the trend shown by **disarmed** thioglycoside donor. Thus, comparison of (ii/v) and (iii/vi) show that with NPOE (a) the secondary-OHs were ignored completely, and (b) the yield of primary-OH glycosidation was substantially higher.

The latter result recalls the higher yield of **41** with NPOE *vis a vis* NPG$_{AC}$ seen in Scheme 8b.

A Reagent Combination for Unparalleled Regioselectivity

The superiority of NPOEs over their disarmed counterparts, which had been seen in Schemes 8b (**38→44**) and 9 (entries ii/iii and v/vi) was an inviting observation that warranted further scrutiny. Unexpected good fortune arose from efforts to find Lewis acid surrogates that could be used to generate I$^+$, needed for n-pentenyl activation, but that would not affect acid labile protecting groups. Experiments showed that lanthanide triflates decomposed *N*-iodosuccinimide (Scheme 10a) and rearranged NPOEs to NPG$_{AC}$ (Scheme 10b) (*54*).

However, further study showed remarkable contrasts between the ytterbium and scandium salts. Thus Scheme 10b also shows that Yb(OTf)$_3$/NIS triggered glycosidation of ROH with the NPOE but not with the NPG$_{AC}$ (*55*).

The advantage of this fortuitous observation is hypothesized in Scheme 10c. Thus, diol **56** could be treated with **excess** NPOE **53**, so as to optimize regioselective monoglycosidation, leading to **57**, in full confidence that any

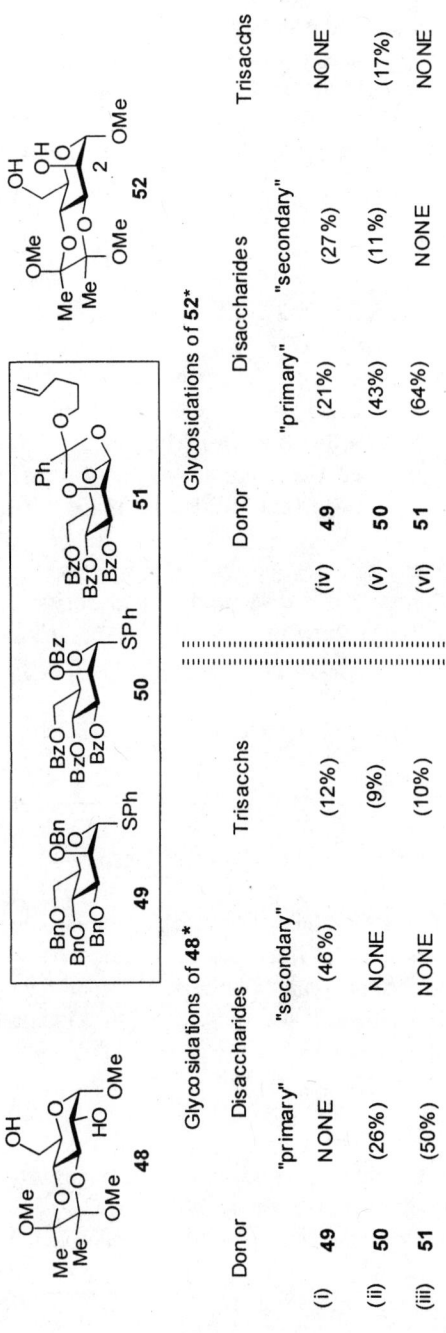

Scheme 9. *Primary versus secondary glycosidations*

NPG$_{AC}$ **54**, produced by rearrangement, would not be activated by Yb(OTf)$_3$, thereby guaranteeing that there would be no further glycosidation of **57**.

On the other hand, since Sc(OTf)$_3$ activates both NPOEs and NPGs$_{AC}$, its use would lead to the double glycosidation product **58**.

The hypothetical situation in Scheme 10c was reduced to practice with a variety of diols. The studies in Scheme 10d show that the specific use of Yb(OTf)$_3$/NIS affords **exquisite** regioselective glycosidation of diols **59, 60, 61** and **62** with NPOE. By contrast, Sc(OTf)$_3$/NIS gave substantial amounts of double glycosidation products in all cases (*56*).

n-Pentenyl Orthoesters in Synthesis of LAM Sub-Domains of *Mycobacterium tuberculosis*

The PIM Moiety[57]

Returning to LAM, **3** (Scheme 2), the biosynthetic chronology of the molecule involves phosphatidyl inositol **[PI]** → phosphatidyl inositol mannoside(s) **[PIMs]** → lipomannan **[LM]** → **[LAM]** (*16*). Therefore our synthetic strategy has been to assemble the various domains according to the biosynthetic pathway (*58*). Our first target was therefore the **PhosphatidylInositolMannoside (PIM)** domain, and our procedure (Scheme 11) (*57*) took advantage of the RDAS lessons learned from the synthesis of **47** in Scheme 8. In so doing, we were also mindful of future requirements to add the "mannan core" of LAM. Thus, regioselective mannosylation of diol **66** using three equivalents of NPOE **67a** promoted by Yb(OTf)$_3$/NIS, afforded *pseudo*disaccharide **68a** exclusively in 96% yield along with NPG$_{Ac}$ **69a** (arising from excess **67**). Separation of **68a** was easily achieved.

In keeping with the findings in Scheme 8, the proper MATCH for the O2-OH of **68** is an armed donor (*48*). Accordingly, NPG$_{ALK}$ **43** was prepared from **67c** by the general procedure described in Scheme 7a, and was coupled, using Sc(OTf)$_3$/NIS, to give compound **70a** in 94% yield. (Notably, glycosidation with an NPOE gave only ~40%).

The α/α component of **70a** was separated very easily after debenzoylation, and subsequent standard transformations, and removal of the allyl protecting group, afforded the monoester **71**. The phosphoglycerolipid moiety was attached using the phosphoamidite **72** and 1-H-tetrazole followed by oxidation with MCPBA, and product **73** was then hydrogenolyzed to obtain the desired AC$_3$ PIM$_2$ product **74** (*57*).

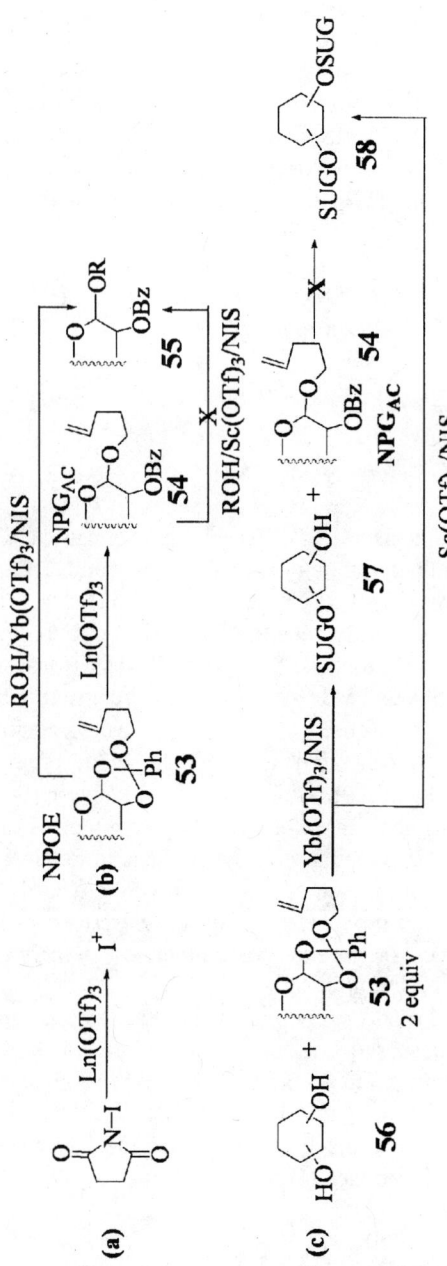

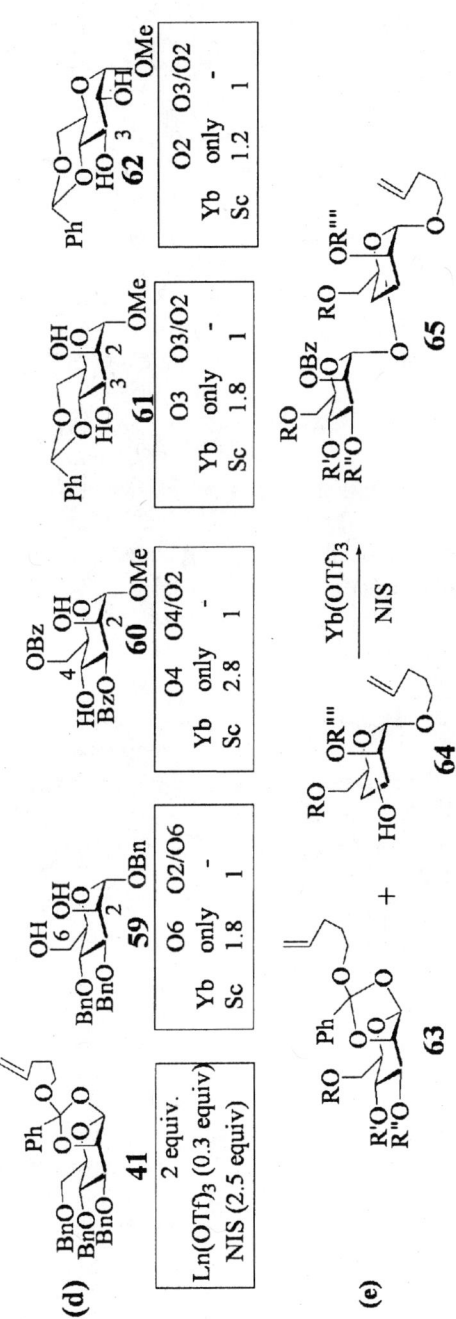

Scheme 10. *Lanthanide triflates and n-pentenyl donors s.*

The Dodecasaccharide Lipomannan Component (59)

For elaboration of the mannan backbone of LAM **3**, a PIM *pseudo*trisaccharide precursor was needed, that was differentiated at the primary-OHs of the O6 and O2 mannosides. These in turn could be obtained by using the appropriate NPOE precursor. Experimentation showed that tritylated the silylated derivatives **67b** and **c** were the best choice for obtaining a *pseudo*trisaccharide **70b** using the strategy in Scheme 11. The primary-OH of the "left-hand" mannoside was then liberated in **75** (Scheme 12).

Extending the mannan, α-1,6-backbone took advantage of the fact that NPOE **67d** has two benzoates, the one at O2 being in masked form. Reaction with acceptor **75** therefore added the 2,6-dibenzoylated-mannoside unit of **76a** in 93% yield in step (i), and saponification in step (ii) led to diol **76b**.

Further processing of diol **76b** took advantage of (a) the exquisite regioselectivity offered by Yb(OTf)$_3$ as shown in Scheme 10d (*41*), and (b) the experiments in Scheme 9 (entry vi) showing that NPOEs do not match the C2-OH of mannosides e.g **52**. Items (a) and (b) ensured that diol **76b** could be treated with a generous excess of **67c** to optimize formation of a new dibenzoate in step (i), which upon saponification in step (ii), would gave triol **77a**.

Iterative application of steps (i) and (ii) to **77a** then led to tetraol **77b** and pentaol **77c**, excellent yields being maintained throughout as shown in Scheme 12.

Compound **77c** now has the mannoside α-1-6 backbone. For the target at hand it was now necessary to add the O-2 mannose residues at the five-OH groups. Experiments showed that the best donor for simultaneous delivery of five mannosides proved to be the trichloroacetimidate **78**, prepared as indicated in Scheme 7c. The dodecasaccharide **79** was thereby obtained in 86% yield. Protecting groups adjustments, followed by "end-game" strategies such as used in Scheme 11, then led to lipomannan **80** (*59*).

The Arabinofuranose Domain

The cartoons of LAM in Scheme 1 draw attention to the extensive arabinan array, and our transposition, **3** (Scheme 3), shows the presence of linear and branched furanose components. Furanose donors have received less attention than pyranose counterparts due, presumably, to the infrequency with which they have been isolated from natural sources (*60*). However, this circumstance may be due, in turn, to the great lability of furanosides. Indeed, the recent increase in the number of furanose-containing natural products could be the direct result of modern, less drastic, isolation methods (*61*).

Early attempts (*62*) to prepare n-pentenyl furanosides in our laboratory utilized Fischer methodology, e.g Scheme 13a, which requires stopping the reaction when the kinetically formed n-pentenyl furanoside, **81**, is optimal, and then separating it from pyranoside **82**. The result for n-pentenyl galactofuranoside was only fair, a result that was not encouraging.

The great success with n-pentenyl pyranosyl orthoesters, as evident from the synthesis of **80,** prompted us to test the technology for furanosides. The preparative approach in Scheme 5a would not be appropriate, since formation of furanoside, rather than pyranoside could not be guaranteed.

Methyl furanosides are available by several routes (*63*), and are therefore a good starting point. The desired arabinofuranoside, **83a**, is available commercially; but can also be prepared as indicated in Scheme 13b (*64*). The furanose ring survived perbenzoylation and treatment with HBr to afford the glycosyl bromide **83c**. Standard treatment with lutidine in the presence of pent-4-enol gave orthoester **84**, and the key orthoester diol, **85**, was then obtained and variously protected analogs, **86**, **87** and **88**, in Scheme 13c were obtained routinely.

Rearrangement of the furanosyl orthoesters to the corresponding *n*-pentenyl furanosides occurred with Yb(OTf)$_3$. Of special interest for the work ahead, were the structures with one and two free-OHs, **89b** and **91**. With respect to the latter, the route via triester **92** was abandoned because chemoselective cleavage of the chloroacetates was problematic. The alternative via **90b** worked well; but the direct rearrangement of **85** proved best, there being no evidence of self-condensation, as was feared.

Probing Chemoselectivity in *n*-Pentenyl Furanosyl Derivatives

In view of the high reactivity of furanosides in general, it was of interest to see whether the chemoselective coupling exhibited by pyranosides, as exemplified in Scheme 10c, could be duplicated. A relevant test target was the linear component in the arabinan domain of LAM **3**. Accordingly, NPG acceptor **89b** was treated with 2 equivalents of NPOE **86** under catalysis by Yb(OTf)$_3$/NIS. Diarabinan **93a** was produced in 77% yield, and readily separated from NPG **89a** (which had come from the excess NPOE **86**). Iterating the last two steps, in which the yields were upheld, gave trisaccharide **94** (Scheme 14a).

Synthesis of a Dodecaarabinofuranan[65]

With the success in Scheme 14a, we turned our attention to strategies for the branched arabinans (Scheme 13b). The same acceptor, **89b**, was coupled with

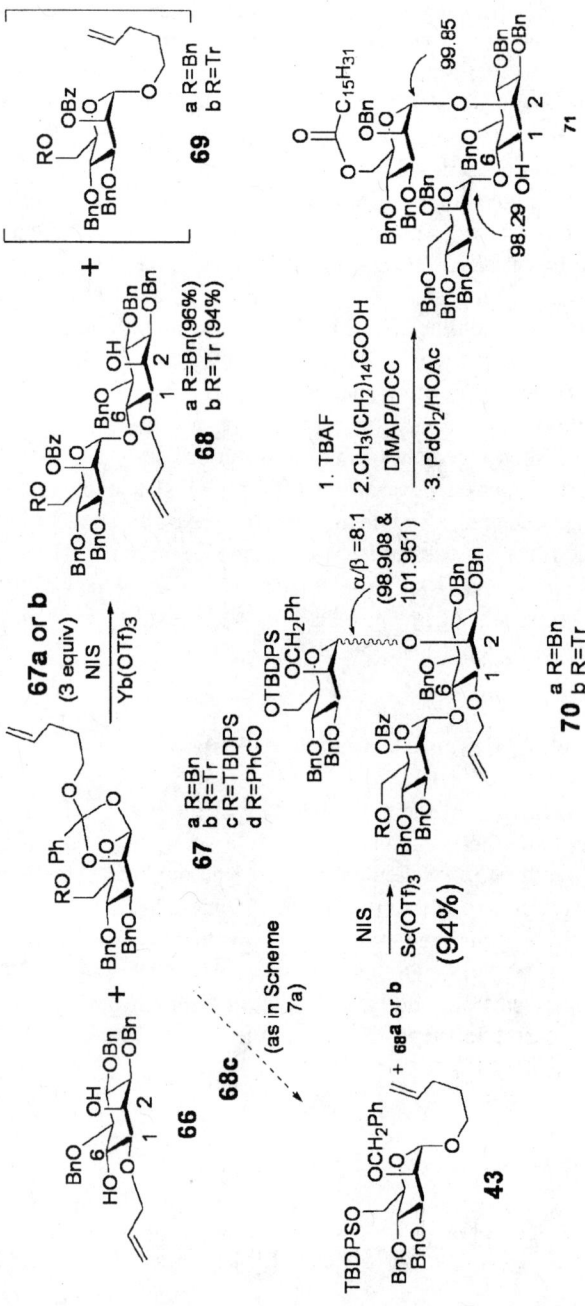

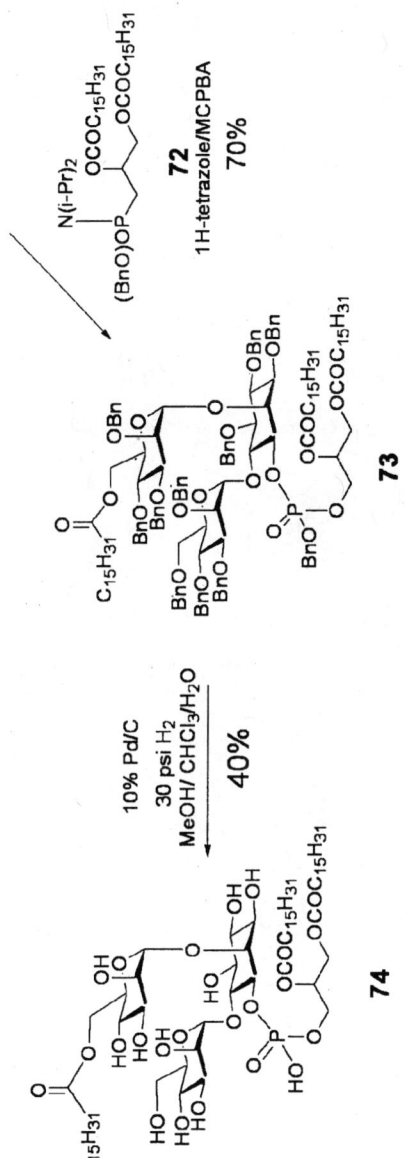

Scheme 11. Synthesis of a dimannosylated phosphoinositide

224

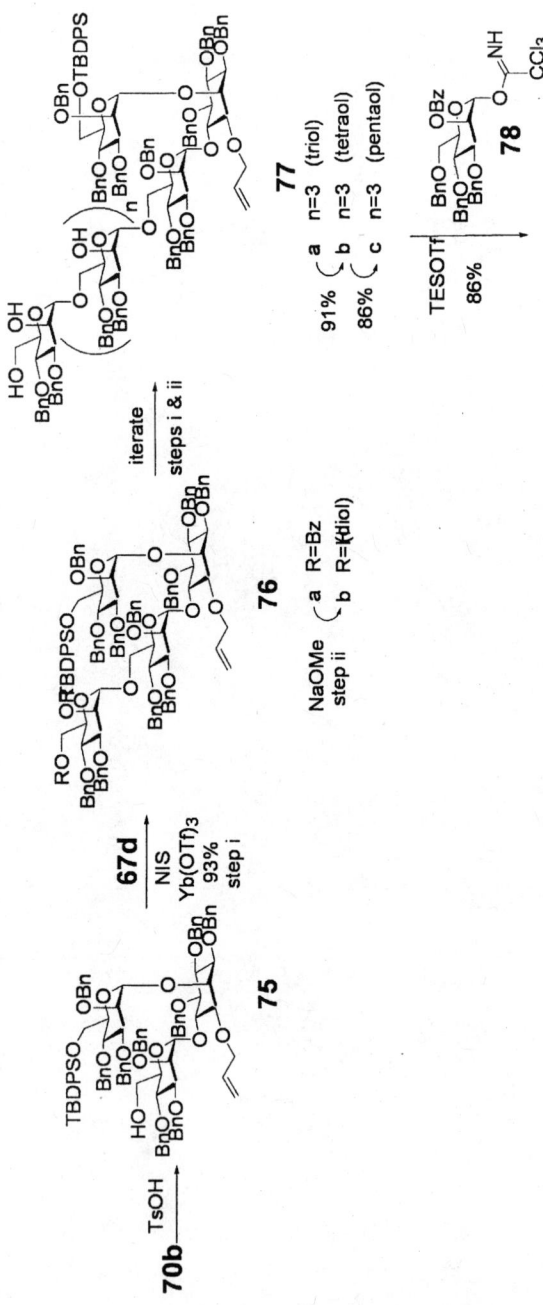

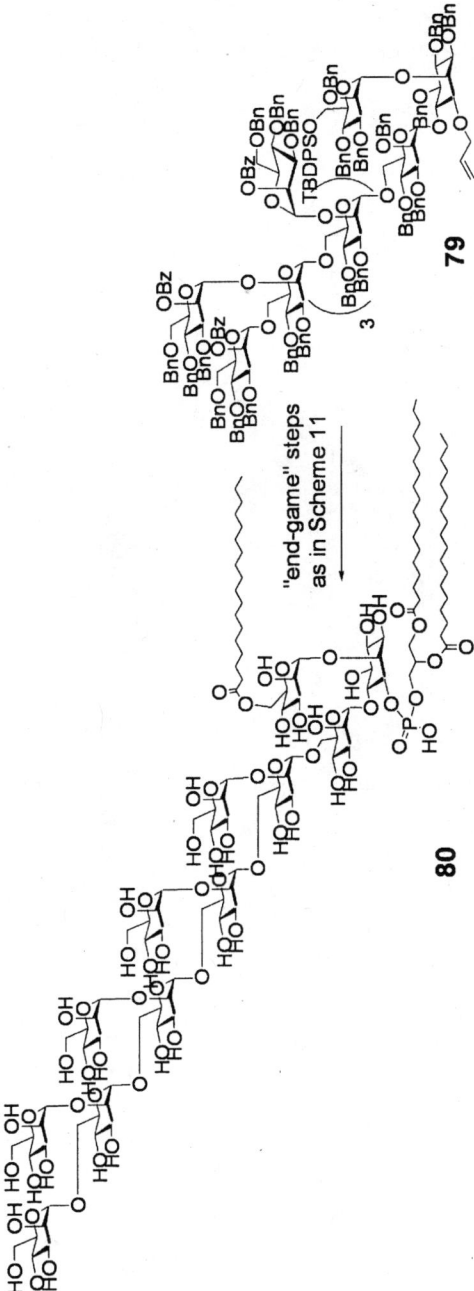

Scheme 12. Synthesis of a dodecasaccharide lipomannan of M. tuberculosis.

(a) Arabinose → 81 + 82 (TsOH/DMSO, pentenol)

(b) Arabinose → 83 (MeOH, HCl) → 84 (pentenol, lutidine) → 85 (NaOMe)

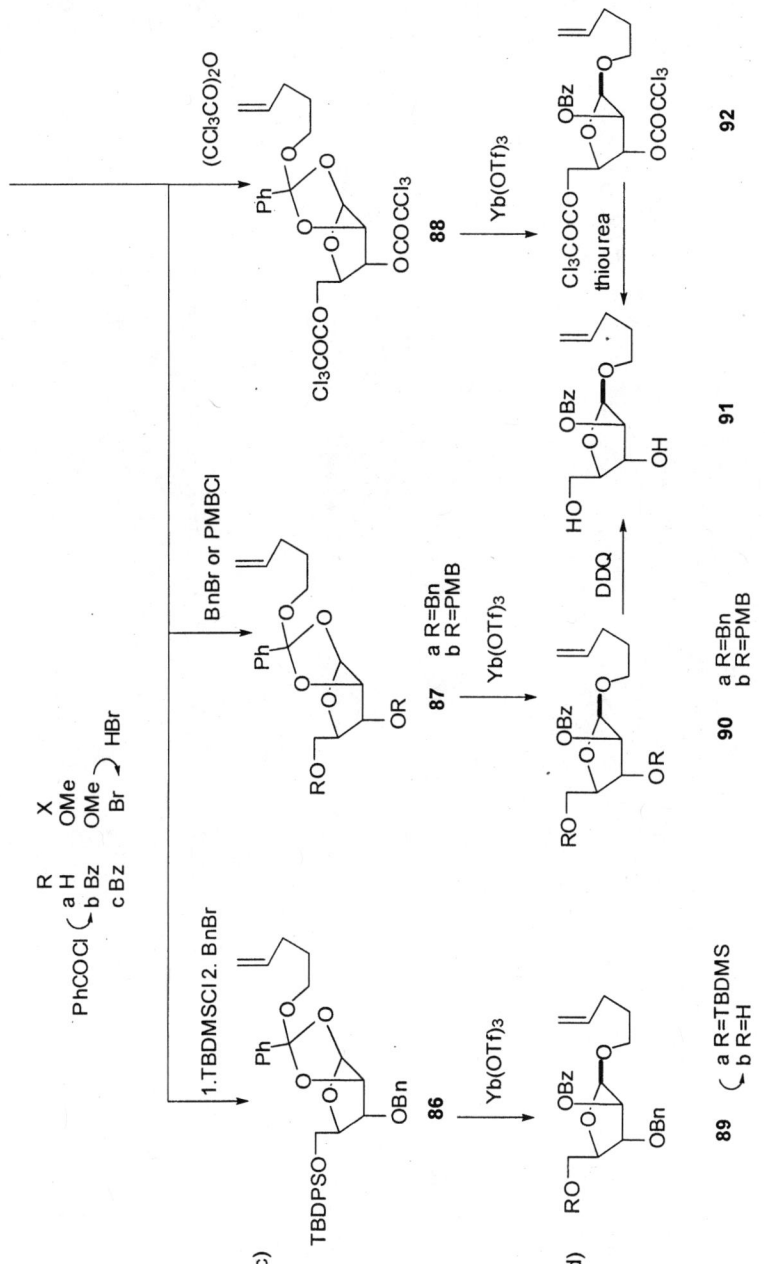

Scheme 13. Preparation of n-pentenyl furanosyl donors

228

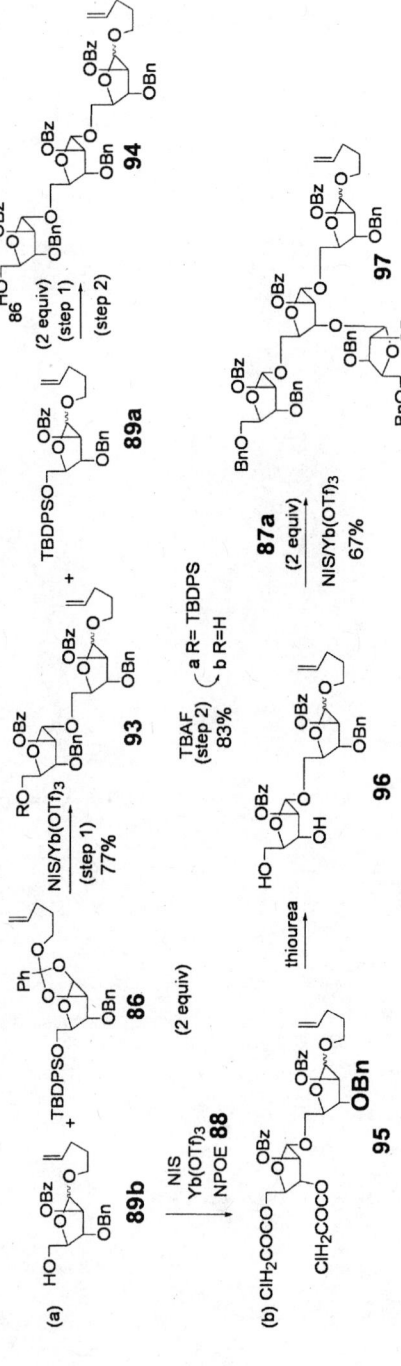

Scheme 14. Chemoselectivity of n-pentenyl furanosyl derivatives

donor **88**, which produced the disaccharide **95**. Dechloroacetylation to diol **96** occurred without event, and both hydroxyls were successfully furanosylated, to give **97** in 67% yield.

The double glycosidation, **96→97**, was most encouraging, such double glycosidations could be applied to provide the branched units of **3** (Scheme 2) using larger arabinan donors. But *n*-pentenyl glycofuranosides, especially where disarmed, would be less responsive than their NPOE counterparts, and so would need a stronger promoter than $Yb(OTf)_3$.

Accordingly, we did a preliminary test. Attempts to couple donor **94a**, with acceptor **98** under the agency of TESOTf/NIS gave, surprisingly, disaccharide **99a** (Scheme 15).

Clearly, the distal furanose moiety of **94a** had been cleaved at the point indicated in Scheme 15, and transferred to acceptor **98**. Repetition with omission of NIS led to the same result indicating that the pentenyl moiety was not involved. A purely acid-transfer process was confirmed when a comparable glycosidation occurred with the methyl glycoside **100** and acceptor **98** to give **99b**.

The lesson from Scheme 15 was that n-pentenyl furanosides are not reactive enough to serve as donors. Accordingly, NPG **97** was oxidatively hydrolysed to glycose, **101a**, which was then converted into the trichloroacetimidate **101b** (Scheme 16).

A test acceptor diol was generated by extending the previously synthesized trisaccharide **94** with the dichloroacetylated NPOE **88** to give **102a**, and thence **102b** (Scheme 16).

Treatment of the latter with three equivalents of trichloroacetimidate **101b** gave a mixture in 75% yield; but the desired dodecasaccharide **104** proved to be the minor component. The major product was the octasaccharide **103**.

The formation of **104** as the minor coupling product in Scheme 16 suggested that the free-OH in the major product **103** was too hindered for the approach of a second equivalent of donor **101b**. The contrast with the 67% yield of **97** (Scheme 14b), taught us that there would be better success with a "smaller" donor. Indeed, reaction of the same diol, **102b**, with excess NPG **86**, and subsequent desilylation afforded the hexafuranosyl acceptor diol **105** in 62% yield (Scheme 17a).

For a test donor, the diol **91** was double glycosylated with NPG **86** to give trisaccharide NPG **106a**, and the derived diol **106b** again double glycosylated, this time with NPG **87a**, to the pentafuranoside **107a,** from which the trichloroacetaminde **107c** was obtained in the usual way (Scheme 17b).

As indicated in Scheme 17c, acceptor **105** and donor **107c**, coupled under the influence of $Yb(OTf)_3$, gave the 26-mer arabinofuranoside **108** in 78% yield.

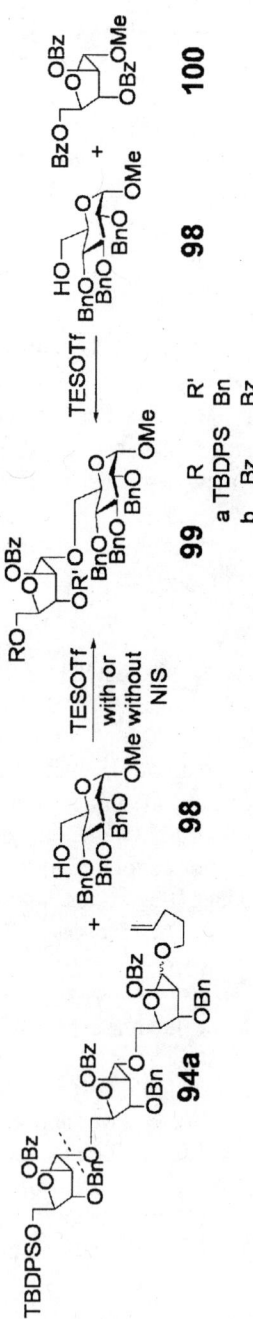

Scheme 15. *Acid lability rears its ugly head*

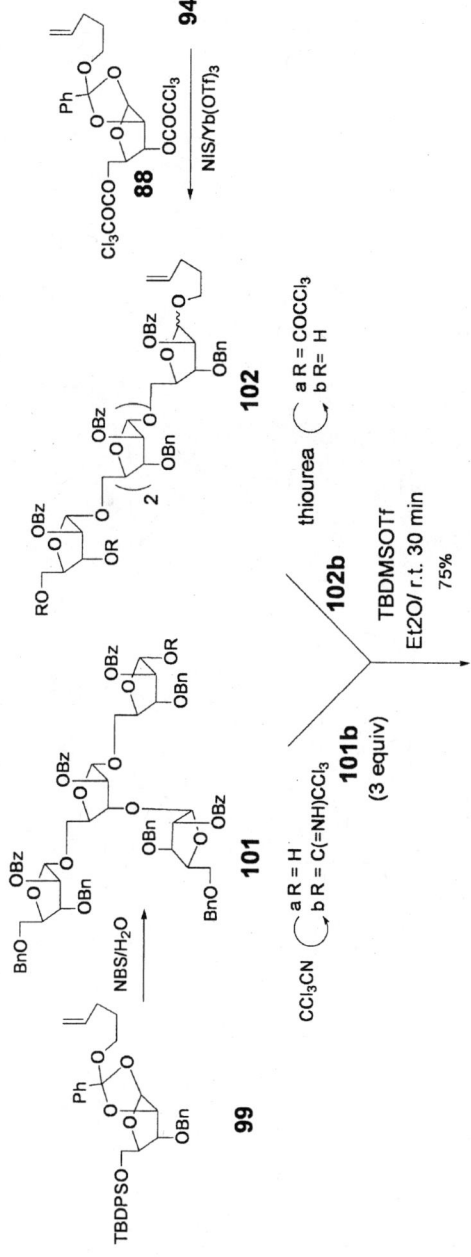

Scheme 16. A strategy for convergent assembly. Continued on next page.

232

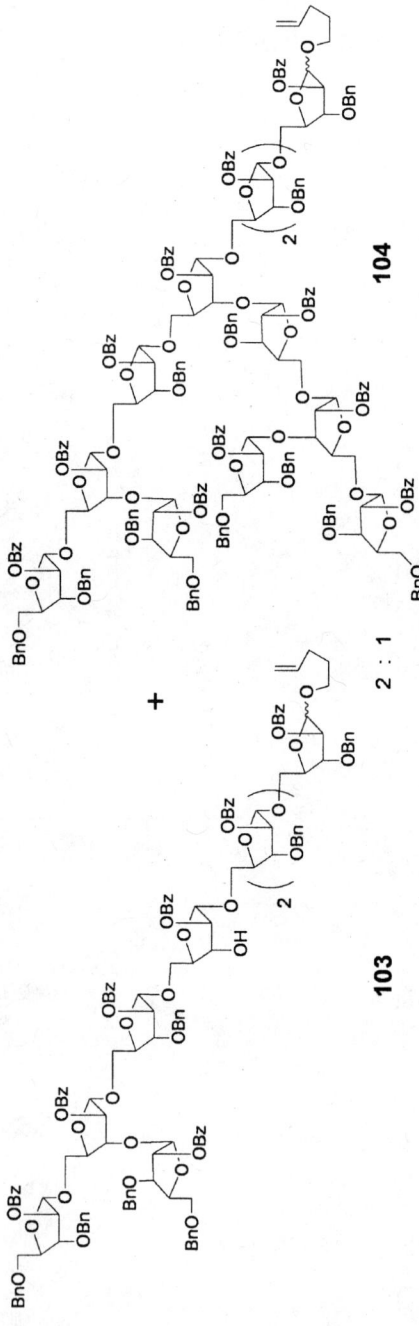

Scheme 16. Continued.

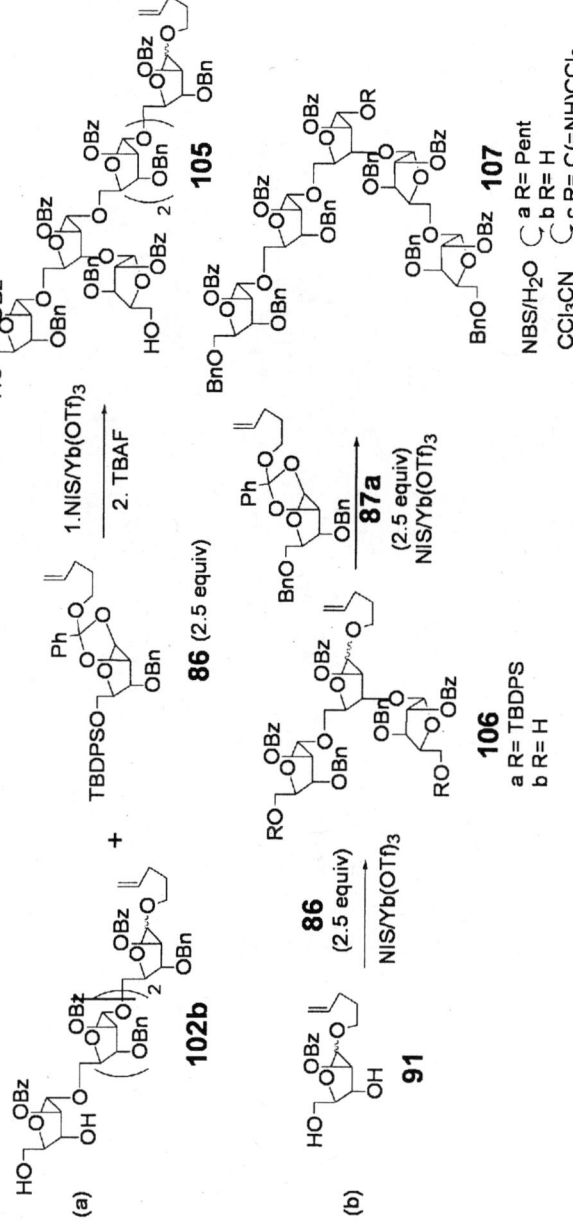

Scheme 17. Synthesis of a 26-mer-arabinofuranan. Continued on next page.

234

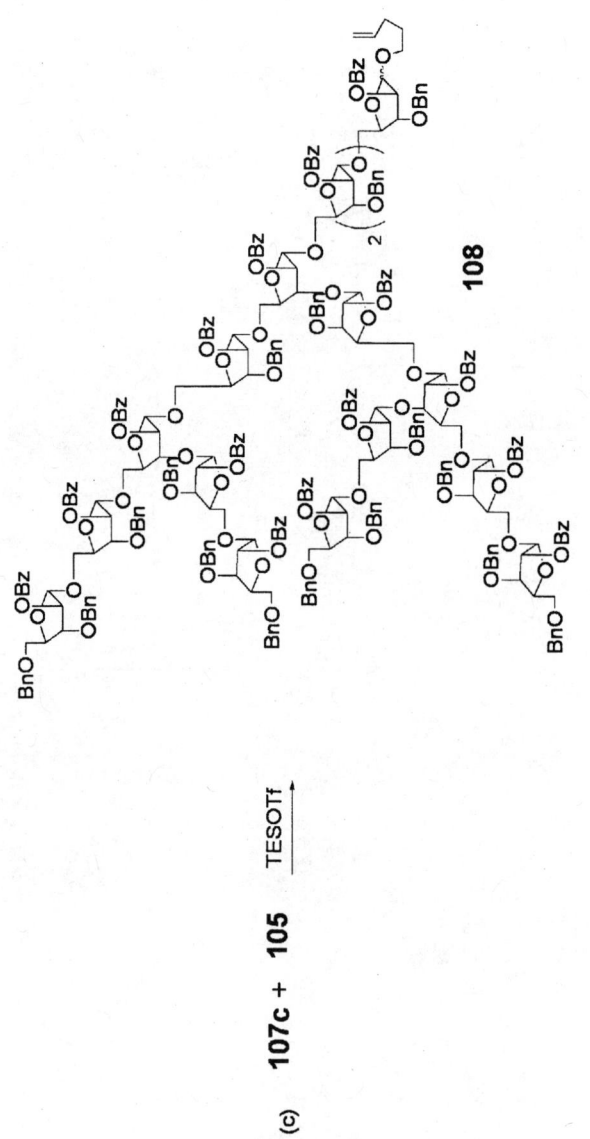

(c) **107c + 105** $\xrightarrow{\text{TESOTf}}$ **108**

Scheme 17. Continued.

Conclusion

We believe that the MATCH that has resulted in the rapid syntheses of the LAM components described above, is the direct result of Reciprocal Donor Acceptor Selectivity (RDAS) that emanated from the fortuitous observations outlined in Scheme 8.

The basis for RDAS is under investigation, and our progress, which is beyond the scope of this article, has been outlined elsewhere[27]. However, we can identify two factors that are obvious from the above account: (i) NPOEs and NPGs$_{Ac}$ display the same preference for a given diol-OH, but NPOEs give much higher yields; (ii) the fortuitous discovery that NIS, activated by Yb(OTf)$_3$, is specific for NPOEs, whereas NPGs$_{AC}$ (and NPGs$_{ALK}$) require Sc(OTf)$_3$. The combination of items (i) and (ii) permits highly selective glycosylation by NPOEs. Indeed this MATCH can be so exquisite that the mismatched-OHs do not need to be protected. The traditional protection/deprotection practices, to ensure that only one-OH is presented to the donor, are therefore substantially reduced.

The NPOE methodology has also been extended for the first time to furanoses. The success of this venture may be judged by the fact that the NPOE of arabinofuranose is the sole source for preparing a dodecaarabinan consisting of linear and branched components.

A particularly attractive feature of the n-pentenyl furano derivatives, NPOEs or NPGs, is the ability to liberate the anomeric centers oxidatively by use of NBS in aqueous medium. The extreme sensitivity of furanosides to acid catalysed hydrolysis is therefore circumvented. The glycofuranoses so formed can then be converted into thioglycosisdes or trichloroacetimidates to provide other types of donors.

Acknowledgements

We are grateful to the National Science Foundation (CHE-0243436) for support of the work described above.

References

1. World Health Organization. Tuberculosis. Fact Sheet No 104, 2004.
2. Dye, C.; Scheele, S.; Dolin, P.; Pathania, V.; Raviglione, M. C. Consensus statement. Global burden of tuberculosis: Estimated incidence, prevalence, and mortality by country. WHO global surveillance and monitoring project. *JAMA*, **1999**, *282*, 677-686.

3. Kochi, A. The global tuberculosis situation and the new global control strategy of the world health organization. *Tubercle*, **1992**, *72*, 1-6.
4. (a) Consensus Statement. Global burden of tuberculosis: estimated incidence, prevalence and mortality by country. W.H.O. Global Surveillance and Monitoring Project. Dye, C.; Scheele, S.; Dolin, P.; Pathania, V.; Raviglione, M. C. *JAMA*, **1999**, *282*, 677-686. (b) *New York Times* August 19, 2006, page 2.
5. Murray, J. F. *Bull Int. Union Tuberc Lung Dis.*, **1991**, *66*, 199-204. Brudney, K.; Dobkin, J. *Am. Rev. Respir. Dis.*, **1991**, *144*, 746-749.
6. Dormandy, T. "The White Death" New York University Press, New York, 2000.
7. Zumla, A.; Squire, S. B.; Chintu, C.; Grange, J. M. The tuberculosis pandemic: implications for health in the tropics. *Trans. Royal Soc. Tropical Med. Hygiene*, **1999**, *93*, 113-117.
8. Guerin, C. The history of BCG. In: Rosenthal SR ed. Boston: Little, Brown & Co.; 1957, 48-53.
9. Colditz, G. A.; Brewer, T. F.; Berkey, C. S. et al. Efficacy of BCG vaccine in the prevention of tuberculosis. Meta-analysis of the published literature. *JAMA*, **1994**, *271*, 698-702.
10. Brennan, P. quoted in article by Maureen Rouhi in Chemical and Engineering News. May 17, 1999.
11. Koch, R *Berliner Klinische Wochenschrift*. **1882**, *19*, 221-230; *Review Inf. Dis.* **1982**, *4*, 1270-1274.
12. Hayashi, Y.; Ebina, T.; Suzuki, F.; Ishida, N. *Microbiol. Immunol.*, **1981**, *25*, 305-316.
13. Mukai, M.; Kibota, S.; Morita, S.; Akanuma, A. *Cancer*, **1995**, *75*, 2276-2280.
14. Kobayashi, M.; Herndon, D. N.; Pollard, R. B.; Suzuki, F. *Immunol. Lett.*, **1994**, *40*, 199-205.
15. Lesprit, P.; Zagdanski, A. M.; de la Blanchardiere, A.; Rouveau, M.; Decarpes, J. M.; Frija, J.; Lagrange, P.; Modai, J.; Molina, J. M. *Medicine*, **1997**, *76*, 423-431.
16. Riviere, M.; Moisand, A.; Lopez, A.; Puzo, G. *J. Mol. Biol.*, **2004**, *344*, 907-918.
17. Turnbull, W. B.; Shimizu, K. H.; Chatterjee, D.; Homans, S. W.; Truemann, A. *Angew. Chem. Int. Ed.*, **2004**, *43*, 3918-3922.
18. Kremer, L.; Gurcha, S. S.; Bifani, P.; Hitchen, P. G.; Baulard, Al.; Morris, H. R.; Dell, A.; Brennan, P. J.; Besra, G. S. *Biochem. J.*, **2002**, *363*, 437.
19. Chatterjee, D., *Curr. Open. Biol.*, **1997**, *1*, 579.
20. Porcelli, S. A.; Besra, G. S. In Intracellular Pathogens in Membrane Interactions and Vacuole Biogenesis. *Gorvel, J. P. (ed)*, **2004**, 230-249. New York: Kluwer Academic Plenum Publishers.
21. Trost, B. M. *Science*, **1983**, *219*, 245-250.

22. Spijker, N. M.; van Broeckel, C. A. A. *Angew. Chem., Int. Ed. Engl.* **1991**, *30*, 1-80.
23. Masamune, S.; Choy, W.; Petersen, J. S.; Sita, L. R. *Angew. Chem., Int. Ed. Engl.* **1985**, *24*, 1-76.
24. Paulsen, H. In *"Selectivity a Goal for Synthetic Efficiency"* Bartmann, W.; Trost, B. M. Eds. 1984. Paulsen, H. *Angew. Chem., Int. Ed. Engl.* **1982**, *21*, 155-173.
25. *Chemical & Engineering News,* September 19, 2005, page 38.
26. (a) Isbell, H. S. *Ann. Rev. Biochem.* **1940**, *9*, 65-92. (b) Frush, H. L.; Isbell, H. S. *J. Res. NBS*, **1941**, *27*, 413.
27. Bert Fraser-Reid, K. N. Jayaprakash, J. Cristóbal López, Ana M. Gómez, Clara Uriel, in press.
28. Veeneman, G. H.; van Boom, J. H. *Tetrahedron Lett.* **1990**, *31*, 275-278.
29. Mootoo, D. R.; Konradsson, P.; Udodong, U.; Fraser-Reid, B. *J. Am. Chem. Soc.* **1988**, *110*, 5583-5584.
30. Veeneman, G. H.; van Leeuwen, S. H.; van Boom, J. H. *Tetrahedron Lett.* **1990**, *31*, 1331-1334.
31. Kanie, O.; Ito, Y.; Ogawa. T. *J. Am. Chem. Soc.* **1994**, *116*, 12073-12074. Demchenko, A. V.; De Meo, C. *Tetrahedron Lett.* **2002**, *43*, 8819-8822.
32. Haines, A. H. *Advan. Carbohydr. Chem. Biochem.*, **1976**, *33*, 11.
33. Richardson, A. C.; Williams, J. M. *Tetrahedron*, **1967**, *23*, 1641.
34. Flowers, H. M.; *Carbohydr. Res.* **1982**, *00*, 418. Flowers, H. M. *Carbohydr. Res.* **1975**, *39*, 245.
35. Angyal, S. J.; Irving, G. C.; Rutherford, V. D.; Tate, M. E. *J. Chem. Soc.*, **1965**, 6662. Angyal, S. J.; Tate, M. E. *J. Chem. Soc.*, **1965**, 6949.
36. Kanie, O.; Hindsgaul, O. In "Solid Support Oligosaccharide Synthesis and combinatorial Carbohydrate Libraries" Ed. Seeberger, P. H. Wiley Interscience, New York 2001, Chapter 12.
37. Fraser-Reid, B.; Udodong, U. E.; Wu, Z.; Ottosson, H.; Merrett, R.; Rao, C. S.; Robert, C.; Madsen, R. *Synlett*, **1992**, 927-942.
38. Fraser-Reid, B.; Grimme, S.; Piacenza, P.; Mach, M.; Schlueter, U. *Chem. Eur. J.* **2003**, *9*, 4687-4692.
39. Norberg T. In "Modern Methods in Carbohydrate synthesis" Ed. Khan, S. H and O'Neill, R. A. Harwood Academic Publishers, Amsterdam 1996, Chapter 4.
40. Mayer, T. G.; Schmidt, R. R. *Eur. J. Org. Chem.* **1999**, 1153-1159.
41. (a) Jayaprakash, K. N.; Radhakrishnan, K. V.; Fraser-Reid, B. *Tetrahedron Lett.* **2002**, *43*, 6953-6955. (b) Jayaprakash, K. N.; Fraser-Reid, B. *Synlett* **2004**, 301-305.
42. Jayaprakash, K. N.; Fraser-Reid, B. *Org. Lett.* **2004**, *6*, 4211-4214.
43. Anilkumar, G.; Jia, Z. J.; Kraehmer, R.; Fraser-Reid, B. *J. Chem. Soc. Perkin I* **1999**, 3591.

44. Douglas, N. L.; Ley, S. V.; Lucking, U.; Warriner, S. L. *J. Chem. Soc. Perkin I* **1998**, 51-65.
45. "Solid Support Oligosaccharide Synthesis and combinatorial Carbohydrate Libraries" Ed. Seeberger, P. H. Wiley Interscience, New York 2001.
46. Zhang, Z.; Ollmann, I. R.; Ye, X-S.; Wischnat, R.; Baasov, T.; Wong, C.-H. *J. Am. Chem. Soc.* **1999**, *121*, 734-753.
47. Plante, O. J.; Palmacci, E. R.; Seeberger, P. H. *Science* **2001**, *291*, 1523-1527.
48. Anilkumar, G.; Nair, L. G.; Fraser-Reid, B. *Org. Lett.* **2000**, *2*, 2587-2590.
49. Fraser-Reid, B.; López, J. C.; Gómez, A. M.; Uriel, C. *Eur. J. Org. Chem.* **2004**, 1387-1395.
50. Fraser-Reid, B.; López, J. C.; Radhakrishnan, K. V.; Mach, M.; Schlueter, U.; Gómez, A. M.; Uriel, C. *Can. J. Chem.* **2002**, *124*, 1075-1087.
51. Fraser-Reid, B.; López, J. C.; Radhakrishnan, K. V.; Nandakumar, M. V.; Gómez, A. M.; Uriel, C. *Chem. Commun.* **2002**, 2104-2105.
52. Fraser-Reid, B.; Anilkumar, G.; Nair, L. G.; Radhakrishnan, K. V.; López, J. C.; Gómez, A. M.; Uriel, C. *Aust. J. Chem.* **2002**, *55*, 123-130.
53. Uriel, C.; Agocs, A.; Gómez, A. M.; López, J. C.; Fraser-Reid, B. *Org. Lett* **2005**, *7*, 4899-4902.
54. Jayaprakash, K. N.; Radhakrishnan, K. V.; Fraser-Reid, B. *Tetrahedron Lett.* **2002**, *43*, 6953-6955.
55. Jayaprakash, K. N.; Fraser-Reid, B. *Synlett* **2004**, 301-305.
56. Jayaprakash, K. N.; Fraser-Reid, B. *Org. Lett* **2004**, *6*, 4211-4214.
57. Jayaprakash, K. N.; Lu, Jun.; Fraser-Reid, B. *Bioorg. Med. Chem. Lett,* **2004**, *14*, 3815-3819
58. Brennan, P.. *Tuberculosis* **2003**, *1*, 1.
59. Jayaprakash, K. N.; Lu, Jun.; Fraser-Reid, B. *Angew. Chem. Int. Ed. Engl.* **2005**, *44*, 5754-5763.
60. For a recent summary of developments in furanosyl donors see: Lowary, T. L. In Glycoscience: Chemistry and Biology. Fraser-Reid, B.; Tatsuta, K.; Thiem, J. Eds.; *Springer, Heidelberg*, **2001**, *3*, 1696.
61. Previato, J. O.; Gorin, P. A.; Mazurek, M.; Xavier, M. T.; Fournet, B.; Wieruszesk, J. M.; Mendoca-Previato, L. *J. Biol. Chem.* **1990**, *265*, 2518. Lederkremer, R. M.; Lima, C.; Ramirez, M. I.; Ferguson, M. A. J.; Homans, S. W.; Thomas-Oates, J. E. *J. Biol. Chem.* **1991**, *265*, 19611.
62. Arasappan, A.; Fraser-Reid, B. *Tetrahedron Lett.*, **1995**, *36*, 7967.
63. Ness RK, and Fletcher, HG, Jr.. *J. Amer. Chem Soc.* **1958**, *80*: 2007.
64. Lu, J, Fraser-Reid, B. *Org. Lett.* **2004**, *6*, 3051.
65. Lu, J and Fraser-Reid, B. *Chem. Commun.* **2005**, 862.

Chapter 11

Lipopolysaccharide Antigens of *Chlamydia*

P. Kosma[1,*], H. Brade[2], and S. V. Evans[3]

[1]Department of Chemistry, University of Natural Resources and Applied Life Sciences, A–1190 Vienna, Austria
[2]Leibniz Center for Medicine and Biosciences, D–23845 Borstel, Germany
[3]Department of Biochemistry and Microbiology, University of Victoria, Victoria, British Columbia V8W 3P6, Canada

Chlamydiae contain in their outer leaflet family- and species-specific carbohydrate epitopes which may be exploited for diagnostic and therapeutic purposes. On the basis of a series of semisynthetic neoglycoconjugates chlamydia-specific and cross-reactive epitopes were defined using a panel of monoclonal antibodies. In addition, crystal structures of the Fab fragments complexed to synthetic Kdo ligands revealed the molecular basis for the binding of the charged oligosaccharides. Thus terminal Kdo units are recognized by a highly conserved binding motif whereas proximal Kdo units are either bound via induced-fit conformational rearrangements in the binding site or via specific binding imposed by stacking interactions.

Chlamydiae are obligate intracellular bacterial parasites which are responsible for a broad spectrum of acute and chronic diseases in humans and animals (*1*). According to estimates by the World Health Organization 2.8 million Americans are affected - in the context of sexually transmitted diseases - whereas several million of cases of blindness in Africa and Asia are caused by infections from *Chlamydia trachomatis*. Moreover, chronic diseases such as arthritis, atherosclerosis, asthma and neurodegenerative diseases have been linked to chlamydial infections. A primary infection does not lead to a lasting protective immunity and in many cases, asymptomatic infections by *Chlamydia* remain unrecognized. The prevalence of chlamydial episodes increases with age and seroepidemiological studies indicate that 50-75% of the population is being exposed to the microorganism.

Treatment of chronic chlamydial infections is possible by administration of macrolide and other antibiotics, but evidence has been accumulated concerning the emergence of antibiotic-resistant strains. The development of effective vaccines, however, would provide a highly desirable preventive tool against chlamydial infections. Despite considerable efforts in recent years, no proven vaccine has been developed thus far. Many approaches have focused on the major outer membrane protein (MOMP) as the target candidate antigen and some reports have appeared on the effectiveness of these protein as well as of DNA vaccines (*2*).

In this chapter, the carbohydrate antigenic determinants located in the outer membrane being specific for chlamydial epitopes will be characterized, the synthesis of related neoglycoconjugates and their binding modes with a series of monoclonal antibodies will be described in molecular detail. These studies should help to improve our understanding of the manifold interactions of this bacterial parasite with the host cells and pave the way for the generation of rationally designed and potentially protective antibodies.

Chlamydia and *Chlamydia*-induced Diseases

Chlamydiaceae constitute a unique family within the bacterial kingdom. According to a recent reclassification (*3*), the family *Chlamydiaceae* comprises two genera – *Chlamydia* and *Chlamydophila* with several species which have specifically adapted to survival in intracellular environments.

Chlamydia trachomatis is the leading cause of preventable blindness and the most common sexually transmitted bacterial species. Sequelae of infections by *C. trachomatis* include (neonatal) conjunctivitis and trachoma of the eye (serotypes A-C), lymphogranuloma venereum (serotypes L1-L3) cervicitis, pelvic inflammatory disease, salpingitis, epididymitis, prostatitis (serotypes D-K) as well as urethritis and reactive arthritis.

The genus *Chlamydophila* harbours two human-pathogenic species, *C. psittaci* and *C. pneumoniae* leading to respiratory tract infections. Infections by *C. psittaci* are transmitted by birds and do not have a high incidence rate, but may induce severe pulmonary infections with a mortality rate of ~5%. The second important species *C. pneumoniae* (formerly known as TWAR) was described in 1989 and since then has been linked to atypical pneumonia, bronchitis and asthma (*4*). In addition, the bacterium has been associated with the development of atherosclerotic and cardiovascular disease as well as multiple sclerosis and Alzheimer's disease. *C. psittaci* and the species *C. pecorum* cause infections relevant in veterinary medicine.

Chlamydiae undergo a unique developmental cycle (48-72 h) during the infection process (*5*). The infectious, spore-like particles termed elementary bodies (EB, 200-400 nm in size) adher to the membrane of the host cell and by active endocytosis become incorporated into a phagosome. By unknown mechanisms the elementary bodies inhibit the fusion of phagosomes into lysosomes. Subsequently EBs differentiate into the metabolically active reticulate bodies (RB). Following replication, the reticulate bodies reorganize into the infectious elementary bodies forming inclusion granules, which are eventually either released by reverse endocytosis or which lead to lysis of cells and inclusions (Figure 1).

The immune response to chlamydial species involves cell-mediated reactions, local secretion of IgA-antibodies and humoral recruitment of IgM, IgA and IgG antibodies. The most important antigenic determinants reside in the outer leaflet of the chlamydial outer membrane and comprise the major outer membrane protein (MOMP) and lipopolysaccharide (LPS).

Structure and Synthesis of *Chlamydia*-specific Carbohydrate Epitopes

In contrast to the conserved inner-core structure of enterobacterial LPS, Chlamydiae contain a highly truncated glycolipid which is composed of lipid A and Kdo residues only (*6,7*). The chemical structure of chlamydial lipid A reflects a potential downregulation of the immune response by the host organism, since it contains unusual long-chain fatty and (*R*)-3-hydroxyalkanoic acids (*8,9*) and no more than five fatty acyl chains in the lipid A backbone (Figure 2). A unique feature of chlamydial lipid A is the absence of hydroxy-fatty acids at the 3 and 3' positions. The low endotoxic activity of chlamydial LPS has been known and recently, synthetic lipid A derivatives were shown to involve the pentaacyl species only in the Toll-like receptor 4 mediated signaling pathway (*10-12*).

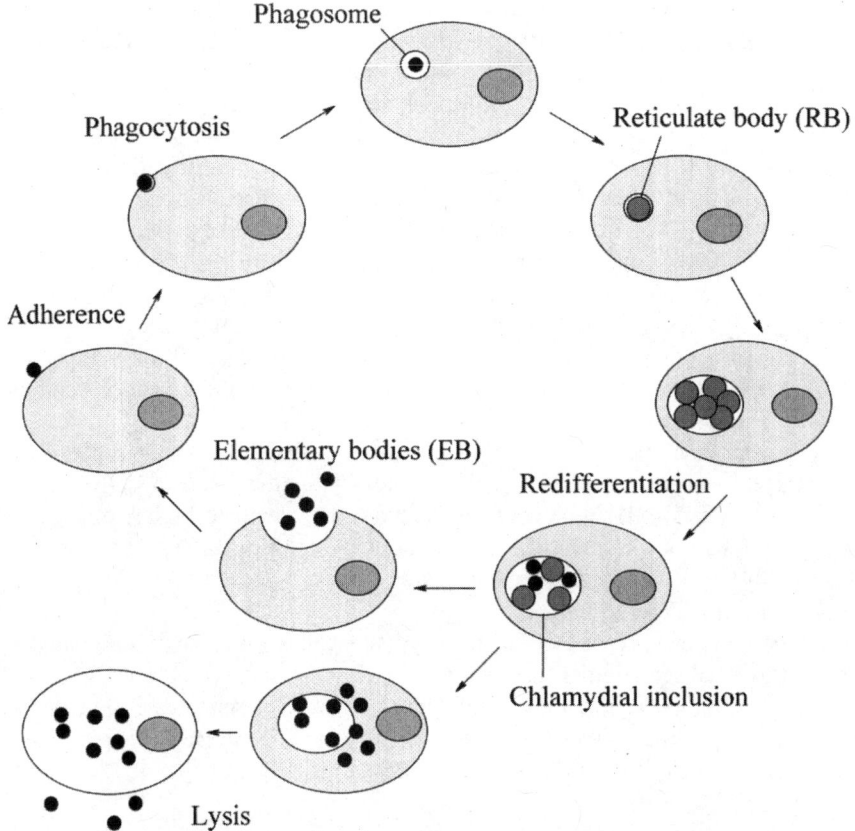

Figure 1. Developmental cycle of Chlamydia.

All *Chlamydiaceae* share a common carbohydrate epitope **1** – formerly called the genus-specific epitope - attached to the lipid A anchor, which resembles the deep rough mutant LPS structures of *Enterobacteriaceae* (*13*). For *Chlamydia*, however, the 3-deoxy-D-*manno*-octulosonic acid (Kdo) region constitutes an immunodominant epitope of three Kdo units of the sequence α-Kdo-(2→8)-α-Kdo-(2→4)-α-Kdo-(2→6)-Lipid A, wherein the (2→8)-linkage confers *Chlamydia*-specificity (*14*). Additionally, species-specific Kdo oligosaccharides are present in the LPS of *C. psittaci* which have been isolated from recombinant strains expressing the respective Kdo transferase and which were structurally characterized by NMR (*15*). Thus, a Kdo trisaccharide of the sequence α-Kdo-(2→4)-α-Kdo-(2→4)-α-Kdo **2** and a branched Kdo tetrasaccharide α-Kdo-(2→4)-[α-Kdo-(2→8)]-α-Kdo-(2→4)-α-Kdo **3** were found linked to lipid A (Figure 3). These oligosaccharides are assembled by

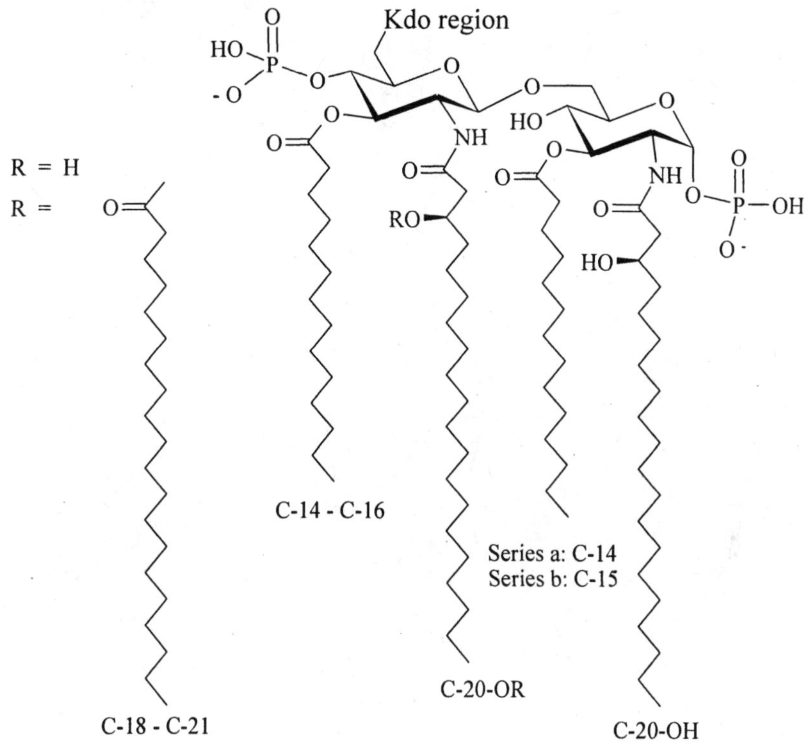

Figure 2. Structure of lipid A species from C. trachomatis serotypes L_2, E and F.

multifunctional Kdo transferases, which most likely may only accomodate a single substrate binding site (*16*).

For the generation of monoclonal antibodies recognizing Kdo epitopes, a series of allyl glycosides comprising α-(2→4)- as well as α-(2→8)- linked Kdo disaccharides and the related tri- and tetrasaccharide derivatives have been prepared and converted into neoglycoproteins (*17*). As an example, the synthesis of the branched Kdo tetrasaccharide (*18*) is outlined (Figure 4). Coupling of the 7,8-diol acceptor derivative **4** with the readily available Kdo bromide donor **5** under Helferich conditions proceeded in remarkable regioselectivity to give the α-(2→8)-linked trisaccharide derivative **6** in 59% isolated yield. The trisaccharide was then elaborated into the donor derivative **7** via hydrogenolysis, acetylation and treatment with $TiBr_4$. Finally, condensation with 7,8-carbonate derivative **8** afforded the branched tetrasaccharide derivative **9** (32% yield) together with the β-linked isomer and trisaccharide glycal ester as by-products. Deprotection of **9** under alkaline conditions provided the Kdo tetrasaccharide allyl glycoside **10** as sodium salt. For the generation of immunoreagents and

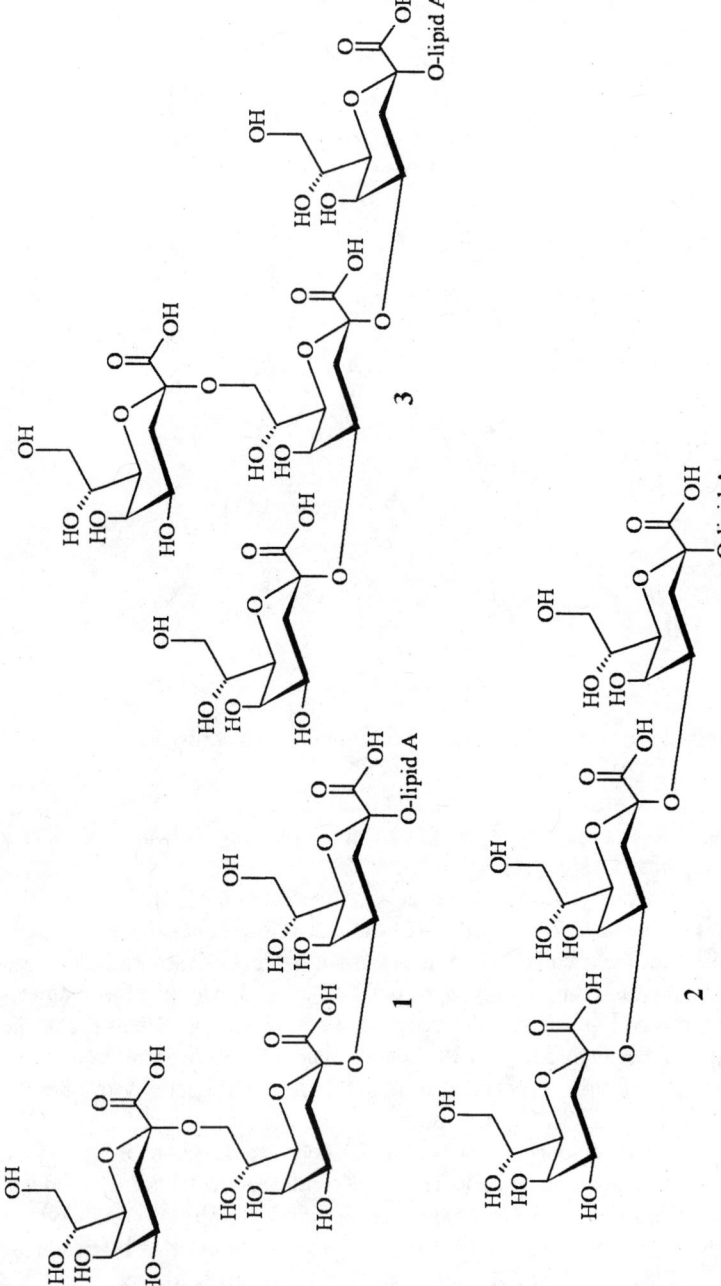

Figure 3. Chemical structure of the Chlamydia-specific Kdo trisaccharide epitope 1 and the C. psittaci specific oligosaccharides 2 and 3.

affinity supports, however, acidic conditions have to be avoided, since the ketosidic linkages are sensitive to acid hydrolysis and furthermore, interresidue lactone formation would be induced. Thus, the allyl group was elongated by UV-mediated addition of cysteamine (19,20), and the spacer derivatives were isolated in pure form by chromatography on ion-exchange resins. Activation of the amino group by the action of thiophosgene and subsequent coupling with bovine serum albumin under slightly alkine conditions furnished the corresponding neoglycoconjugate 12.

By using these and other related neoglycoproteins, a panel of well-defined monoclonal antibodies has been obtained (21). These antibodies may be classified into cross-reactive antibodies recognizing Kdo mono- and α-(2→4)-linked Kdo disaccharides present in enterobacterial LPS and into *Chlamydia*-specific antibodies reacting with family-specific as well as species-specific epitopes.

Antigenic structures of chlamydial epitopes

Synthetic neoglycoconjugates have been used in immunization protocols to generate murine monoclonal antibodies and to characterize their binding epitopes in EIA and EIA-inhibition assays. As can be seen from Table 1, mAb S25-23 - a high affinity antibody with a K_D of 350 nM (22) - requires the complete trisaccharide sequence for binding. MAb S25-2 has a more relaxed binding specificity and binds also the α-(2→8)-linked disaccharide part with reduced affinity. Furthermore, this antibody displays cross-reactivity with the α-

Table 1. Specificity of monoclonal antibodies for Kdo epitopes

	Concentration (ng/ml) mAb yielding $OD_{405} > 0.2$			
Neoglycoconjugate[a]	S25-2	S25-23	S45-18	S69-4
α-Kdo	>1000	>1000	n.d.[b]	>1000
α-Kdo-(2→8)-α-Kdo	32	>1000	n.d.	>1000
α-Kdo-(2→8)-α-Kdo-(2→4)-α-Kdo 2	0.5	500	>1000	
α-Kdo-(2→8)-α-Kdo(C-1$_{red}$)-(2→4)-α-Kdo	125	125	n.d.	n.d.
α-Kdo-(2→4)-α-Kdo	1000	>1000	250	>1000
α-Kdo-(2→4)-α-Kdo-(2→4)-α-Kdo	500	>1000	4	125
α-Kdo-(2→4)-α-Kdo-(2→4)-α-Kdo [α-Kdo-(2→8)]⌐	n.d.	n.d.	n.d.	63

[a]BSA-conjugates were coated on microtiter plates at ligand concentrations of 2 pmol/ml.
[b]n.d. Not determined. Source: Adapted from References 21 and 24 (by permission of Oxford University Press and Maney Publishing, respectively).

246

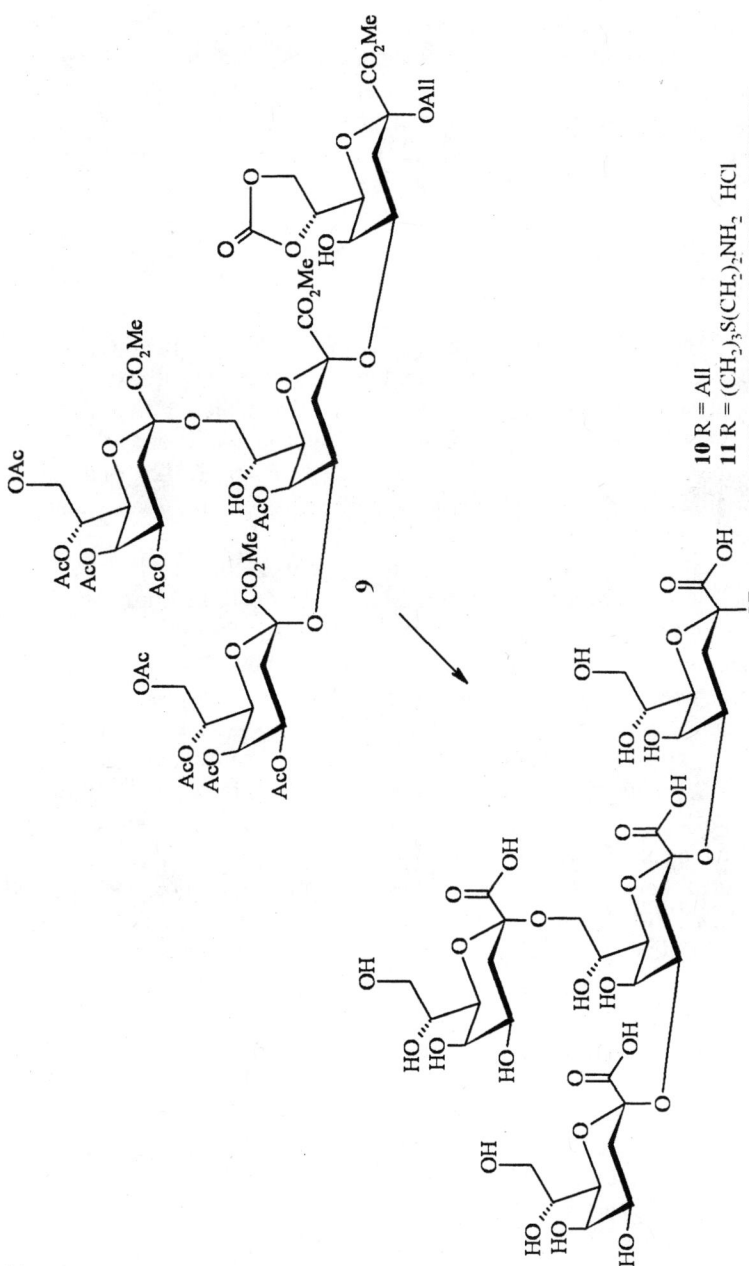

Figure 4. Synthesis of the C. psittaci specific Kdo tetrasaccharide 10.

10 R = All
11 R = (CH$_2$)$_3$S(CH$_2$)$_2$NH$_2$·HCl
12 R = (CH$_2$)$_3$S(CH$_2$)$_2$NH(C=S)NH-BSA

-(2→4)-linked disaccharide unit, albeit with significantly reduced affinity. A second group of antibodies which has been obtained using the α-(2→4)-linked Kdo trisaccharide as the immunizing agent (23) afforded the (2→4)-specific antibody S45-18 (24). Recently, another related mAb, termed S69-4 has been generated using neoglycoconjugate 12 as the immunogen, which does not bind to any of the α-(2→8)-linked compounds (25). Furthermore, mAb S69-4 displays an enhanced binding for the branched tetrasaccharide, but still binds to the α-(2→4)-interlinked trisaccharide moiety within the tetrasaccharide epitope. The antibody allows for selective staining of chlamydial inclusions from *C. psittaci* and thus for a serological differentiation of *C. psittaci* from other chlamydial species.

For further definition of the epitope specificities of the antibodies, di- and trisaccharide analogues have been prepared containing single carboxyl-reduced Kdo units, in order to evaluate the contribution of the acidic substituents to the binding (26). Whereas reduction of the terminal and proximal carboxylates abolished the trisaccharide reactivity completely, mAb S25-23 and S25-2 tolerate modification of the internal Kdo-residue (entry 4, Table 1). A schematic view on the different epitopes is given in Figure 5.

Similar antibody specificities have been detected in human sera from patients with *Chlamydia*-induced urogenital and respiratory tract infections and may thus be employed for serological detection of chlamydial infections where a direct antigen test is not applicable and also for the follow-up of antichlamydial treatments. There is, however, limited evidence for neutralizing properties of these antibodies. In one case, a monoclonal antibody CP-33 has been described with reactivity towards the family-specific LPS epitope conferring protection in vitro against *C. pneumoniae* strain TW-183 (27). This antibody recognizes the α-Kdo-(2→8)-α-Kdo-(2→4)-α-Kdo trisaccharide, albeit in a different conformational arrangement, which may be due to the steric influence of the neighboring phosphorylated lipid A backbone (Figure 1). Thus, a detailed analysis and improved modelling of the bioactive structure of the ligands more closely related to the native location of the antigens in the outer membrane is needed.

Crystal structures of antibody-ligand complexes

Up to now only a very small number of crystal structures of Fab-fragments complexed to carbohydrate ligands has been reported (28,29), none of them comprising charged sugar residues. Based on the detailed data of the epitope characterization, Fab-fragments and single chain antibodies of selected specificities have been subjected to crystallization studies. Recently, the crystal structures of S25-2 and S45-18 alone and in complex with a variety of Kdo ligands have been solved at atomic resolution (30). Amino acid sequence

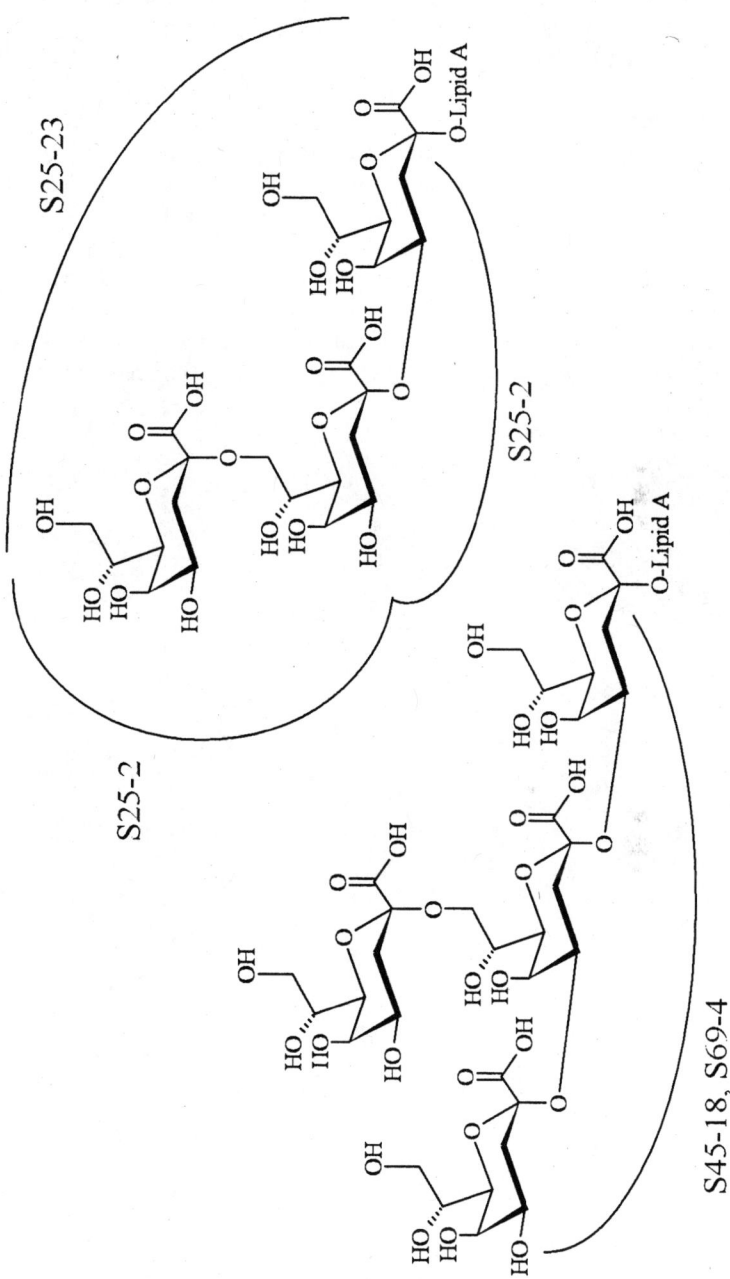

Figure 5. Major Chlamydia-specific and cross-reactive epitopes

data of these antibodies indicate that a large portion corresponds to germline segments with nearly identical sequences in the light chains and subtle but significant differences in the H3 domain between the antibodies. The major interactions result from a network of hydrogen bonds towards O-7, O-5 and O-4 of Kdo as well as a salt bridge extending from tyrosine H33 and arginine H52 to the carboxyl group of the terminal Kdo unit (Figure 6). These binding motifs of terminal Kdo units are almost identical in all antibody complexes studied. Major differences arise in the binding mode seen for the internal Kdo units. Within the binding site of S25-2 relaxed binding is oberserved for the α-Kdo- 2→8)-α-Kdo disaccharide and the cross-reactive α-Kdo-(2→4)-α-Kdo unit, which is accomplished by the very same amino acid residues interacting with different parts of the ligands (Figure 7). In particular, whereas tyrosine H33 forms hydrogen bonds with O-7 of the 8-substituted Kdo (Figure 7, left), this

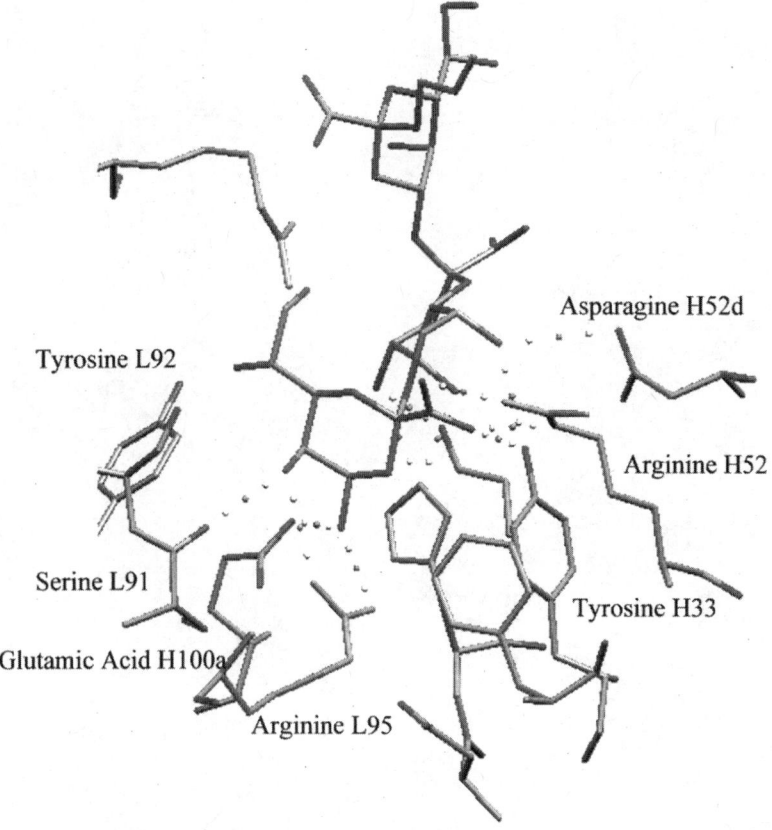

Figure 6. Binding environment of α-Kdo-(2→8)-α-Kdo-(2→4)-α-Kdo bound to the Fab-fragment of mAb S25-2 (adapted from pdb-file 1Q9Q).

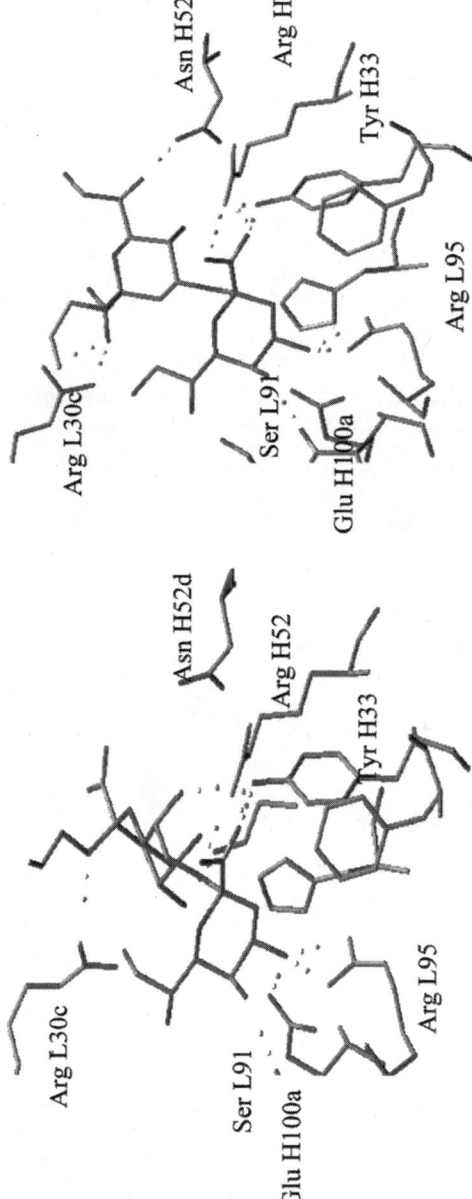

Figure 7. Structure of α-Kdo-(2→8)-α-Kdo (left) and α-Kdo-(2→4)-α-Kdo (right) disaccharide bound to S25-2 (Adapted from pdb-files 1Q9R and 1Q9T).

hydrogen bond is present in a similar fashion, albeit extending to O-5 of the 4-substituted Kdo. Thus, identical amino acid residues in the binding site may nevertheless contact different parts of a cross-reactive epitope. Flexibility is also seen in arginine L30c which interacts with the proximal Kdo unit in case of the trisaccharide, but moves downward to provide a salt bridge with the proximal carboxylic group of the cross-reactive α-Kdo-(2→4)-α-Kdo disaccharide (Figure 7, right). The crystal structures of the α-Kdo-(2→8)-α-Kdo disaccharide as well as the α-Kdo-(2→8)-α-Kdo-(2→4)-α-Kdo trisaccharide complex also corroborate the serological data obtained with carboxyl-reduced analogues carrying the C-1 hydroxymethyl group at the 8-substituted Kdo, since this carboxylic group of the internal Kdo is not observed in salt bridge or hydrogen bond formation (Figure 7 and Figure 8). In contrast, binding of mAb S45-18 is characterized by a rigid interaction with the α-Kdo-(2→4)-α-Kdo-(2→4)-α-Kdo trisaccharide due to hydrophobic interactions exerted by a mutated phenylalanine H97 residue towards all three Kdo units (Figure 9). Similar to the binding mode of the cross-reactive α-Kdo-(2→4)-α-Kdo disaccharide with mAb S25-2, O-5 of the internal Kdo is hydrogen-bonded to tyrosine H33 in the complex with S45-18. A second hydrogen bond extends to from O-5 of Kdo to lysine H52d. As seen in the complex obtained with a bisphosphorylated pentasaccharide, isolated from a recombinant *C. psittaci* strain expressing the chlamydial Kdo transferase, there is no contact of the lipid A backbone to the protein (Figure 10). This may be explained by the choice of the immunizing agent, which had been a neoglycoconjugate containing the Kdo trisaccharide only.

Acknowledgments

Financial support of this work by the Austrian Science Fund (grants P 13843 and P 17407) and DFG (SFB 470-C1) is gratefully acknowledged.

References

1. Campbell, L. A.; Kuo, C. C. *Nat. Rev. Microbiol.* **2004**, *2*, 23-32.
2. Christiansen, G.; Birkelund, S. *Best Pract. Res. Clin. Obstet. Gynaecol.* **2002**, *16*, 889-900.
3. Everett, K. D. E.; Bush, R. M.; Andersen, A. A. *Int. J. Syst. Bacteriol.* **1999**, *49*, 415-440.
4. Grayston, J. T.; Aldous, M. B.; Easton, A.; Wang, S.-P.; Kuo, C.-C.; Campbell, L. A.; Altman, J. *J. Infect. Dis.* **1993**, *168*, 1231-1235.
5. Moulder, J. W. *Microbiol. Rev.* **1991**, *55*, 143-190.
6. Nano, F. E.; Caldwell, H. D. *Science* **1985**, *228*, 742-744.

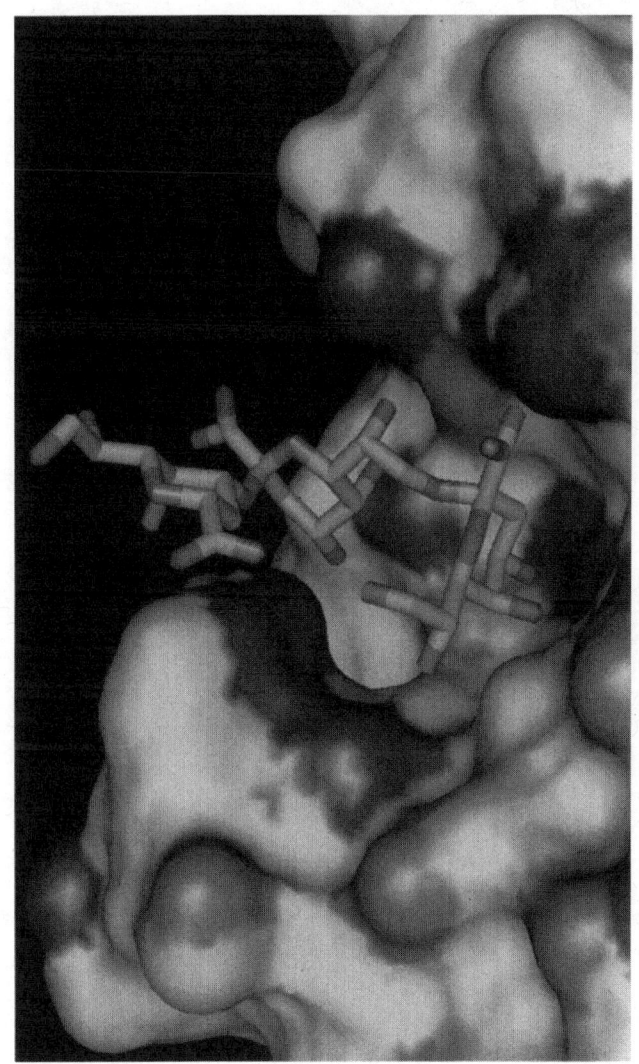

Figure 8. Structure of α-Kdo-(2→8)-α-Kdo-(2→4)-α-Kdo trisaccharide bound to mAb S25-2.

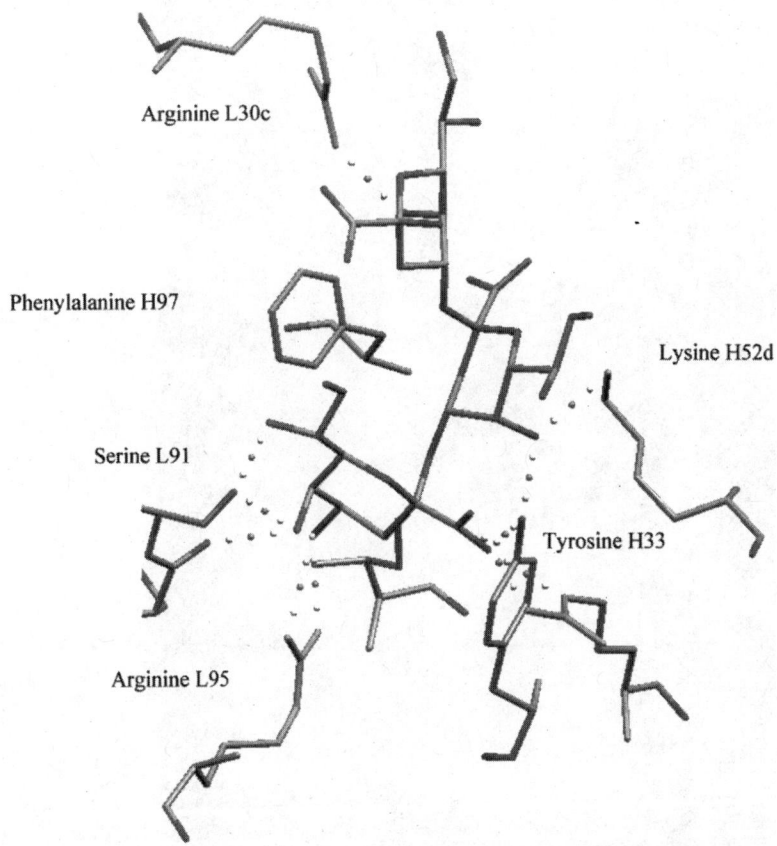

Figure 9. Binding environment of α-Kdo-(2→8)-α-Kdo-(2→4)-α-Kdo bound to the Fab-fragment of mAb S25-2(Adapted from pdb-file 1Q9Q).

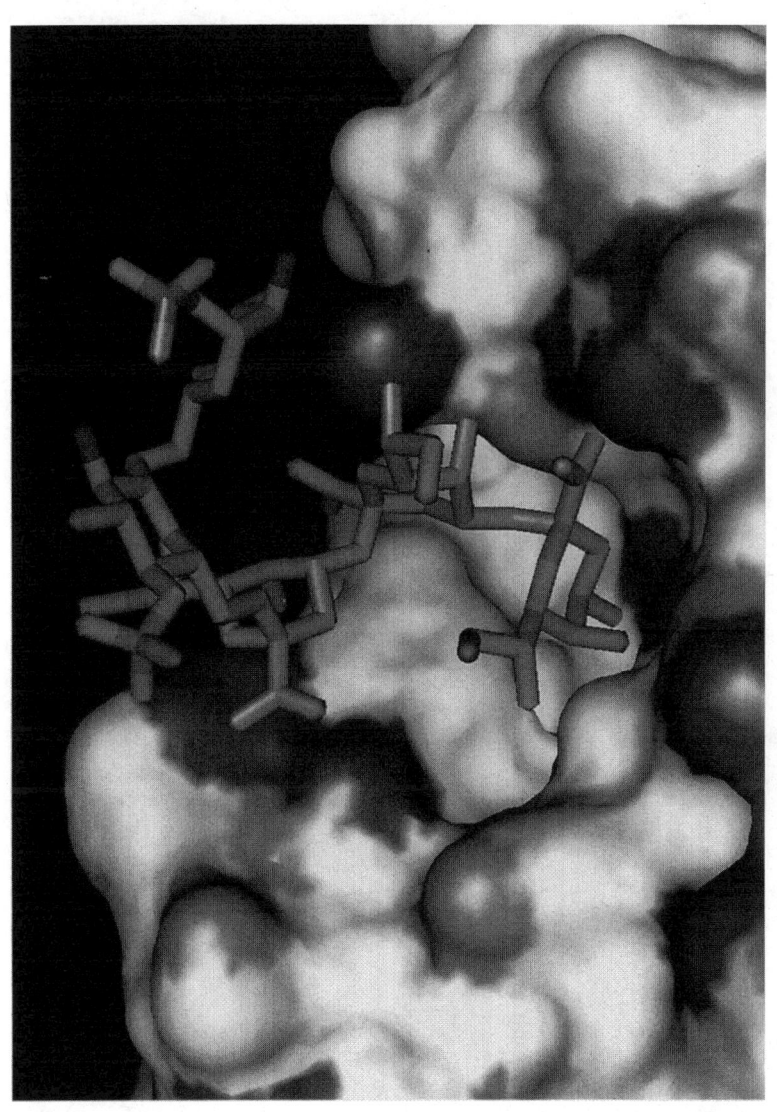

Figure 10. Binding environment of α-Kdo-(2→4)-α-Kdo-(2→4)-α-Kdo-(2→6)-β-4P-GlcN-(1→6)-α-GlcN-1P bound to the Fab-fragment of mAb S45-18 (Adapted from pdb-file 1Q9W).

7. Brade, H. *Endotoxins in Health and Disease*; Brade, H.; Opal, S. M.; Vogel, S. N.; Morrison, D. C., Eds.; Marcel Dekker Inc.: New York, Basel, 1999; pp 229-242.
8. Qureshi, N.; Kaltashov, I.; Walker, K.; Doroshenko, V.; Cotter, R. J.; Takayama, K.; Sievert, T. R.; Rice, P. A.; Lin, J.-S.; Golenbock, D. T. *J. Biol. Chem.* **1997**, *72*, 10594-10600.
9. Rund, S.; Lindner, B.; Brade, H.; Holst O. *J. Biol. Chem.* **1999**, *274*, 16819-16824.
10. Heine, H.; Müller-Loennies, S.; Brade, L.; Lindner, B.; Brade, H. *Eur. J. Biochem.* **2003**, *270*, 440-450.
11. Prebeck, S.; Brade, H.; Kirschning, C. J.; da Costa, C. P.; Dürr, S.; Wagner, H.; Miethke, T. *Microbes Infection* **2003**, *5*, 463-470.
12. Zamyatina, A.; Sekljic, H.; Brade, H.; Kosma, P. *Tetrahedron* **2004**, *60*, 12113-12137.
13. Nurminen, M.; Leinonen, M.; Saikku, P.; Mäkelä, H. *Science* **1983**, *220*, 1279-1281.
14. Brade, H.; Brade, L.; Nano, F. E. *Proc. Natl. Acad. Sci. USA* **1987**, *84*, 2508-2512.
15. Rund, S.; Lindner, B.; Brade, H.; Holst O. *Eur. J. Biochem.* **2000**, *267*, 5717-5726.
16. Mamat, U.; Baumann, M.; Schmidt, G.; Brade, H. *Mol. Microbiol.* **1993**, *10*, 935-941.
17. Kosma, P. *Endotoxins in Health and Disease*; Brade, H.; Opal, S. M.; Vogel, S. N.; Morrison, D. C., Eds.; Marcel Dekker Inc.: New York, Basel, 1999; pp 257-282.
18. Kosma, P.; Reiter, A.; Hofinger, A;, Brade, L; Brade, H. *J. Endotoxin Res.* **2000**, *6*, 57-69.
19. Lee, R. T.; Lee, Y. C. *Carbohydr. Res.* **1974**, *37*, 193-201.
20. Fu. Y.; Baumann, M.; Kosma, P.; Brade, L.; Brade, H. *Infect. Immun.* **1992**, *60*, 1314-1321.
21. Brade, L.; Zych, K.; Rozalski, A.; Kosma, P.; Bock, K.; Brade, H. *Glycobiol.* **1997**, *7*, 819-827.
22. Müller-Loennies, S.; MacKenzie, C. R.; Patenaude, S. I.; Evans, S.; Kosma, P.; Brade, H.; Brade, L.; Narang, S. *Glycobiol.* **2000**, *10*, 121-130.
23. Kosma, P.; Schulz, G.; Unger, F. M. *Carbohydr. Res.* **1989**, *190*, 191-201.
24. Brade, L.; Rozalski, A.; Kosma, P.; Brade, H. *J. Endotoxin Res.* **2000**, *6*, 361-368.
25. Müller-Loennies, S.; Gronow, S.; MacKenzie, R.; Brade, L.; Kosma, P.; Brade, H. *Glycobiol.* submitted.
26. D'Souza, F. W.; Kosma, P.; Brade, H. *Carbohydr. Res.* **1994**, *262*, 223-244.
27. Peterson, E. M.; De La Maza, L. M.; Brade, L.; Brade, H. *Infect. Immun.* **1998**, *66*, 3848-3855.

28. Cygler, M.; Rose, D. R.; Bundle, D. R. *Science* **1991**, *253*, 442-446.
29. Villeneuve, S.; Souchon, H.; Riottot, M.-M. ; Mazie, J.-C. ; Lei, P.-S.; Glaudemans, C. P. J. ; Kovac, P. ; Fournier, J.-M. ; Alzari, P. M. *Proc. Natl. Acad. Sci. USA* **2000**, *97*, 8433-8438.
30. Nguyen, H. P.; Seto, N. O. L.; MacKenzie, C. R.; Brade, L.; Kosma P.; Brade, H.; Evans, S. V. *Nature Struct. Biol.* **2003**, *10*, 1019-1025.

Chapter 12

Synthetic Carbohydrate-Based Antitumor Vaccines

Rebecca M. Wilson[1], J. David Warren[1], Ouathek Ouerfelli[2], and Samuel J. Danishefsky[1–3,*]

[1]Laboratory for Bioorganic Chemistry and [2]Organic Synthesis Core Laboratory, Sloan-Kettering Institute for Cancer Research, 1275 York Avenue, New York, NY 10021
[3]Department of Chemistry, Columbia University, 3000 Broadway, New York, NY 10027

Our laboratory has a longstanding program devoted to the preparation and immunological evaluation of fully synthetic carbohydrate-based antitumor vaccine constructs. We present herein a brief historical account of the evolution of this research program, and describe key achievements, including the synthesis of the Globo-H-KLH construct, which is currently in clinical trials. Remarkable advances in carbohydrate and glycopeptide assembly techniques have allowed for the synthesis of increasingly sophisticated, structurally complex vaccine constructs. In this setting, we have successfully accomplished the synthesis of a highly complex unimolecular pentavalent vaccine construct, in which five prostate and breast cancer-associated carbohydrate antigens are displayed on a single polypeptide backbone.

Introduction

A distinguishing characteristic of malignantly transformed tumor cells is the display of aberrant levels and types of cell surface carbohydrates. These carbohydrates are commonly expressed as glycoproteins, mucins, and glycosphingolipids on the cell surface. Interestingly, antibodies that recognize these glycoconjugates are occasionally found in the sera of human cancer patients, suggesting the possibility of an immune response to the tumor state. Further evaluation reveals that antibody formation can be evoked by suitable glycoconjugates, but not by the oligosaccharides alone. However, having been elicited in this way, most antibodies are particularly sensitive to the structure of the carbohydrate domain. Given these observations, it has been postulated that vaccination with tumor-specific glycoconjugates could provide immunological protection against micrometastases and circulating tumor cells. Clearly, the possibility of inducing an immune response to cancer could have an enormous impact on the *in vivo* diagnosis and treatment of the disease (*1*).

Our general approach to the design and evaluation of carbohydrate-based anticancer vaccines is illustrated in Figure 1 (*2*). Thus, through the process of glycal assembly, we seek to prepare homogeneous, fully synthetic constructs consisting of one or more tumor associated carbohydrate antigens, which are then appended through a linker to an appropriate immunogenic carrier protein, such as keyhole limpet hemocyanin (KLH) (*3*). These vaccine constructs are injected into mice in the hopes of eliciting an antibody response. Finally, the antibodies thus obtained are harvested and evaluated *in vitro* for their ability to bind cells known to overexpress the antigens in question.

Results and Discussion

Synthesis of the Monomeric Globo-H-KLH Construct.

A number of known tumor-associated carbohydrate-based antigens are presented in Figure 2. One of the first carbohydrate antigens synthesized in our laboratory was the Globo-H hexasaccharide. First isolated from the human breast cancer cell line MCF-7 (*4, 5*), overexpression of this antigen was later found to be associated with a number of other cancer types, including colon, lung, ovarian, and prostate (*6*). Globo-H is expressed on the cell surface as a glycolipid and also possibly as a glycoprotein.

The synthesis of the Globo-H-KLH conjugate, which made use of our highly convergent glycal assembly protocol, commenced with the preparation of the ABC trisaccharide fragment **8** (Figure 3) (*7-9*). Thus, the cyclic carbonate **1**, which would ultimately serve as the B-ring of the hexasaccharide, was treated with DMDO to afford α-epoxide **2**. The latter was coupled with **3**, itself derived

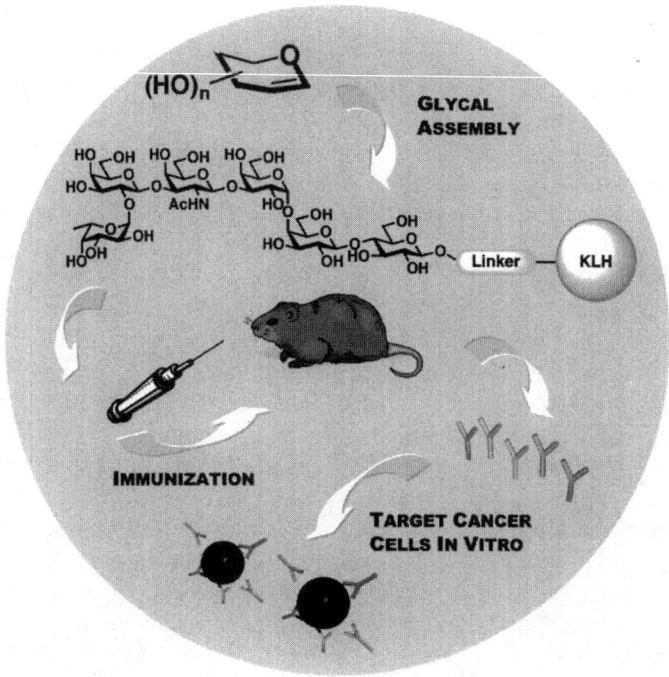

Figure 1. Design of a synthetic carbohydrate-based anticancer vaccine.

from D-glucal, to give rise to the AB disaccharide **4**. Through a series of functional group manipulations, the C_4' axial hydroxyl functionality of the B ring was selectively unmasked to afford **5**. This free hydroxyl group would serve as the site of connection of the B and C rings. Glycal **6**, which would constitute the C ring, was derived from D-galactal. Treatment of **6** with DMDO afforded the α-epoxide which was opened to the requisite β-anomeric fluoride upon exposure to TBAF. Finally, benzyl protection of the resultant alcohol provided the coupling precursor **7**. With both components in hand, we were prepared to attempt the coupling between the glycal acceptor **5** and donor **7**. In the event, glycosylation proceeded with moderate selectivity under Mukayama-Nicolaou conditions to afford the trisaccharide as a 4.5:1 (α:β) anomeric mixture at the B-C ring junction. Subsequent removal of the C-ring PMB protecting group, under oxidative conditions, provided the ABC trisaccharide **8**.

The synthesis of the DEF trisaccharide commenced again with epoxide **2** (Figure 4). Thus, the zinc chloride-mediated coupling of **2** and **9** proceeded regioselectively to afford the disaccharide **10**. At this point, we hoped to differentiate the C_2' and C_4 hydroxyl groups in a fucosylation reaction. We postulated that fucosylation would proceed preferentially at the C_2' equatorial

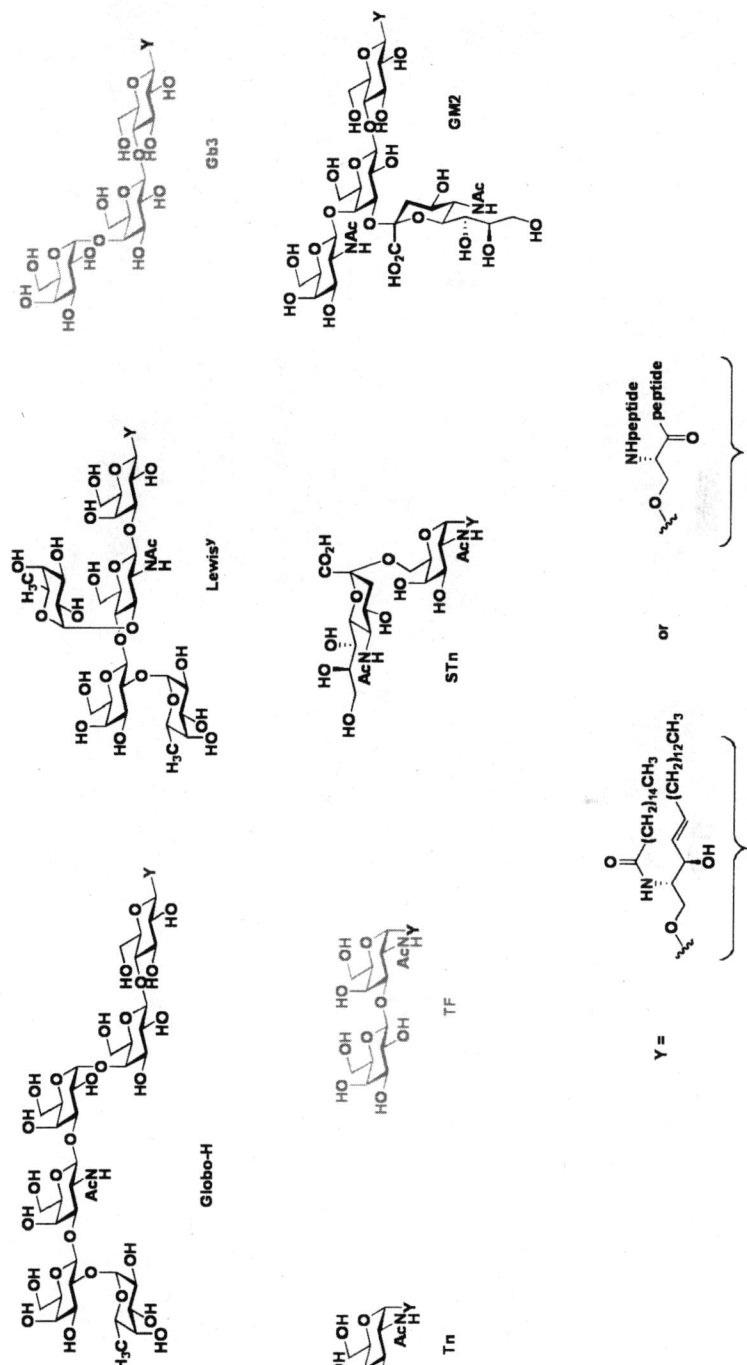

Figure 2. Representative tumor-related carbohydrate-based antigens.

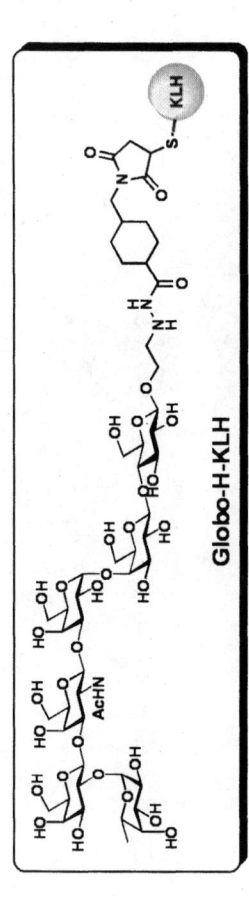

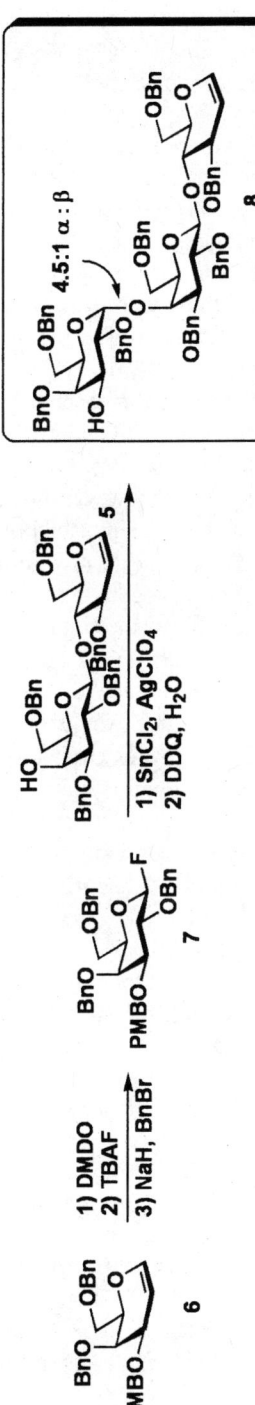

Figure 3. Synthesis of the Globo-H glycal: ABC trisaccharide.

hydroxyl. Indeed, exposure to Mukayama-Nicolaou fucosylation conditions with **11** afforded the desired trisaccharide **12** as the predominant product (47% yield). Although moderate amounts of the undesired regioisomer were observed (~ 8%), the convenience of the overall route was deemed to compensate for this compromise in selectivity. Acetylation of the C_4 hydroxyl followed by iodosulfonamidation of the glycal afforded the DEF trisaccharide **13**. Although it is possible to achieve direct azaglycosidation of iodo sulfonamide donors, we had previously observed that such "direct rollover" couplings are not feasible with severely hindered substrates. As such, we turned to a two-step coupling protocol involving the intermediacy of an activated thioglycoside. Thus, treatment of **13** with lithium ethanethiolate afforded the thioglycoside coupling partner, **14**.

The stage was now set for the coupling of the ABC and DEF trisaccharide domains. Thus, methyl triflate-promoted coupling of thioglycoside **14** and glycal **8** proceeded in 60% yield to afford a hexasaccharide adduct as a 6:1 ratio of stereoisomers at the newly formed C-D ring junction (Figure 5). However, upon further structural evaluation, it was determined that the major stereoisomeric product of the coupling reaction was in fact the undesired α- anomer. This result was particularly surprising given our prior experience with analogous systems. We tentatively attributed this unexpected product distribution to the failure of the sulfonamide to participate in the activation of the donor. Thus, certain structural features of **14** may have forced the intermediacy of an active onium species in lieu of the cyclic sulfonamido active species, which would have dictated glycosidation from the requisite β-face. If this were the case, then presumably by making small changes in the nature of the DEF donor system, we would be able to encourage formation of the desired cyclic sulfonamide intermediates, which could lead to improvements in the product distribution.

One straightforward opportunity for variation would be in the disposition of the C_4 oxygen in the future D-ring. Earlier in the synthesis of the DEF trisaccharide, this functionality had existed as a free alcohol (*cf.* **12**). Indeed, attempts to convert the trisaccharide (**12**) to the coupling partner **16** were successful. In the event, methyl triflate-mediated coupling of **16** with **8** proceeded in 80% yield to afford the hexasaccharide **18** as a 10:1 mixture of isomers at the C-D ring junction, *with the requisite β-anomer as the major adduct*. This remarkable result serves to underscore the potential impact that small changes in the nature of the donor ring may have on such coupling reactions. Indeed, by replacing an acetate group with a free hydroxyl, it was possible to effect a substantial reversal in the directionality of the coupling reaction of the DEF and ABC sectors of the Globo-H glycal.

The final phase of the Globo-H-KLH monomer synthesis commenced with the global deprotection of the hexasaccharide **18** (Figure 6). The fully deprotected intermediate was then peracetylated and subjected to epoxidation followed by epoxide opening with allyl alcohol to afford, following removal of the acetate protecting groups, the allyl glycoside **19**. The newly appended allyl moiety would serve as the functional handle for attachment to the KLH carrier protein. Thus, ozonolysis of the terminal olefin was followed by appendage of the linker molecule through reductive amination. Finally, conjugation to KLH provided the Globo-H-KLH monomeric vaccine construct.

Preliminary *in vivo* investigations of the Globo-H-KLH construct confirmed its immunological potential (*10*); mice vaccinated with Globo-H-KLH were found to produce high titer IgM and IgG responses to the Globo-H antigen. The antibodies thus obtained were found to be reactive to Globo-H-positive cell line, MCF-7, but not to Globo-H-negative cell line, B78.2. Furthermore, the Globo-H antibodies thus raised were capable of inducing complement-mediated lysis of the MCF-7 cells.

Based on the results of these and other preclinical evaluations, the Globo-H-KLH conjugate was advanced to Phase I clinical trials for breast and prostate cancers (*11, 12*). The results of both trials served to establish the safety and lowest functional immunogenic dose of the Globo-H-KLH construct. Importantly, the immunogenicity of the vaccine was demonstrated for each patient regardless of the individual tumor burden level, indicating that the efficacy of the antibody response is somewhat independent of the cancer stage. We now have an approved protocol for Phase II/III clinical trials of the Globo-H-KLH vaccine, which are planned to commence in 2006.

Having established the validity of the carbohydrate antitumor vaccine concept both at the levels of synthesis and apparent chemical safety, we next sought to prepare increasingly complex constructs that would more closely mimic the glycoprotein arrangements typically found on the surface of transformed cells. Importantly, it has been reported that carbohydrate antigens are often displayed in triplicate on adjacent threonine and serine residues (*13*). Furthermore, such clusters are apparently the preferred targets of monoclonal antibodies. With these considerations in mind, we successfully prepared a number of "clustered" vaccine constructs consisting of a single antigen displayed in triplicate on a serine or threonine polypeptide backbone (*cf.* Tn(c), TF(c), and STn(c), Figure 7). Importantly, such clustered vaccines were found to exhibit enhanced levels of antibody induction when compared to their non-clustered counterparts. Several such clusters are currently being evaluated in preclinical and clinical settings (*14-19*).

Unimolecular Multivalent Vaccine Constructs.

In light of the encouraging immunogenicity observed with our first- and second-generation vaccine constructs in both preclinical and clinical settings, we hoped to develop increasingly robust vaccine candidates. Since transformed cells are known to harbor varying degrees of heterogeneity with regard to the type and

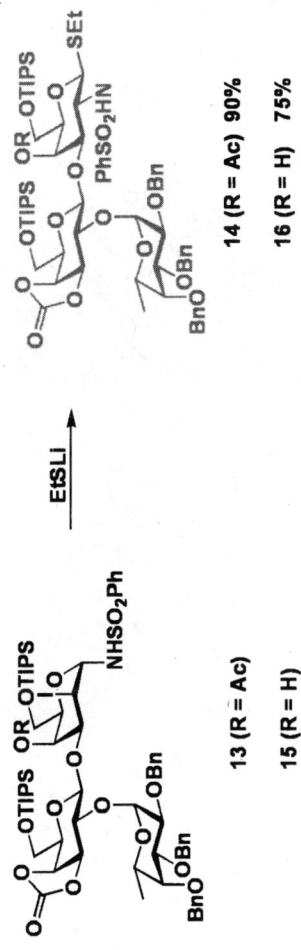

Figure 4. Synthesis of the Globo-H glycal: DEF trisaccharide.

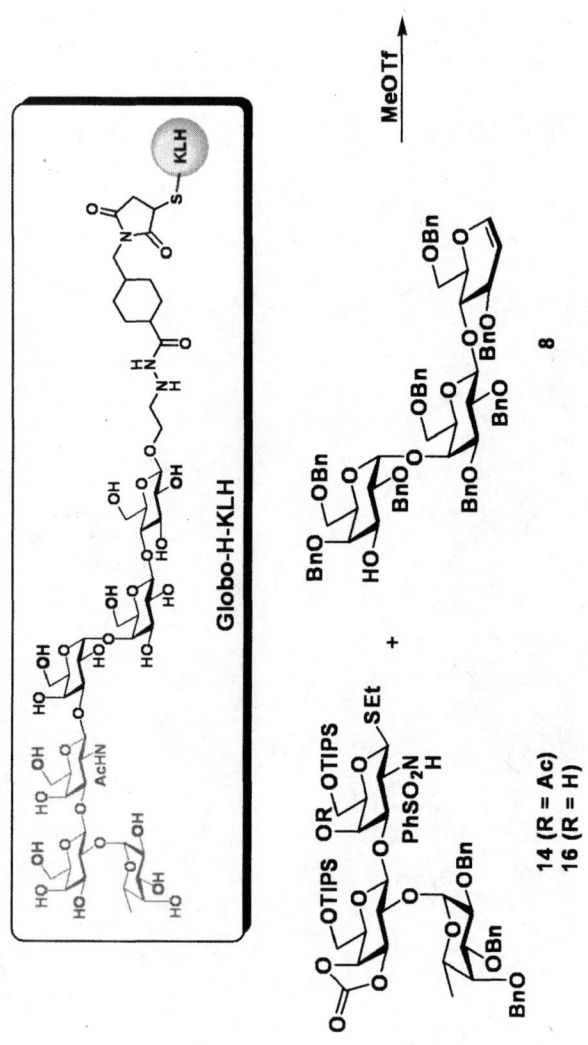

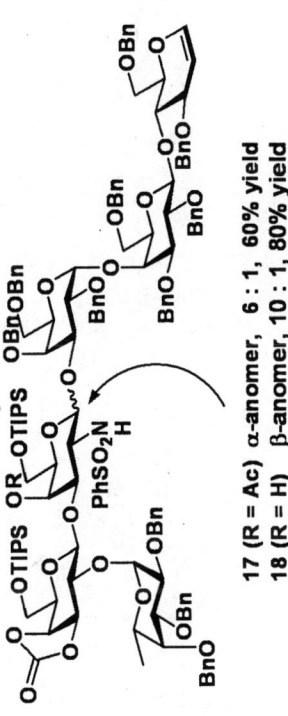

Figure 5. Synthesis of Globo-H glycal: coupling of ABC and DEF domains.

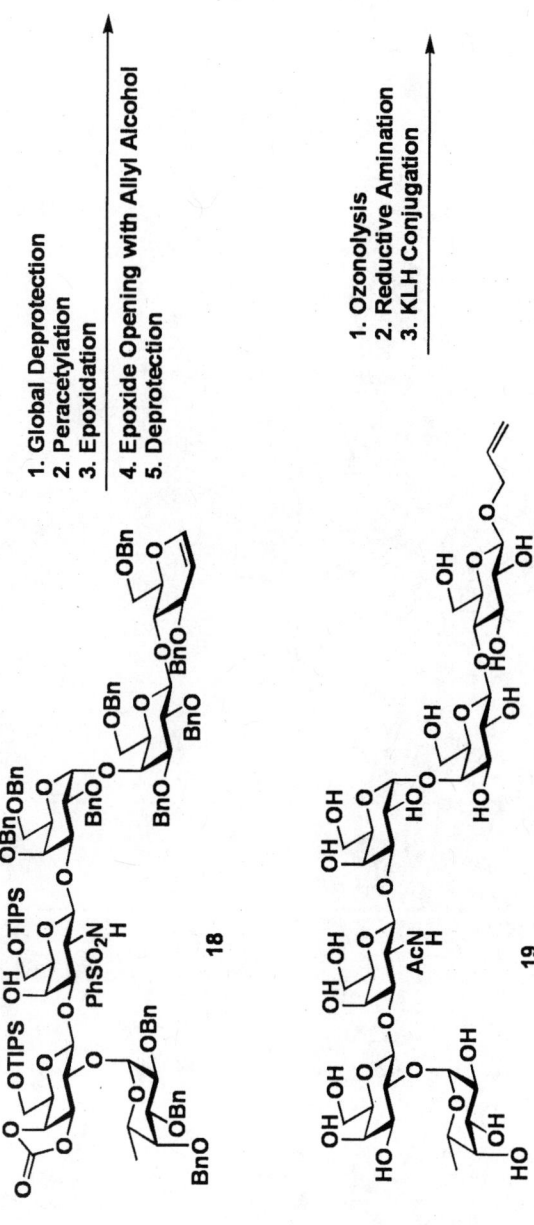

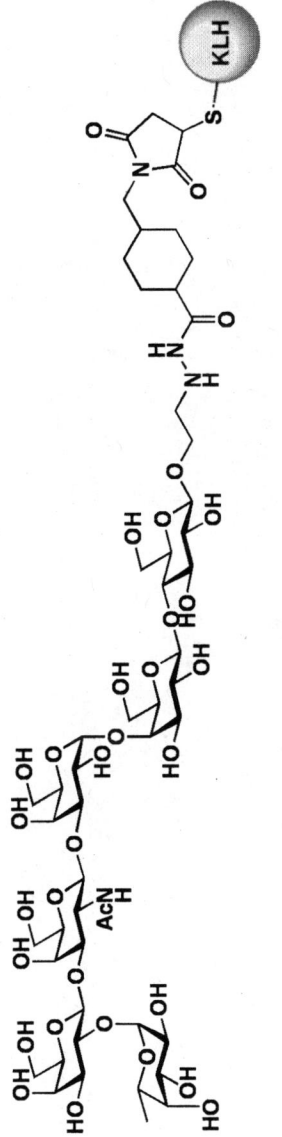

Figure 6. Completion of Globo-H monomer synthesis and conjugation.

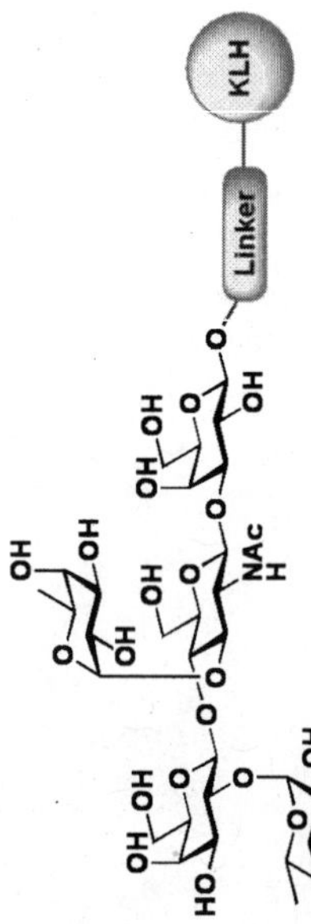

Lewis^y

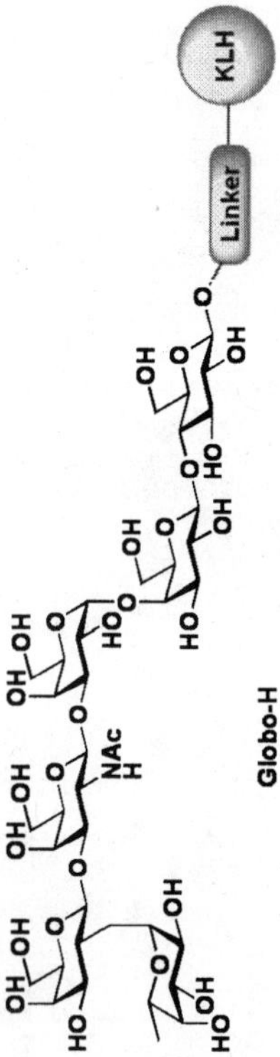

Globo-H

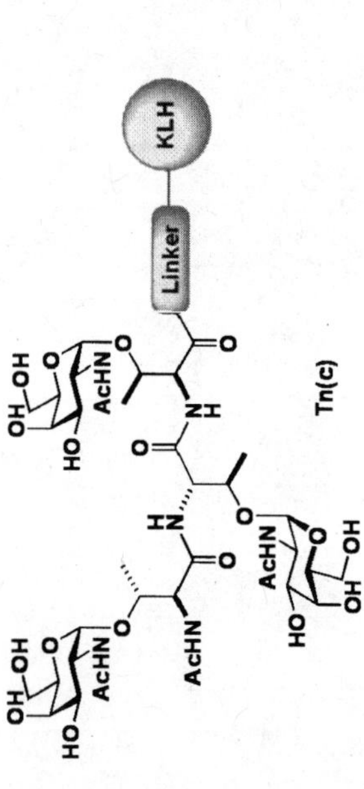

Tn(c)

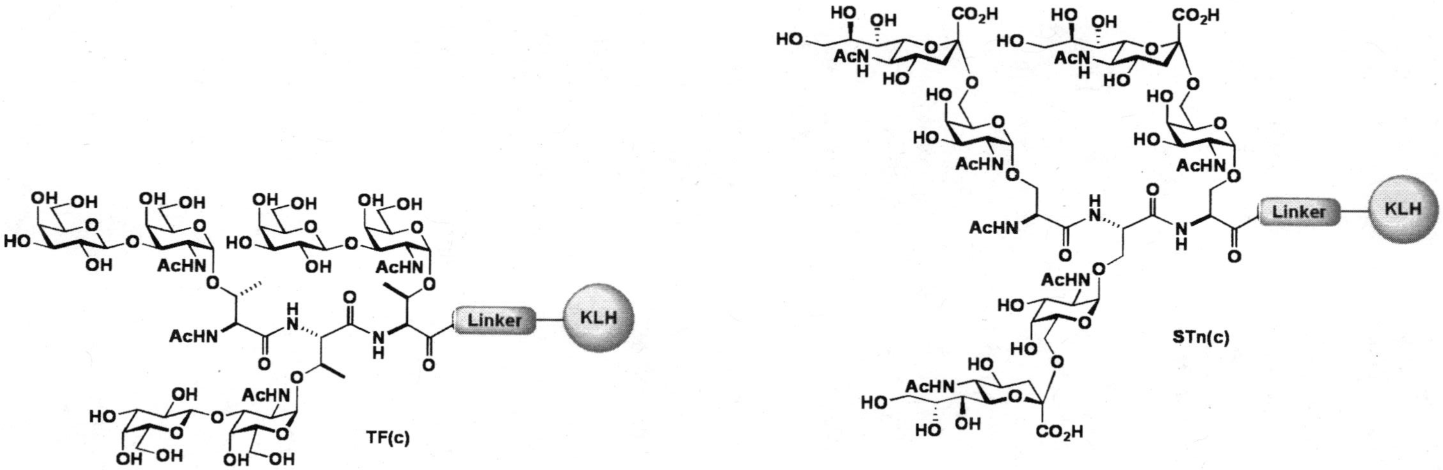

273

Figure 7. Conjugate vaccines in use.

distribution of antigens expressed on their cell surfaces, the development of a truly potent and broadly useful carbohydrate-based vaccine should take this into account (20). The variety and quantity of antigen expression may fluctuate as a function of the stage of cellular development. Thus, even within a particular cancer type, there is a great deal of antigen heterogeneity. We speculated that the inclusion of additional carbohydrate antigens closely associated with a particular cancer type could well increase the percentage of tumor cells targeted. One can imagine two possible implementations of this multiantigenic vaccine. The first, we term the polyvalent vaccine approach and the second is the unimolecular, multivalent vaccine approach (Figure 8).

According to the polyvalent method, mixtures of several different monovalent antigen-KLH constructs would be co-injected in the hopes of inducing antibody response to each individual antigen. This strategy has been implemented in preclinical settings, and it has been demonstrated that the monomeric constructs administered in this way are each able to evoke antibody formation at levels comparable to those obtained when the vaccine is administered individually (21). As expected, the overall level of antibody induction is enhanced as a result of the co-administration of different epitopic vaccines.

This approach, however, suffers from a number of significant practical limitations. First, the implementation of the polyvalent vaccine method requires the employment of significantly increased levels of carrier protein (i.e. KLH). The extended use of excess carrier protein may actually lead to a decrease in the immunogenicity of the carbohydrate antigen, due to the fact that prior exposure of the host immune system to the carrier protein can lead to an increased antibody response to the carrier protein itself (22, 23). The result is a decreased T-helper response and, ultimately, diminished induction of antibodies toward the carbohydrate antigen itself. A second consideration arises at the regulatory level; the clinical approval of a polyvalent vaccine mixture would require the nontrivial validation of each individual component. Finally, the synthesis of each carbohydrate-KLH construct requires a low-yielding final conjugation reaction. For these reasons, we prefer a unimolecular, multivalent approach in which several different tumor-associated carbohydrate antigens are displayed on a single polypeptide backbone.

There are a number of potential advantages to this polyvalent strategy. First, only a single conjugation event to carrier protein is required. Clearly, this feature is particularly attractive from a synthetic standpoint, as the final conjugation to KLH is typically low-yielding. Furthermore, from a regulatory standpoint, the parallel registration of various antigens ina single multivalent entity would certainly be more straightforward. Finally, it is at least conceivable that some level of antigenic synergy would be achieved through the display of several different antigens along a single polypeptide.

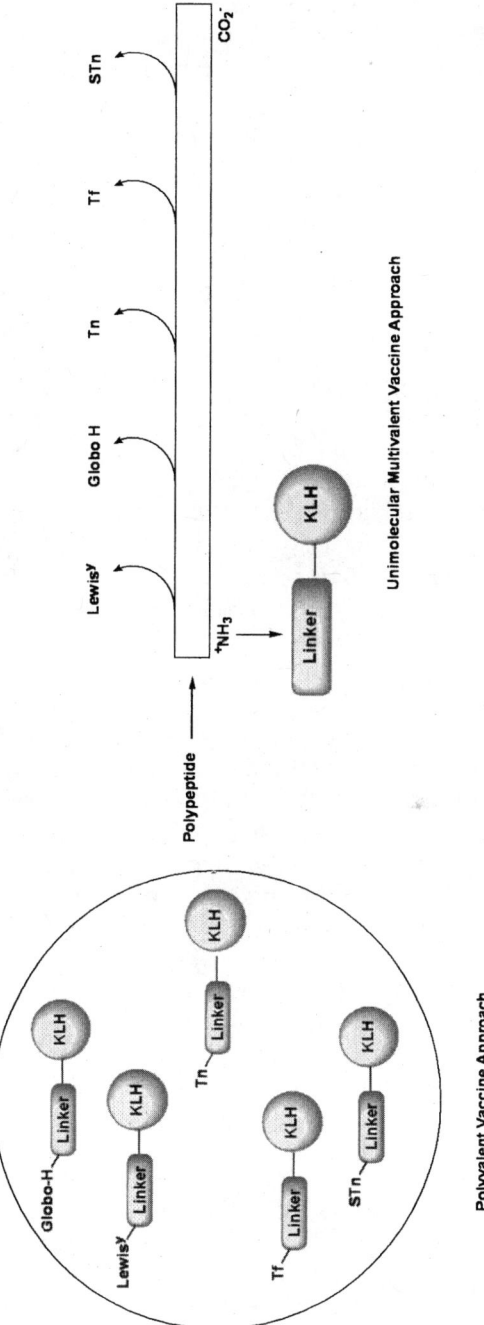

Figure 8. Approaches to multivalent vaccines.

Implementation of the unimolecular multivalent vaccine strategy would require the development of methods for the efficient and selective assembly of glycosylamino acids from complex glycan precursors. Toward this end, three methods have been developed, each of which allows for the introduction of the amino acid moiety in a stereochemically-defined manner to provide a glycosylamino acid in which the carbohydrate is separated from the peptide backbone by a four-carbon linker (Figure 9) (24).

The first method (a) commences with a pentenyl glycoside (25). Thus, ozonolysis of the terminal olefin followed by a Horner-Emmons reaction, as shown, yields the dehydroamino acid. The internal olefin is then subjected to enantioselective reduction, followed by standard functional group management to afford the requisite glycosylamino acid cassette. The stability of the pentenyl glycoside as opposed to the corresponding allyl glycoside allows the glycal to be functionalized at a fairly early stage of the synthesis, thus minimizing the impact of the chemical loss associated with the accompanying glycal epoxidation step. A significant drawback of this method lies in the fact that the stereocenter must be installed through enantioselective reduction of the dehydroamino acid.

According to the second commonly used method (b) (26), a protected allyl glycoside is subjected to olefin cross-metathesis with protected allyl glycine. Reduction of the olefin with concurrent benzyl ester removal provides the glycosylamino acid cassette. A main advantage of this method in comparison with the ozonolytic protocol lies in the fact that the amino acid stereocenter does not have to be installed in the course of the assembly of the glycosylamino acid. Indeed, allyl glycine is commercially available in enantiomerically pure form. However, a shortcoming of this protocol derives from the requirement that the allyl ether be installed from the glycal at the relatively late peracetate stage of the carbohydrate synthesis.

Finally, a third protocol (c) has been developed that allows for the installation of the amino acid directly from glycal epoxide or trichloroacetimidate donors via coupling with hydroxynorleucine (27). This method allows for the introduction of the amino acid residue with the stereocenter in place in a single step; however, the efficiency of the reaction is mitigated somewhat by the moderate stereoselectivity of the glycal epoxidation and the need to prepare hydroxynorleucine, which is not commercially available.

In order to test the viability of the unimolecular multivalent vaccine concept, a trivalent construct was prepared in which the Globo-H, Lewis^y and Tn antigens were displayed on a polypeptide backbone (**20**) (28). Notably, sera obtained from vaccination with the trivalent construct were found to contain antibodies against the individual antigens, as determined by ELISA analysis, and to recognize and bind cells that display these three antigens on their surfaces (Figure 10).

Encouraged by the promising results obtained with the trivalent vaccine, we next sought to prepare a pentavalent construct consisting of five different

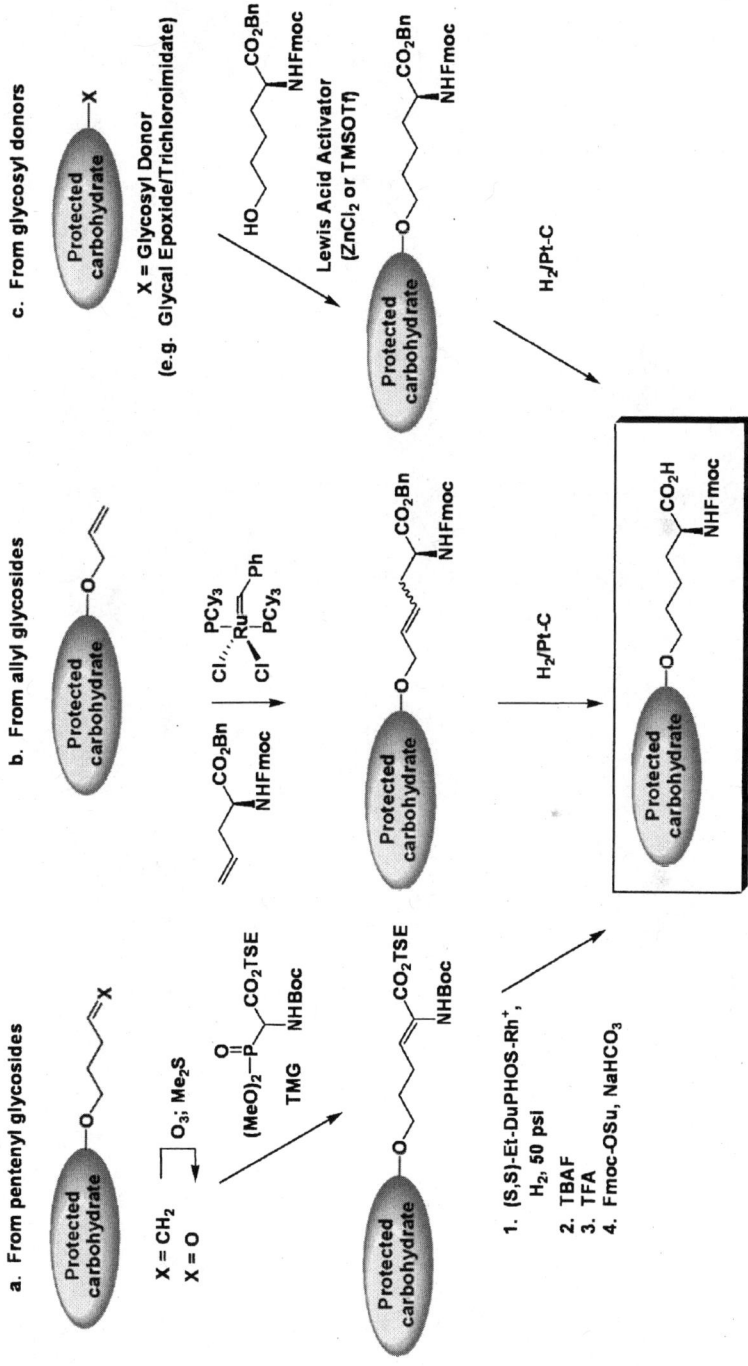

Figure 9. Methods used for the preparation of glycosylamino acids.

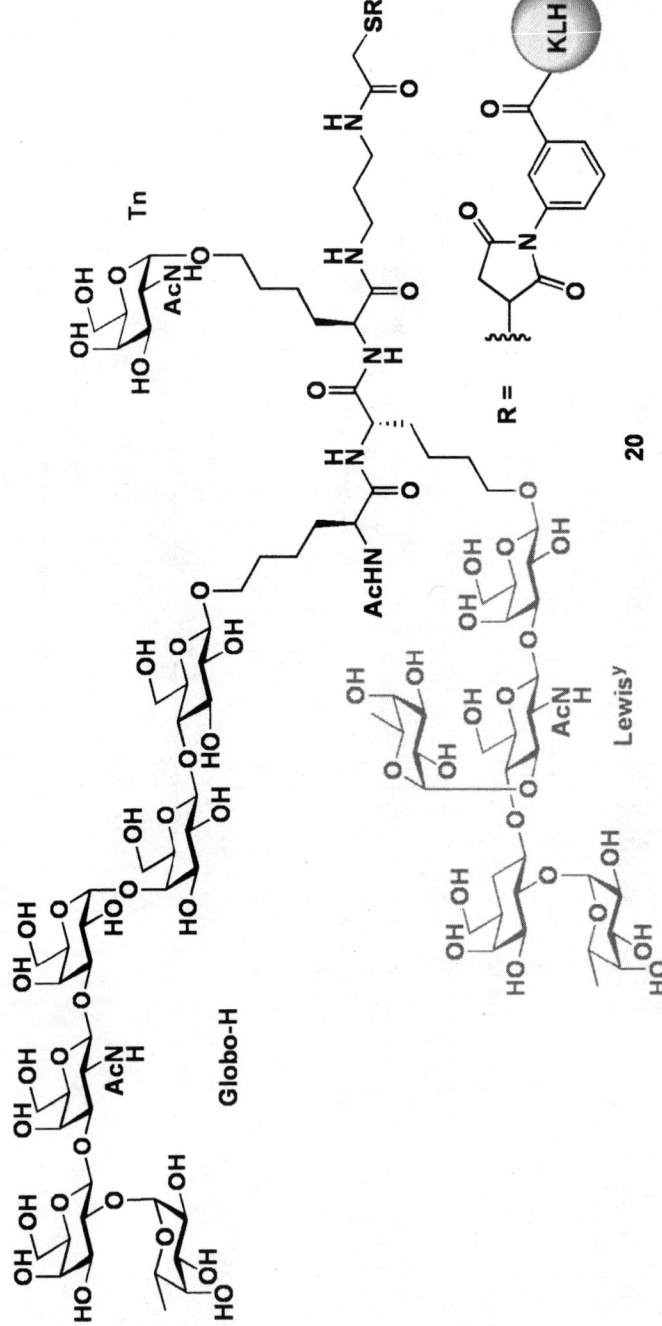

Figure 10. Unimolecular trivalent vaccine construct.

carbohydrate antigens known to be associated with both breast and prostate cancer cell lines. We set as our initial target a pentavalent vaccine construct displaying the Globo-H, Lewis^y, STn, TF, and Tn antigens. This polypeptide would be separately conjugated to KLH (**22**) and the Pam$_3$Cys lipid (**23**) (Figure 11).

The efficient realization of this complex synthetic target would require a slightly modified strategy for glycosylamino acid assembly. The traditional approach to glycosylamino acid synthesis begins with the preparation of the glycodomain through glycal synthesis methods. With the fully assembled glycal in place, the glycosylamino acid is installed through one of the methods described above (Figure 9). Each method for the installation of the amino acid requires an α-selective epoxidation of the appropriately protected glycal. Unfortunately, it is typically difficult to achieve high levels of α-stereocontrol in this types of reaction. Accordingly, we have come to favor a "cassette" approach to glycosylamino acid synthesis, wherein the hydroxynorleucine is appended to an appropriately protected monosaccharide at the outset of the synthesis (Figure 12) (*29*). The glycodomain is then assembled with the amino acid already in place. Although this protocol requires some sacrifice in terms of overall convergence, it also allows one to confront the epoxidation issue at the beginning of the synthesis, avoiding the loss of material associated with coupling of the amino acid to the fully functionalized glycal.

The cassette method has been employed with considerable success in our syntheses of a number of glycosylamino acids. For example, the STn, Tn, and TF glycosylamino acids were each prepared from D-galactal (Figure 13).

We thus prepared the five glycosylamino acids required for the pentavalent vaccine construct (*cf.* **24-28**, Figure 14) (*30*). With these key building blocks in hand, we were now able to assemble the pentapeptide backbone. First, the Tn glycosylamino acid (**28**) was coupled with *tert*-butyl-*N*-(3-aminopropyl) carbamate, which serves as a partial linker. The resulting unit was then elongated to pentapeptide **29** through a series of Fmoc deprotection and coupling reactions, as illustrated in Figure 14. At this point, the next task would be that of replacing the N-terminal Fmoc group (R_1) with an acetyl cap and subsequent removal of the C-terminal Boc functionality (R_2). Installation of a C-terminal thiol linker followed by a two-step global deprotection of the carbohydrate domain afforded glycopeptide **31**, which was conjugated to KLH through derivatization of KLH with maleimide **32**, followed by Michael addition of the thiol to the maleimide handle to afford the KLH-conjugated vaccine **22**, having a glycopeptide to protein ratio of 228:1.

Formation of the Pam$_3$Cys conjugate (**23**) was achieved by functionalization of intermediate **30**. Thus, a series of peptide coupling and deprotection reactions were executed to install alanine and serine amino acids. A global deprotection sequence was followed by treatment with the Pam$_3$Cys pentafluorophenyl ester to afford the Pam$_3$Cys conjugate **23**.

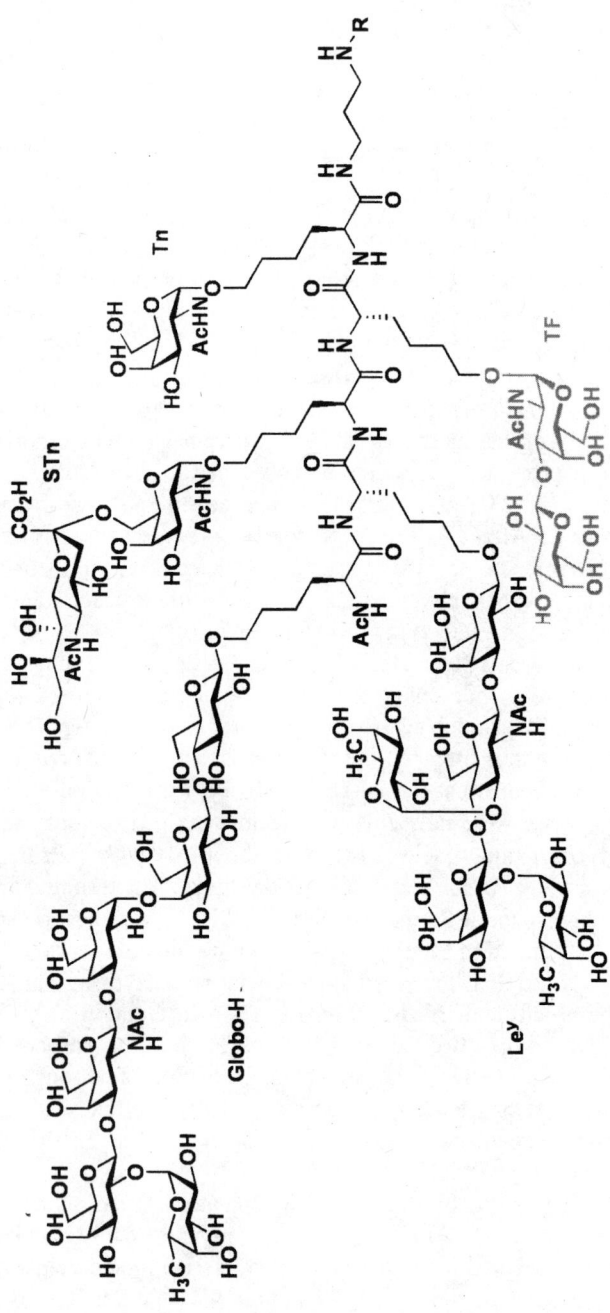

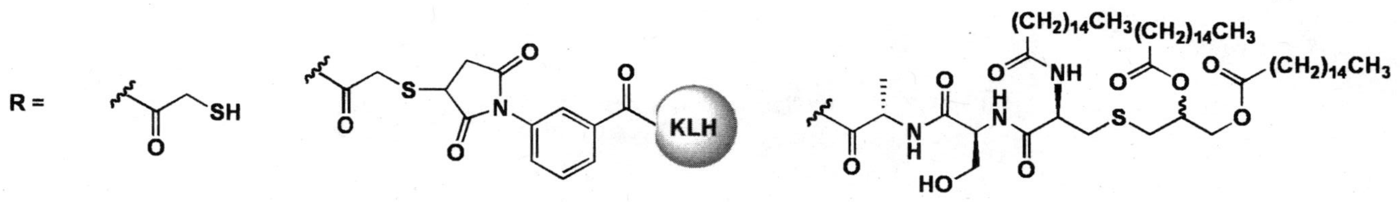

Figure 11. nimolecular pentavalent vaccine construct.

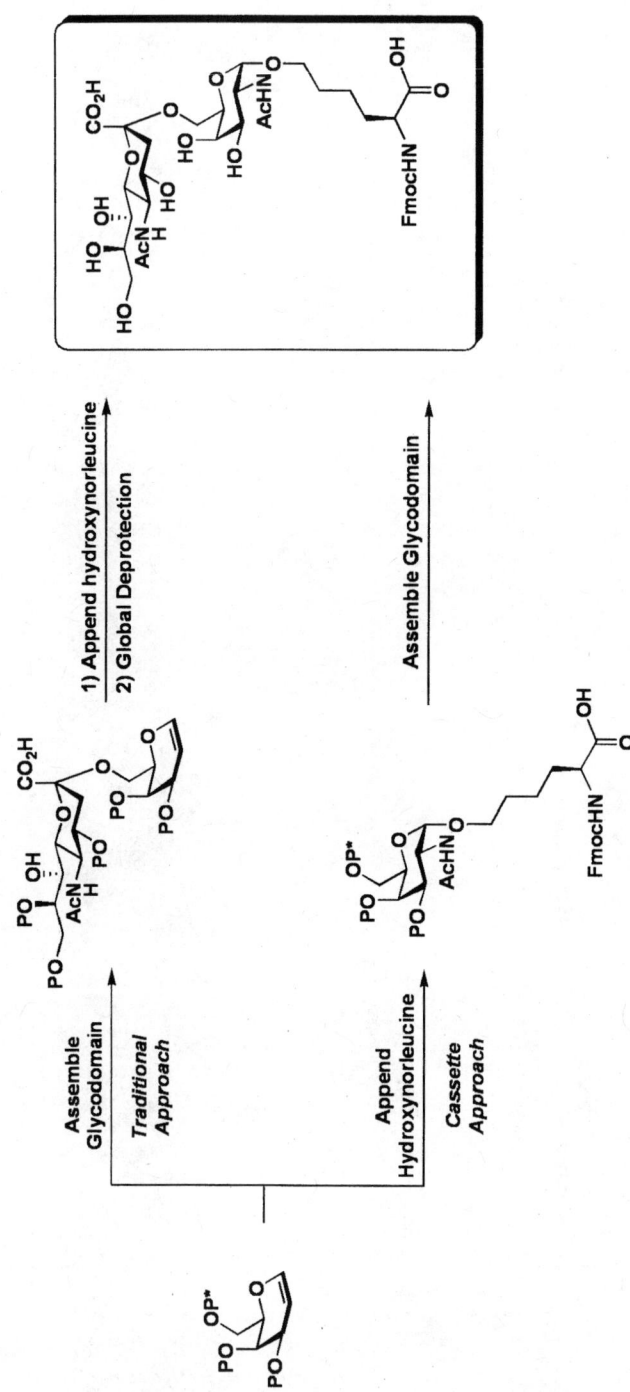

Figure 12. Cassette approach to glycosylamino acid synthesis.

Preliminary biological evaluations of the first-generation pentavalent constructs (**22** and **23**) served to demonstrate the immunological potential of the unimolecular multivalent strategy (*31*). In general, the KLH-conjugated vaccine construct (**22**) demonstrated enhanced immunological activity when compared to the PamCys-conjugate (**23**). Thus, ELISA analysis of **22** revealed that antibodies were formed against each antigen, with the exception of Lewis^y (we note that the failure of Le^y to induce antibodies in this context is not necessarily surprising, given that antigens that are endogenously expressed at high levels, such as Le^y, typically induce a less potent immune response). Moreover, FACS analysis of the pentavalent-KLH conjugate revealed the antibodies induced by this vaccine to be highly reactive to three different antigen-presenting cell lines. Most notably, in each case, the levels of reactivity observed with the **22**-invoked antibodies were higher than those observed with the corresponding monovalent-KLH constructs. These results appear to lend further support for the unimolecular, multivalent approach to vaccine polyvalency (*cf.* Figure 8).

On the basis of these preliminary immunological investigations, we have now prepared two additional unimolecular vaccine constructs (Figure 15). In the first, a modified pentavalent construct (**33**), the Le^y antigen is replaced with the GM2 antigen. The second, a hexavalent vaccine (**34**) incorporates both the GM2 and Le^y antigens. Finally, we are progressing toward the realization of a longstanding goal: the synthesis of a "cluster of clusters" vaccine, in which five different antigens are displayed in triplicate on a single polypeptide backbone (**35**). With this ambitious construct, we hope to reap the combined immunogenic benefits provided by epitope clustering and multivalency in a single vaccine. The results of our efforts to prepare this construct will be reported in due course.

We are now in the process of seeking regulatory approval for bringing fully synthetic monomolecular polyantigenic vaccines forward for clinical evaluation.

Acknowledgements

This research was supported by a the National Institutes of Health (CA28824).

References

1. Le Poole, I. C.; Gerberi, M. A.; Kast, W. M. *Curr. Opin. Oncol.* **2002**, *14*, 641-648.
2. For a general summary of our approach to carbohydrate-based vaccines please see: (a) Danishefsky, S. J.; Allen, J. R. *Angew. Chem. Int. Ed.* **2000**, *39*, 836-863; (b) Keding, S. J.; Danishefsky, S. J. in *Carbohydrate-Based Drug Discovery;* Ed. Wong, C-H.; Wiley-VCH: Weinheim, Germany, 2003,

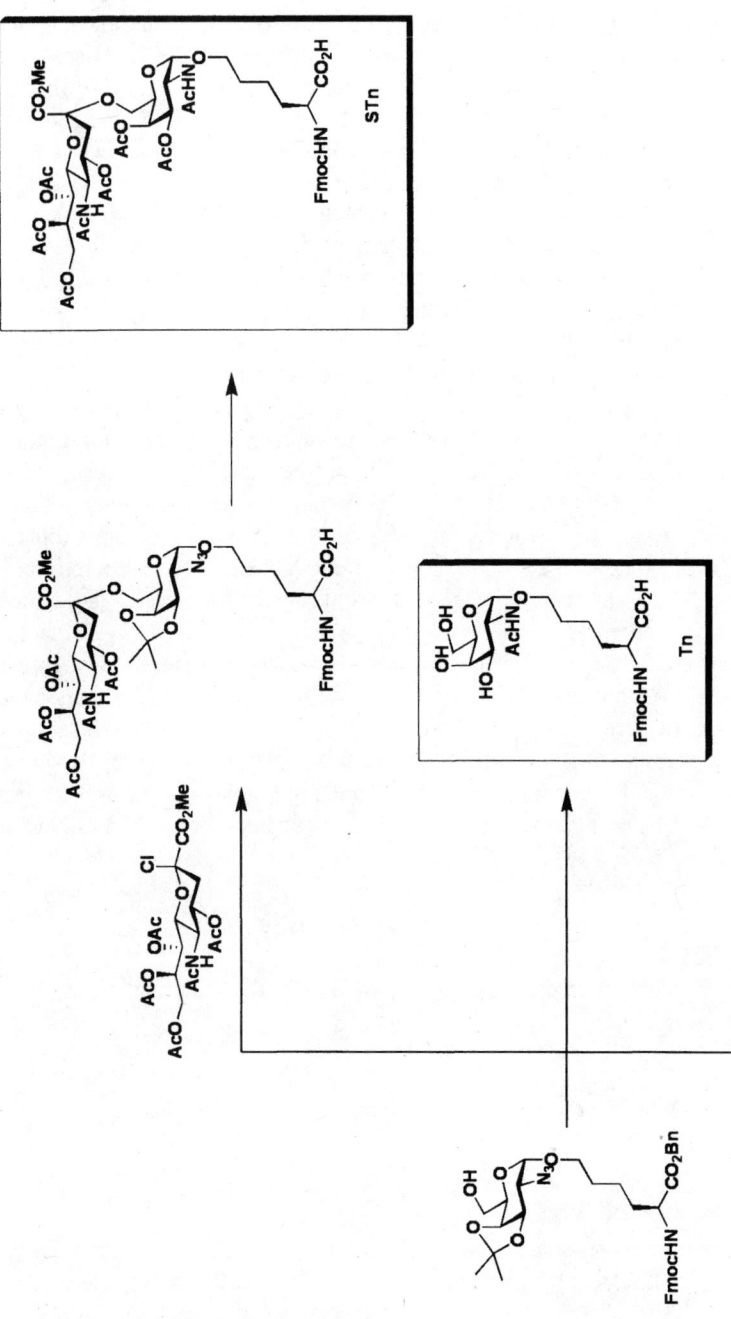

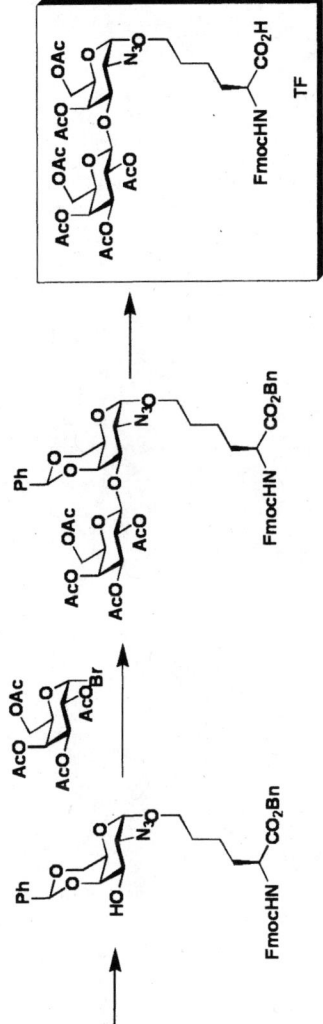

Figure 13. Application of cassette approach to syntheses of STn, Tn, and TF.

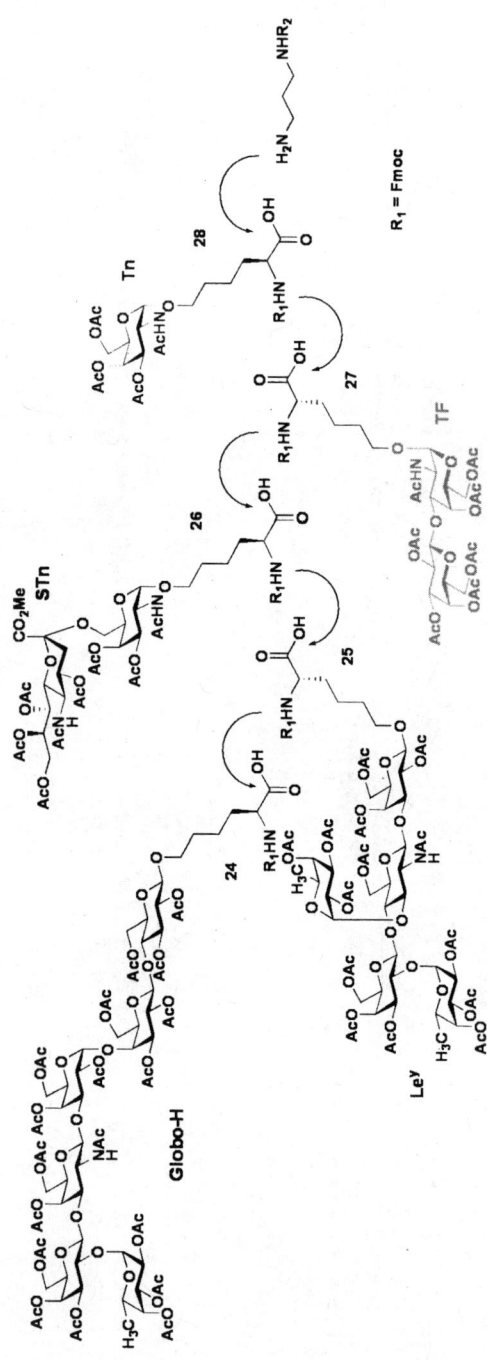

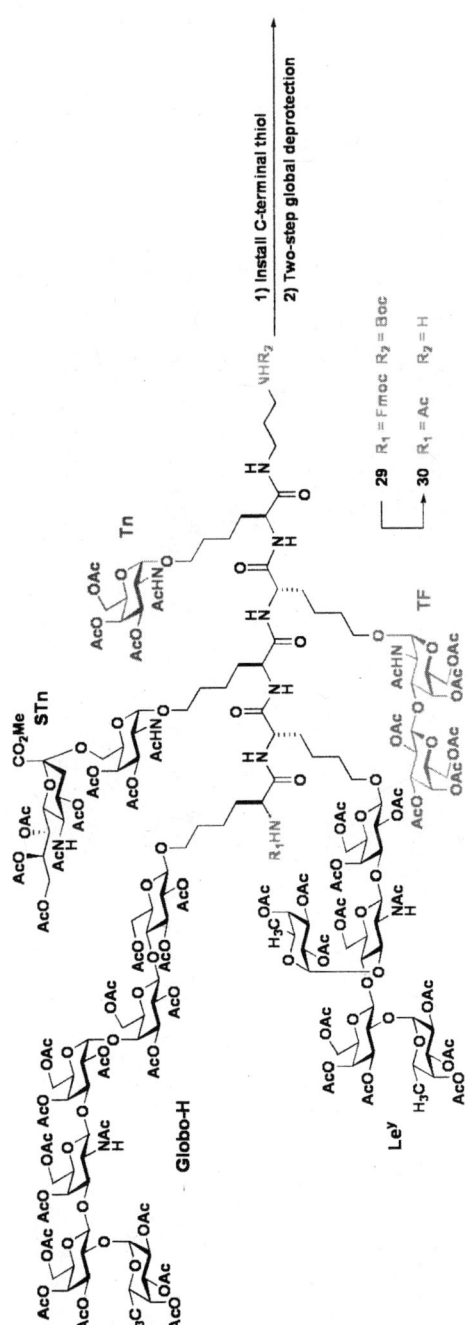

Figure 14. Synthesis of pentavalent vaccine 22. Continued on next page.

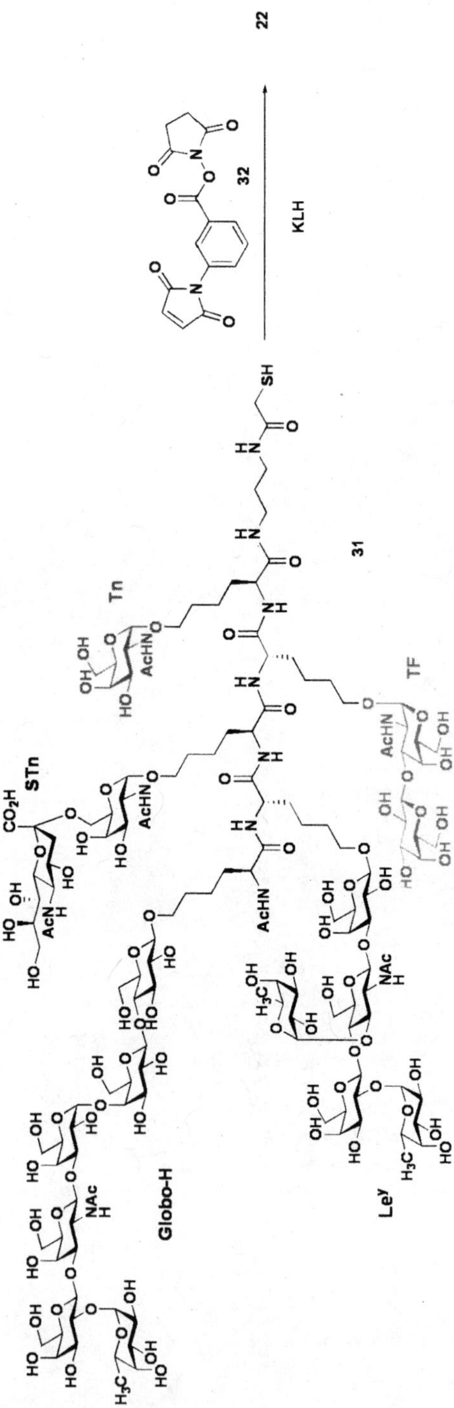

Figure 14. Continued.

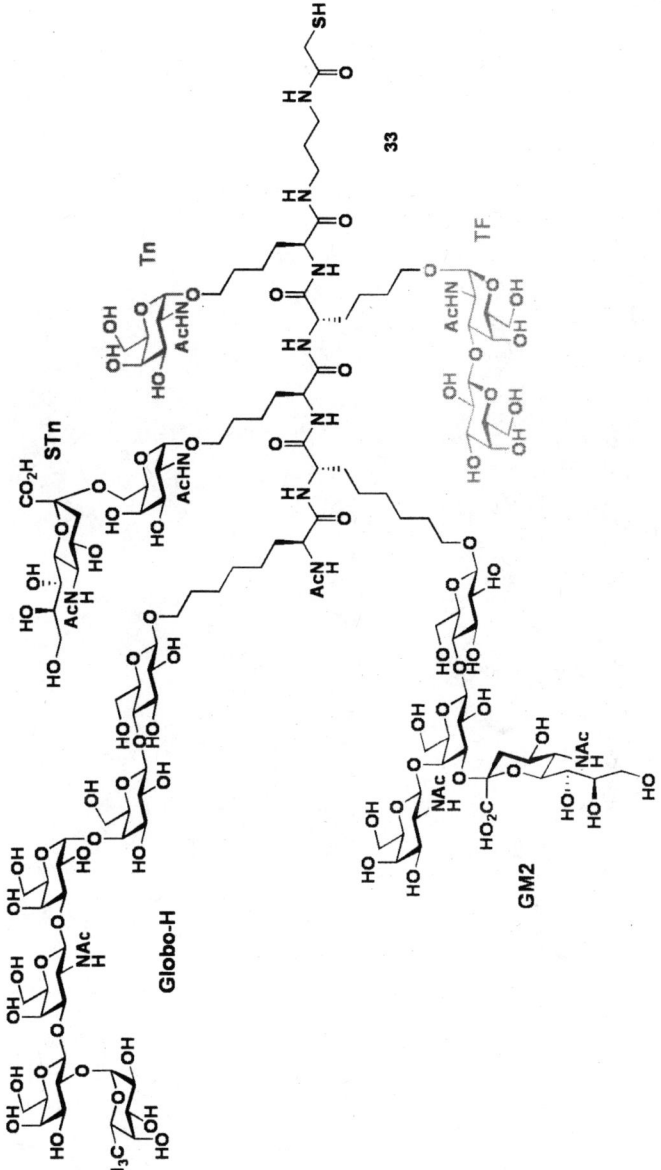

Figure 15. Future work. Continued on next page.

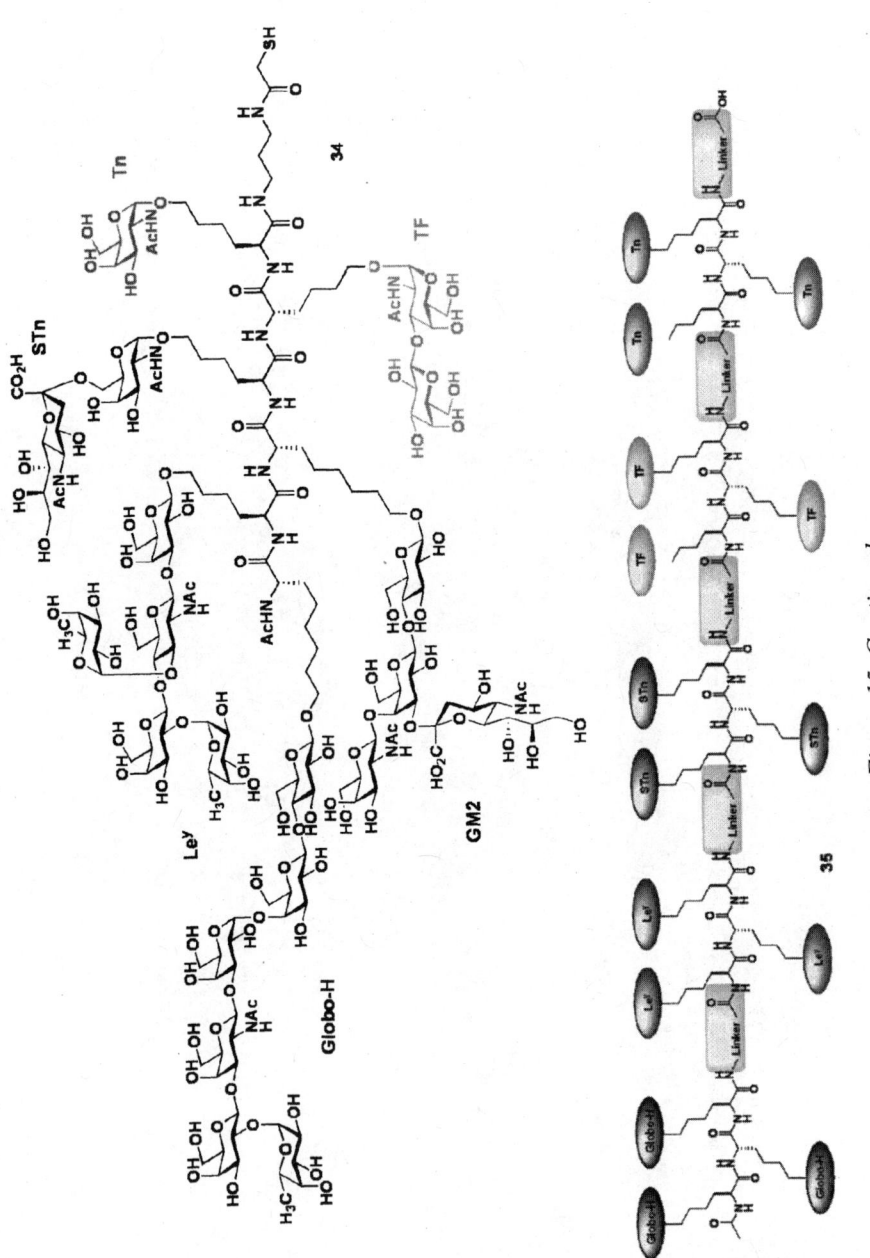

Figure 15. Continued.

Vol. 1, 381-406; (c) Ouerfelli, O; Warren, J. D.; Wilson, R. M.; Danishefsky, S. J. *Expert Rev. Vaccines* **2005**, *4*, 677-685; (d) Warren, J. D.; Geng, X.; Danishefsky, S. J. *Top. Curr. Chem.* **2005**, MS Submitted.
3. (a) Helling, F.; Shang, Y.; Calves, M.; Oettgen, H. F.; Livingston, P. O. *Cancer Res.* **1994**, *54*, 197-203; (b) Helling, F.; Zhang, A.; Shang, A.; Adluri, S.; Calves, M.; Koganty, R.; Longenecker, B. M.; Oettgen, H. F.; Livingston, P. O. *Cancer Res.* **1995**, *55*, 2783-2788.
4. Kannagi, R.; Levery, S. B.; Ishigami, F.; Hakomori, S.; Shevinsky, L. H.; Knowles, B. B.; Solter, D. *J. Biol. Chem.* **1983**, *258*, 8934-8942.
5. Bremer, E. G.; Levery, S. B.; Sonnino, S.; Ghidoni, R.; Canevari, S.; Kannagi, R.; Hakomori, S. *J. Biol. Chem.* **1984**, *259*, 14773-14777.
6. Livingston, P. O. *Semin. Cancer Biol.* **1995**, *6*, 357-366.
7. Bilodeau, M. T.; Park, T. K.; Hu, S.; Randolph, J. T.; Danishefsky, S. J.; Livingston, P. O.; Zhang, S. *J. Am. Chem. Soc.* **1995**, *117*, 7840-7841.
8. Park, T. K.; Kim, I. J.; Hu, S. H.; Bilodeau, M. T.; Randolph, J. T.; Kwon, O.; Danishefsky, S. J. *J. Am. Chem. Soc.* **1996**, *118*, 11488-11500.
9. Allen, J. R.; Allen, J. G.; Zhang, X. F.; Williams, L. J.; Zatorski, I.; Ragupathi, G.; Livingston, P. O.; Danishefsky, S. J. *Chem. Eur. J.* **2000**, *6*, 1366-1375.
10. Ragupathi, G.; Park, T. K.; Zhang, S. L.; Kim, I. J.; Graber, L.; Adluri, S.; Lloyd, K. O.; Danishefsky, S. J.; Livingston, P. O. *Angew. Chem. Int. Ed. Engl.* **1997**, *36*, 125-128.
11. Slovin, S. F.; Ragupathi, G.; Adluri, S.; Ungers, G.; Terry, K.; Kim, S.; Spassova, M.; Bornmann, W. G.; Fazzari, M.; Dantis, L.; Olkiewicz, K.; Lloyd, K. O.; Livingston, P. O. *Proc. Natl. Acad. Sci. USA* **1999**, *96*, 5710-5715.
12. Gilewski, T.; Ragupathi, G.; Bhuta, S.; Williams, L. J.; Musselli, C.; Zhang, X.-F.; Bencsath, K. P.; Panageas, K. S.; Chin, J.; Norton, L.; Houghton, A. N.; Livingston, P. O.; Danishefsky, S. J. *Proc. Natl. Acad. Sci. USA* **2001**, *98*, 3270-3275.
13. Carlstedt, I.; Davies, J. R. *Biochem. Soc. Trans.* **1997**, *25*, 214-219.
14. Kuduk, S. D.; Schwarz, J. B.; Chen, X. T.; Glunz, P. W.; Sames, D.; Ragupathi, G;. Livingston, P. O.; Danishefsky, S. J. *J. Am. Chem. Soc.* **1998**, *120*, 12474-12485.
15. Kagan, E.; Ragupathi, G.; Yi, S. S.; Reis, C. A.; Gildersleeve, J.; Kahne, D.; Clausen, H.; Danishefsky, S. J.; Livingston, P. O. *Cancer Immunol. Immunother.* **2005**, *54*, 424-430.
16. Slovin, S. F.; Ragupathi, G.; Musselli, C.; Olkiewicz, K.; Verbel, D.; Kuduk, S. D.; Schwarz, J. B.; Sames, D.; Danishefsky, S. J.; Livingston, P. O.; Scher, H. I. *J. Clin. Oncol.* **2003**, *11*, 4292-4298.
17. Schwarz, J. B.; Kuduk, S. D.; Chen, X.-T.; Sames, D.; Glunz, P. W.; Danishefsky, S. J. *J. Am. Chem. Soc.* **1999**, *121*, 2662-2673.
18. Sames, D.; Chen, X. T.; Danishefsky, S. J. *Nature* **1997**, *389*, 587-591.

19. Glunz, P. W.; Hintermann, S.; Williams, L. J.; Schwarz, J. B.; Kuduk, S. D.; Kudryashov, V.; Lloyd, K. O.; Danishefsky, S. J. *J. Am. Chem. Soc.* **2000**, *122*, 7273-7279.
20. (a) Zhang, S.; Cordon-Cardo, C.; Zhang, H. S.; Reuter, V. E.; Adluri, S.; Hamilton, W. B.; Lloyd, K. O.; Livingston, P. O. *Int. J. Cancer* **1997**, *73*, 42-49; (b) Zhang, S.; Zhang, H. S.; Cordon-Cardo, C.; Reuter, V. E.; Singhal, A. K.; Lloyd, K. O.; Livingston, P. O. *Int. J. Cancer* **1997**, *73*, 50-56.
21. (a) Ragupathi, G.; Cappello, S.; Livingston, P. O. *Vaccine* **2002**, *20*, 1030-1038; (b) Ragupathi, G.; Koide, F.; Kagan, E.; Bornmann, W.; Spassova, M.; Gregor, P.; Danishefsky, S. J.; Livingston, P.O. *Cancer Immmunol. Immunother.* **2003**, *52*, 608-616.
22. Barington, T.; Gyhrs, A.; Kristensen, K.; Heilmann, C. *Infect. Immun.* **1994**, *62*, 9-14.
23. Sarvas, H.; Makela, O.; Toivanen, P.; Toivanen, A. *Scand. J. Immunol.* **1974**, *3*, 455-460.
24. Our early studies on the best carbohydrate presentation method (i.e. naturally occurring mucin-based, versus nonnaturally occurring hydroxynorleucine-based) indicated that the hydroxynorleucine construct was considerably more immunogenic than the naturally inspired mucin based species (unpublished results).
25. Allen, J. R.; Harris, C .R.; Danishefsky, S. J. *J. Am. Chem. Soc.* **2001**, *123*, 1890-1897.
26. Biswas, K.; Coltart, D. M.; Danishefsky, S. J. *Tetrahedron Lett.* **2002**, *43*, 6107-6110.
27. (a) Keding, S. J.; Endo, A.; Biswas, K.; Zatorski, A.; Coltart, D. M.; Danishefsky, S. J. *Tetrahedron Lett.* **2003**, *44*, 3413-3416; (b) Keding, S. J.; Endo, A.; Danishefsky, S. J. *Tetrahedron* **2003**, *59*, 7023-7031.
28. (a) Allen, J. R.; Harris, C. R.; Danishefsky, S. J. *J. Am. Chem. Soc.* **2001**, *123*, 1890-1897; (b) Ragupathi, G.; Coltart, D. M.; Williams, L. J.; Koide, F.; Kagan, E.; Allen, J.; Harris, C.; Glunz, P. W.; Livingston, P. O.; Danishefsky, S. J. *Proc. Natl. Acad. Sci USA* **2002**, *99*, 13699-13704.
29. Chen, X. T.; Sames, D.; Danishefsky, S. J. *J. Am. Chem. Soc.* **1998**, *120*, 7760-7769.
30. Keding, S. J.; Danishefsky, S. J. *Proc. Natl. Acad. Sci. USA.* **2004**, *101*, 11937-11942.
31. Ragupathi, G.; Koide, F.; Livingston, P. O.; Cho, Y. S.; Endo, A.; Wan, Q.; Spassova, M. K.; Keding, S. J.; Allen, J.; Ouerfelli, O.; Danishefsky, S. J. *J. Am. Chem. Soc.* **2006**, *128*, 2715-2725.

Chapter 13

Synthetic Glycopeptides for the Construction of Anticancer Vaccines

Horst Kunz, Sebastian Dziadek, Sven Wittrock, and Torsten Becker

Institut fuer Organische Chemie, Johannes Gutenberg-Universitaet Mainz, 55099 Mainz, Germany

Glycopeptides of the tandem repeat region of the tumor-associated epithelial mucin MUC1 containing tumor-associated saccharide antigens have been synthesized and conjugated to carrier proteins or T-cell epitope peptides. The T, sialyl T_N, 2,6- and 2,3-sialyl T antigen amino acid building blocks were obtained by biomimetic syntheses starting from the monosaccharide T_N antigen conjugate. To synthesize the MUC1 glycopeptides solid-phase methods with varied anchors were applied. Conjugation of synthetic glycopeptides to bovine serum albumin were achieved by selective reactions at squaric esters. Solid-phase fragment condensation proved successful for constructions of MUC1 glycopeptide-T-cell epitope conjugates. Immunization of mice with a synthetic vaccine consisting of a MUC1 glycopeptide and a T-cell epitope from ovalbumin induced a specific immune response. The elicited antibody only reacted with the glycopeptide. Neither its peptide nor the glycan in a different peptide were recognized. This precise differentiation by antibodies induced by chemically pure vaccines is considered promising for the development of antitumor vaccines.

© 2008 American Chemical Society

Tumor-associated Saccharide Antigens - Early Experiments

Membrane glycoproteins of epithelial cells play key roles in selective biological recognition processes in multicellular organisms. In many cases, the protein portions of these glycoproteins constitute the specific binding sites, for example, in integrin ligands like fibronectin or vitronectin which contain the RGD peptide binding motif (1). In other instances, the carbohydrate portions of glycoproteins are involved in these recognition phenomena as in the differentiation of blood groups (2).

About 30 year ago, cell biological and biochemical investigations had shown that the glycan patterns of glycoproteins on tumor cells are distinctly altered compared to those present on normal cells (3,4). Springer et al. described glycoproteins from epithelial tumor cells carrying the Thomsen-Friedenreich antigen (T antigen) **1** as tumor-associated antigen (5). From cross reactivity of monoclonal antibodies induced by these T-antigen glycoproteins isolated from epithelial tumor cells, Springer and colleagues concluded, that these T antigen glycoproteins are structurally related to asialoglycophorin. On the basis of these structural informations, glycopeptides of the N-terminal portion of glycophorin of M blood group specificity have been synthesized and selectively coupled to bovine serum albumin to give a synthetic vaccine **2** (6). The assumption had been that an immune response selectively elicited against such a structural element present on tumor cells but not on normal cells might help to eradicate malignant cells without destroying healthy cells. Using the synthetic vaccine **2** (Figure 1), an antibody 82-A6 was induced in mice which actually exhibited affinity to epithelial tumor cells, but was not sufficiently tumor-selective (6,7). To understand this at first sight disappointing result, further immunological characterizations of the specificity of antibody 82-A6 were performed. This antibody did not show affinity to glycophorin itself. In glycophorin, the T antigen side chains as in **1** and **2** are covered by 2,3'- and 2,6-linked sialic acids. After removal of these sialic acid residues using N-acetyl-neuraminidases, the thus obtained asialoglycophorins **3** strongly reacted with the antibody 82-A6 (7) giving evidence of its specificity towards the T antigen structure **1** (5-7). On the basis of this specificity toward the saccharide antigen, a number of synthetic vaccines have been constructed by linking T (and T_N) antigen serine or threonine conjugates to proteins (8,9,10). Glycophorin and asialoglycophorin **3**, however, exist in two blood group specificities M and N, which differ in the 131 amino acids containing glycoproteins in only two amino acid positions. One of these positions is the N-terminal amino-acid which is a serine in M blood group specificity **3a**, but a leucine N-blood specificity **3b**.

Surprisingly, antibody 82-A6 was distinctly more reactive towards asialoglycophorin of M specificity **3a**, which has the aminoterminal sequence of the synthetic vaccine **2**, than to asialoglycophorin of N specificity **3b**. This

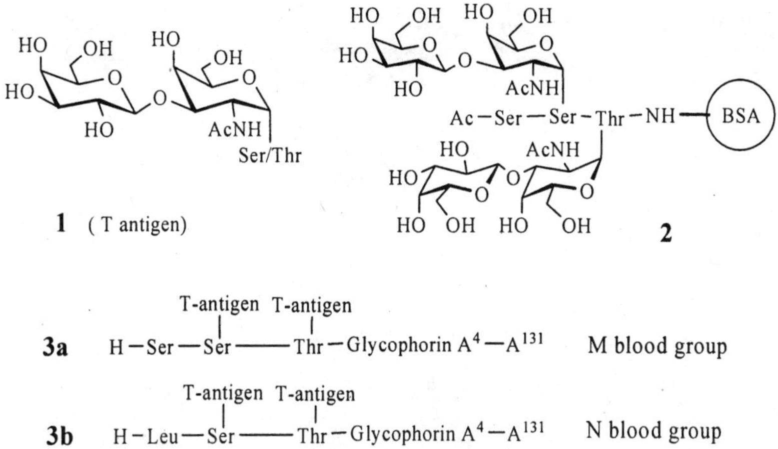

Figure 1. Tumor-associated T antigen structures

observation forced us to conclude that not only the saccharide, but also the amino acid sequence contributes to the epitope recognized by antibody 82-A6 (5-7). We felt it astonishing that an antibody (IgM) elicited by a vaccine **2** representing just a small portion of the N-terminus of asialoglycophorin can distinguish between two large glycoproteins **3** which carry a number of identical T antigen saccharides necessary for recognition, but only are different in the N terminal amino acid.

The conclusion drawn from these results was that for a tumor-selective antigen not only the altered saccharide (here T antigen), but also a tumor selective peptide structure are required. Informations about tumor-relevant peptide structures became available when details of the tumor-associated polymorphic mucin MUC1 had been published (11,12,13,14).

The Tumor-associated Mucin MUC1

Mucin MUC1 is a membrane-bound mucin, expressed on many epithelial tissues and extensively overexpressed on epithelial tumor cells (11-14). Its large extracellular portion contains numerous tandem repeat sequences of 20 amino acids HGVTSAPDTRPAPGSTAPPA (15). Most sites of O-glycosylation are located within the tandem repeats (14). Using antibodies induced by MUC1 isolated from tumor tissues, an immunodominant peptide epitope PDTRPAP was identified (11,12). The tumor-associated form of MUC1 is mainly not characterized by a mutation within the peptide sequence of this heavily glycosylated glycoprotein, but by incomplete formation of the saccharide side chains. This outcome is due to a change in enzyme activities in the mucine O-glycan biosynthesis within the tumor cells (Figure 2).

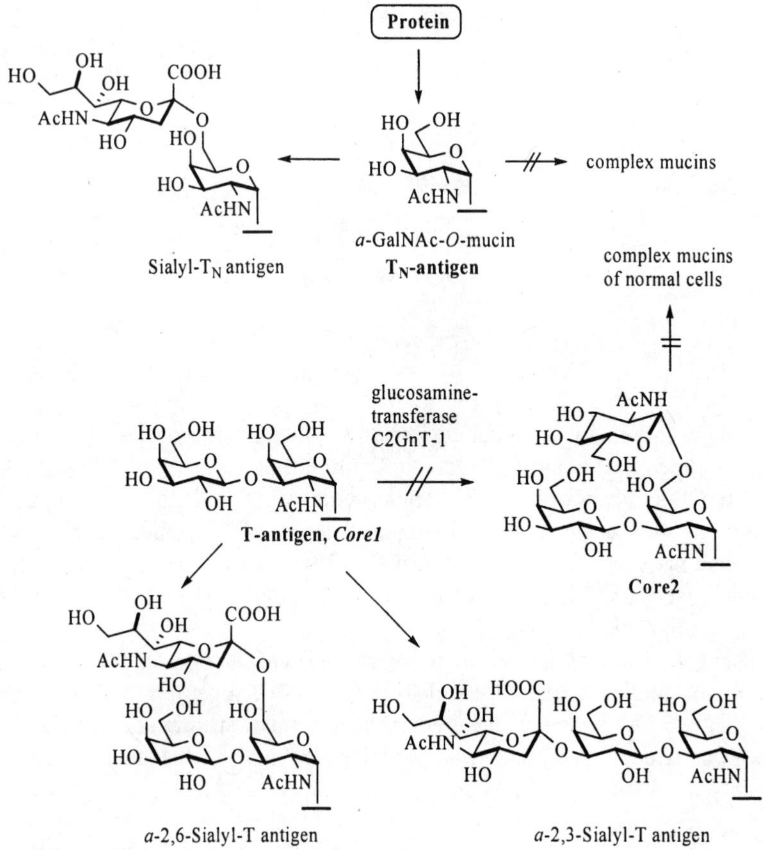

Figure 2. Biosynthesis of mucin O-glycans in malignant epithelial cells

N-Acetyl-galactosamine transferases of overlapping, but different specificity initiate the posttranslational modification of the mucin core protein (16) by addition of N-acetylgalactosamine to serine or threonine residues (Figure 2). Subsequent galactosylation furnishes the T antigen structure, also called core 1 saccharide. In healthy cells, core 1 is transformed into the core 2 structure by β-glucosamine-1,6-transferase C2GnT-1. The core 2 structure is the substrate for efficient alternating galactose and glucosamine transfer processes furnishing the extended complex polylactosamine side chains typical for MUC1 on normal cells. In tumor cells expression of MUC1 is drastically enhanced, however, the expression of the β-1,6-glucosamine transferase C2GnT-1 is strongly downregulated (17). As a consequence, long complex saccharides are not formed. Instead T_N and T antigen accumulate and, consequently, have been identified as tumor associated antigens. As in tumor cells sialyltransferases concomitantly are overexpressed, chain-terminating sialylation of T_N and T antigen form the important tumor-associated antigens sialyl T_N, (2,6)-sialyl T and (2,3)-sialyl T (14,16,17). These short, prematurely sialylated saccharides not only are antigens themselves, they also are responsible for the incomplete coverage of the MUC1 protein backbone resulting in an exposure of these peptide structures to the immune receptors which is blocked in the normal cells by the extended complex saccharides. The now accessible peptide segments within the tandem repeat sequence constitute tumor-selective peptide structure informations. Therefore, it is the hypothesis of this research that synthetic glycopeptides combining tumor-associated saccharide antigens (Figure 2) with peptide sequences of the MUC1 tandem repeat are promising candidates for the development of synthetic antitumor vaccines (18).

Another reason to focus on MUC1 tandem repeat glycopeptides for the construction of tumor-selective antigens resulted from our early syntheses of MUC1 glycopeptides aimed at T_N antigen glycopeptides like **4** (19), and T antigen glycopeptides **5** (20) and **6** (21).

It was revealed by NMR investigations of these compounds that glycopeptides **5** and **6** adopt preferred conformations in water or water containing solutions (20,21) whereas peptides of identical sequences did not prefer a certain conformation. Results obtained by other groups (28,29) suggested that the conformational influence of glycosylation on mucin glycopeptides depends upon the number and the position of glycan side chains in the sequence. It appears justified to conclude from all these observations, that the altered glycosylation pattern of MUC1 in tumor cells not only provides access to cryptic peptide epitopes, but also distinctly influences the spatial presentation of these peptide structures, i.e. it changes the originally rigid, linear conformation of the protein backbone (13b).

Synthetic Glycopeptide Antigens with Sialyl T_N and Sialyl T Structure

The aberrant O-glycan biosynthesis of mucins in epithelial cells results in the formation of T_N and T antigen structures and their sialylated forms as tumor-associated antigens (Figure 2). For reasons of synthetic accessibility, T_N and T antigen glycopeptides had been synthesized at first (5,8-10,18-22). Nevertheless,

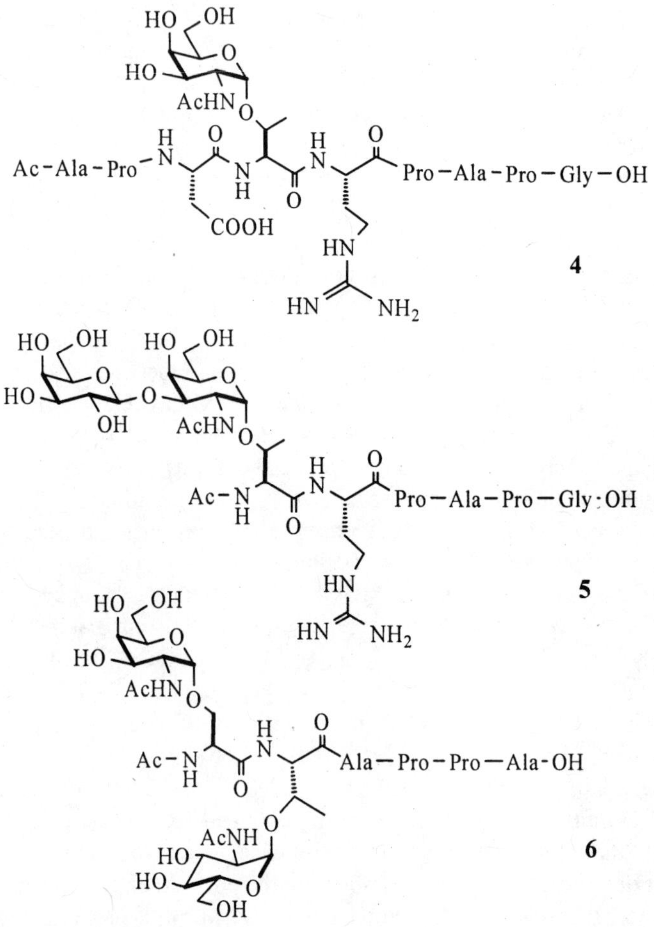

Figure 3. Synthetic glycopeptide partial sequences of MUC1 (19-21)

the indented use of these compounds as antigens in vivo suffers from their rapid clearance in the liver. In addition, due to the distinct overexpression of sialyltransferases, the sialylated T_N and T antigen forms seem to be the more important tumor-associated carbohydrate antigens. Syntheses of glycopeptides with sialyl T_N antigen side chains and peptide sequences of HIV gp120 (24) or MUC1 tandem repeat structure (25) have been described in 1997. In this MUC1 glycopeptide synthesis the glycosylated serine and threonine building blocks were obtained from monosaccharide precursor 7 (26) carrying the fluorenyl-methoxycarbonyl (Fmoc)- and *tert*-butyl ester protecting group combination (25).

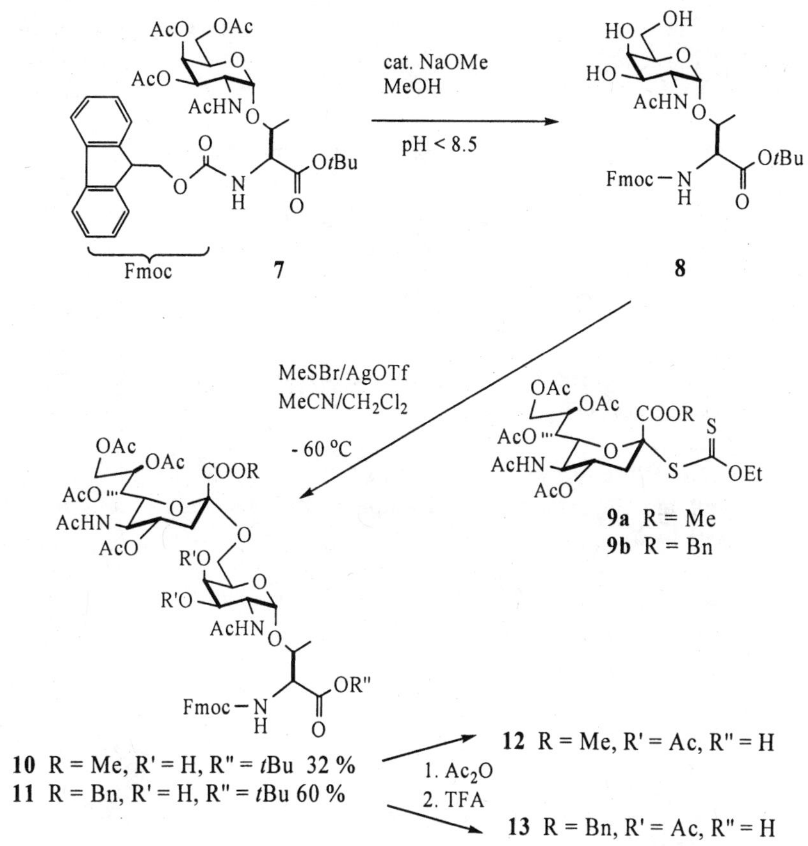

Figure 4. Synthesis of sialyl T_N threonine building block (25,29)

Removal of the O-acetyl groups from the carbohydrate portion to give **8** without affecting the base-sensitive Fmoc group was achieved by a careful transesterification in methanol catalyzed by sodium methanolate at a pH < 8.5. Xanthate of sialylic acid methyl ester **9a** (27) turned out to be the optimal donor for the regio- and stereoselective sialylation of **8** to form the desired sialyl T_N antigen threonine conjugate **10**. In situ formed methylsulfenyl triflate (28) proved to be the best promotor of this reaction (25). Using the xanthate **9b** of the corresponding sialic benzyl ester under these conditions gave an even higher yield of the sialyl T_N threonine derivative **11** (29). O-Acetylation of the 3- and 4-hydroxy groups enhanced the acid-stability of these conjugates, and subsequent acidolysis of the *tert*-butyl ester furnished the Fmoc sialyl T_N threonine building blocks **12** and **13** applicable to solid-phase synthesis of MUC1 tandem repeat glycopeptides (25,29) (Figure 5,6).

By solid-phase syntheses on allylic HYCRON anchor **14** (19) using sialyl T_N building block **12**, glycoundecapeptide **15** was assembled, detached from the resin by palladium(0)-catalyzed cleavage of the allyl ester linkage and careful deprotection reaction procedures (25) (Figure 5).

Particular care was necessary during the saponification of the rather stable sialic methyl ester which was successfully performed with NaOH in water/methanol at pH 11.5 (25) giving the pure glycopeptide **15**.

Treatment of peripheral blood lymphocytes with **15** and similar MUC1 glycopeptides induced only insufficient proliferation indicating a low immunogenicity of glycopeptides having this amino acid sequence. To overcome the low antigenicity of the glycopeptides, a conjugate **16** of a MUC1 tandem repeat sialyl T_N glycopeptide of different amino acid sequence with a part of the tetanus toxoid T-cell epitope was constructed via a fragment condensation on solid-phase (29).

Owing to the use of sialyl T_N threonine building block **13** with benzyl ester protection of the sialic carboxylic function its critical deprotection could be carried out by hydrogenolysis (29).

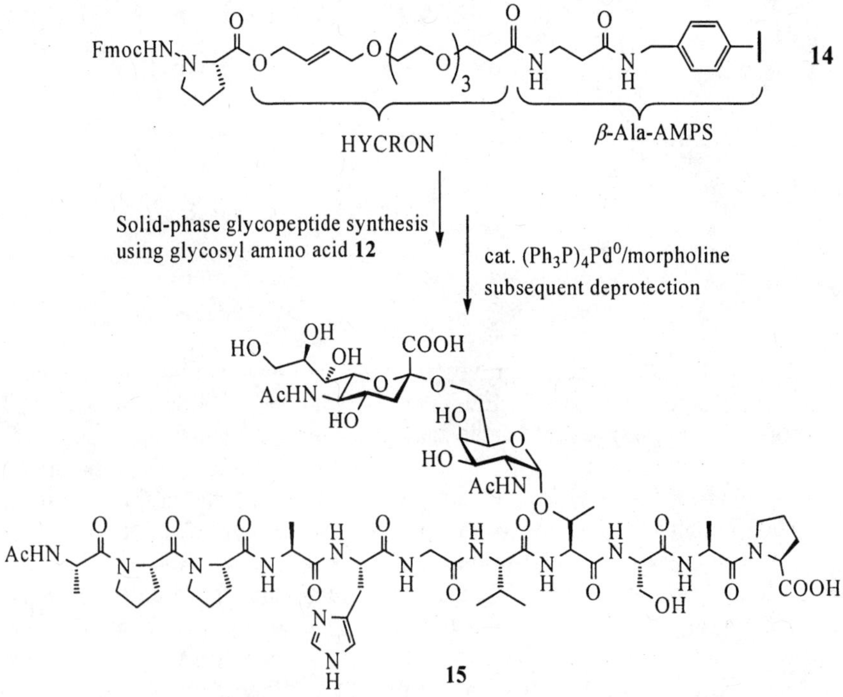

Figure 5. Sialyl T_N glycopeptide sequence of MUC1

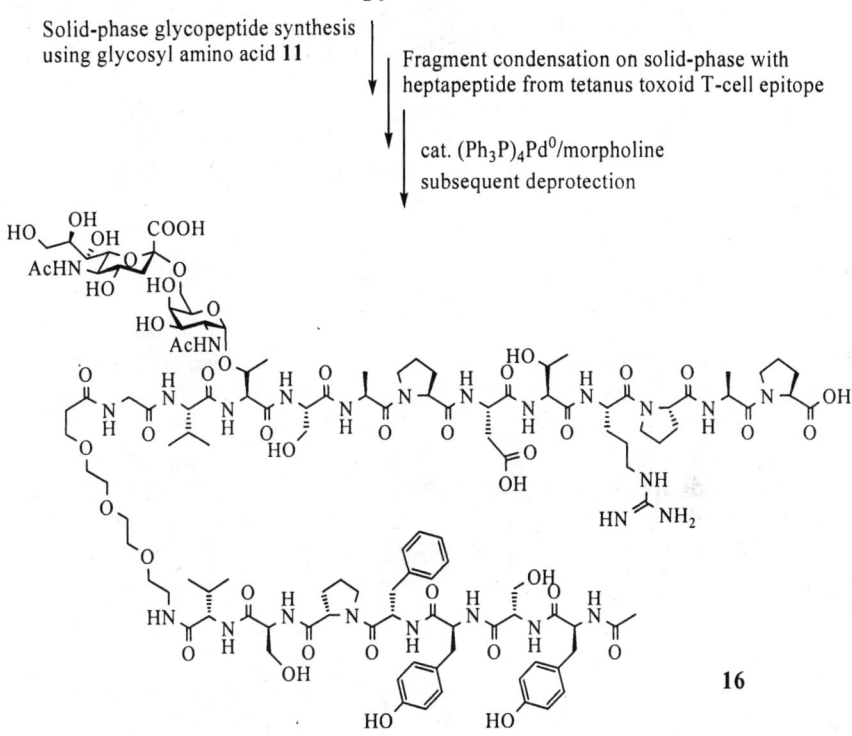

Figure 6. Synthetic vaccine consisting of MUC1 glycopeptide and a T-cell epitope (29)

Immunological evaluation of **16** showed that conjugate **16** induces proliferation of cytotoxic T cells (CD8$^+$ cells) while its partial structures, i.e. the MUC1 sialyl T$_N$ glycododecapeptide in **16** or the unglycosylated dodecapeptide of identical sequence only stimulated proliferation of CD3$^+$ T cells (29).

The observed induction of proliferation of cytotoxic T cells by the MUC1-sialyl T$_N$ glycopeptide construct **16**, however, is considered promising for the further development of antitumor vaccines according to this general hypothesis. Next steps in this development concern the prove of selectivity of the induced immune response towards the tumorassociated MUC1 structures. This purpose needs to couple the synthetic MUC1 glycopeptide antigens to carrier proteins in order to obtain synthetic vaccines. In addition, the synthetic program should also include the other important sialylated tumor-associated saccharide antigens (Figure 2) as well as MUC1 tandem repeat glycopeptides of the complete 20 amino acid sequence.

A unifying biomimetic synthesis of sialyl T and sialyl T_N antigen amino acid building blocks

The biosynthesis of the tumor-associated saccharide antigens proceeds from the monosaccharide T_N conjugate. Imitating the biosynthesis, the synthetic strategy to arrive at all five tumor-associated mucin saccharide antigen serine and threonine conjugates also begins with the T_N antigen conjugate **7** (30). Figure 4 shows the access to the sialyl T_N antigen threonine building blocks **9** and **10** (25,29). The syntheses of the T- and the regioisomeric sialyl T-antigen conjugates is illustrated in Figure 7 (30,31).

As the starting T_N threonine conjugate contains the Fmoc/*tert*-butyl ester protecting group combination in the amino acid portion, all subsequent protecting group manipulations and glycosylation reactions must take care of the base-sensitivity of the Fmoc group as well as of the acid-sensitivity of the *tert*-butyl ester. Introduction of 4,6-benzylidene protection using the corresponding dimethyl acetals proved successful under controlled acid catalysis to give the monofunctional glycosyl acceptors **17**. Their galactosylation without affecting the acid-sensitive *tert*-butyl ester was best achieved under Helferich conditions (30) using galactosyl bromides **18** to furnish the T antigen conjugates **19**. Removal of the methoxy-benzylidene acetal from **19a** and subsequent sialylation of the intermediate using xanthate **9b** of sialic acid benzyl ester formed the 2,6-sialyl T structure **20** which without protection of the 4-OH group was treated with TFA to yield 2,6-sialyl-T threonine building block **21** (30a,31a).

The synthesis of the regioisomeric 2,3-sialyl T antigen conjugate required modification of the strategy in three details. More acid-stable 4,6-benzylidene protection was applied in **17b** and 6-O-benzyl galactosyl bromide was used for Helferich glycosylation to furnish the T antigen threonine conjugate **19b** without affecting the acid-sensitive *tert*-butyl ester. The removal of the O-acetyl groups from **19b** needs particular attention.

It was achieved by transesterification at pH 8.5 without elimination of the Fmoc group to give **22**. Regio- and stereoselective sialylation was best accomplished by xanthate **9b** and promotion with methylsulfenyl triflate. In order to establish a sufficient acid-stability of the glycoconjugates, benzylidene acetal of **23** was cleaved with 80% aq. acetic acid at 80 °C. Subsequent O-acetylation giving **24** and cleavage of the *tert*-butyl ester yielded the desired 2,3-sialyl T-antigen threonine building block **25** (30b,31a). Together with the sialyl T_N (from **6**) and the T antigen (from **19a**) conjugates all five important tumor-associated saccharide structures are thus accessible by this synthetic strategy (30,31).

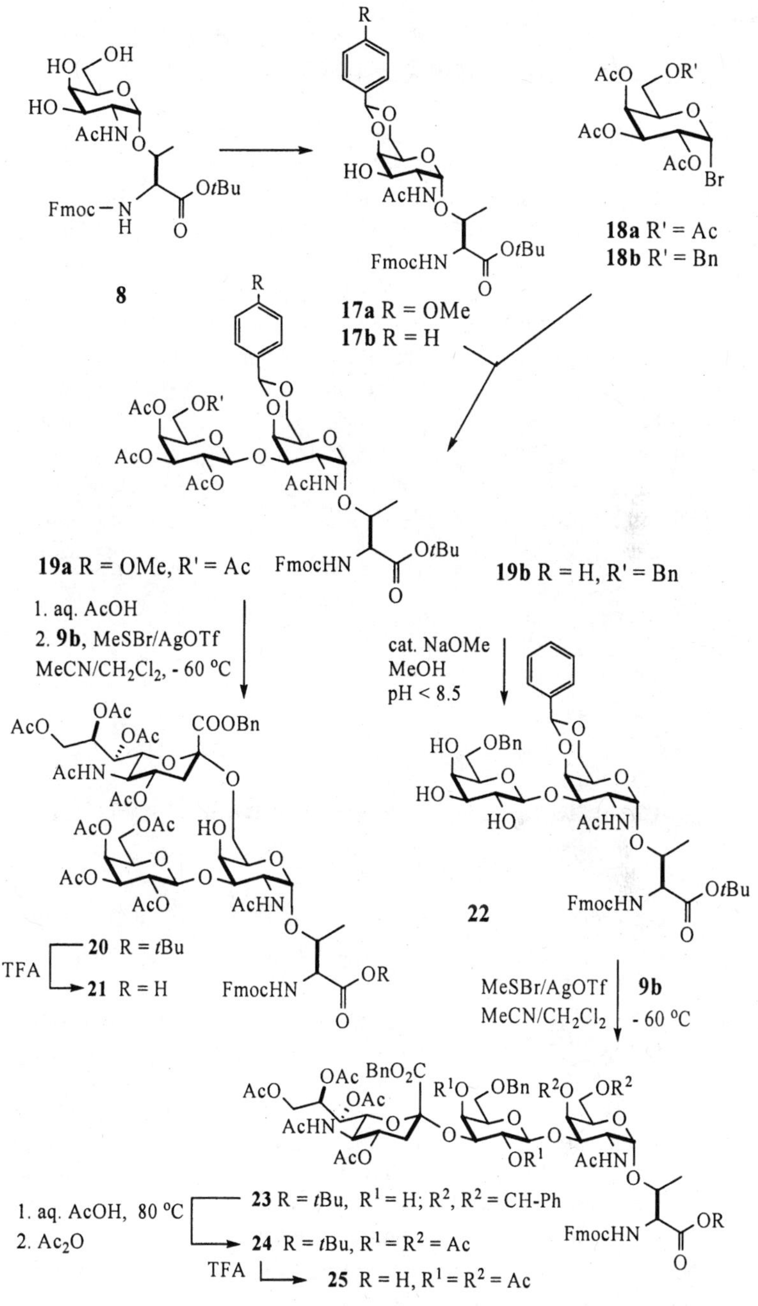

Figure 7. Synthesis of tumor-associated sialyl T threonine conjugates

Formation of synthetic vaccines by conjugation of MUC1 tandem repeat glycopeptides to BSA

Having established a sufficient acid-stability of the Fmoc protected sialyl T_N (**12, 13**) and sialyl T antigen threonine building blocks (**21, 25**), solid-phase synthesis of the corresponding MUC1 tandem repeat glycopeptides could be carried out on the basis of trityl ester linked substrates like **26** according to Fmoc strategy (32).

Uronium salt coupling reagents (33) were used in the solid-phase syntheses. Detachment of the glycopeptides was carried out by acidolysis with TFA in the presence of triisopropylsilane as the cation scavenger. Final careful deprotection gave MUC1 tandem repeat glycopeptides **27-29** in scales of 20-100 mg.

Among the array of MUC1 glycopeptides obtained by this methodology, the sialyl T_N dodecapeptide **30** equipped with a N-terminal polar spacer amino acid was also formed. By its reaction with diethyl squarate (34) one arrives at the squaric monoamide ester **31** which turned out useful for the coupling of the MUC1 glycopeptide antigen to the lysine side chains of bovine serum albumin (BSA).

The synthetic vaccine **32** (32) displays the tumor-associated MUC1 glycopeptide in an exactly specified structure. It is not only applicable as a synthetic vaccine for immunization, but also as a specific tool adhering to the microtiter plates for probing the obtained sera and antibodies.

A synthetic vaccine from combination of a MUC1 glycopeptide and a T-cell epitope

After the promising experiences in the induction of a T-cell response with the construct **16** consisting of the tumor-associated MUC1 glycopeptide and a portion of a T-cell epitope (29), a related strategy was pursued in order to demonstrate the specificity of an induced humoral immune reaction. As transgenic mice which express a receptor specific for the ovalbumin T-cell epitope (35) were available, the MUC1 glycopeptide antigen was combined with a T-cell epitope from ovalbumin (36). The construction of the fully synthetic vaccine was achieved by a fragment condensation of protected glycopeptide antigen **33** with the solid-phase linked T-cell epitope peptide **34** (Figure 10).

After condensation the complex construct was released from the Wang resin (PHB-linker) and subsequently deprotected (see, Figure 8) to furnish the fully synthetic vaccine **35**. The structure and purity of the vaccine **35** was confirmed by 500 MHz ^{1}H-NMR spectroscopy and analytical HPLC showing no impurity.

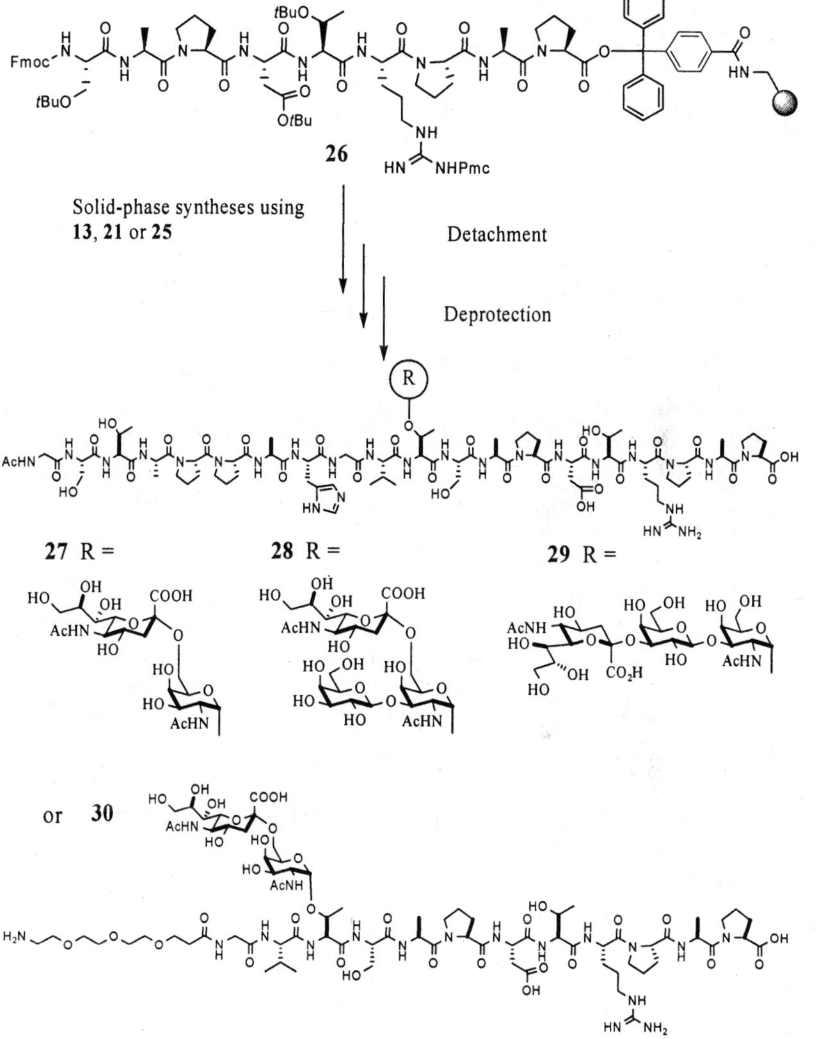

Figure 8. Solid-phase synthesis of sialyl T and sialyl T_N antigen glycopeptides according to lit. (32). Release: TFA, triisopropylsilane, water; deprotection: 1) H_2/Pd, MeOH; 2) NaOMe, pH 9.5; 3) aq. NaOH, pH 11.5, then AcOH.

Figure 9. Conjugation of tumor-associated MUC1 glycopeptides to BSA (32)

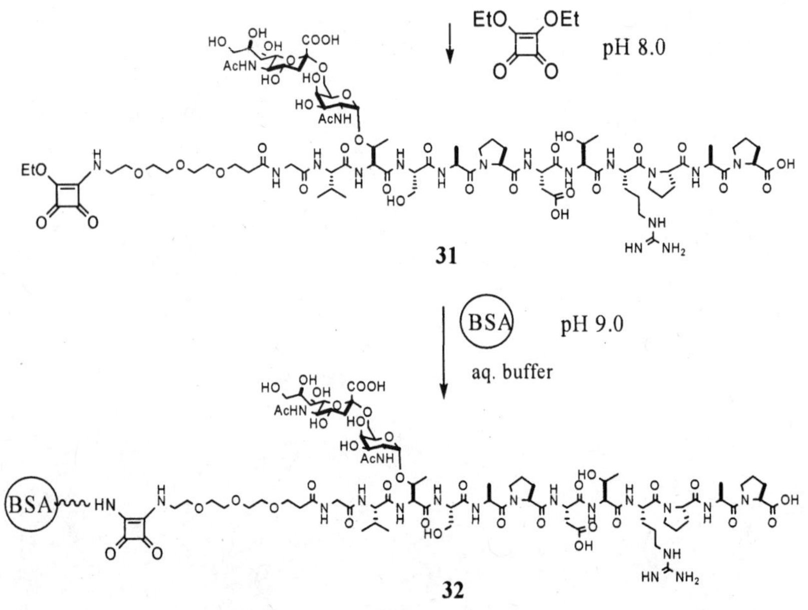

Using the vaccine **35**, transgenic mice (DO 11.10) expressing a receptor specific for the ovalbumin T-cell epitope on their T-cells, were immunized in combination with complete Freund's adjuvant (CFA). After 21 days booster immunizations were carried out with **35** (10 µg) in combination with incomplete Freund's adjuvant (IFA) (**36**). The detection of MUC1 specific antibodies in the sera of the mice was carried out by ELISA five days after each immunization. One of the three immunized mice showed a particularly strong immune response indicating the establishment of an immunological memory.

The specificity of the antibody induced with the synthetic vaccine **35** was verified in a neutralization assay. To this end the MUC1 glycopeptide-BSA conjugate **32** adhered to the microtiter plates served as the binding ligand. After addition of the mouse serum, the bound mouse antibody was detected with a biotinylated secondary anti-mouse antibody. Streptavidin-mediated horseradish peroxidase produced a green color, which was photometrically detected (Figure 11, open circles).

However, after pre-incubation of the induced serum with the synthetic MUC1 sialyl T_N glycopeptide the antibody was completely neutralized (filled squares, Figure 11). It is noteworthy that neither the unglycosylated peptide contained in **36** nor the sialyl T_N antigen threonine in context of another peptide were able to neutralize the antibody induced by **35**. These findings give evidence

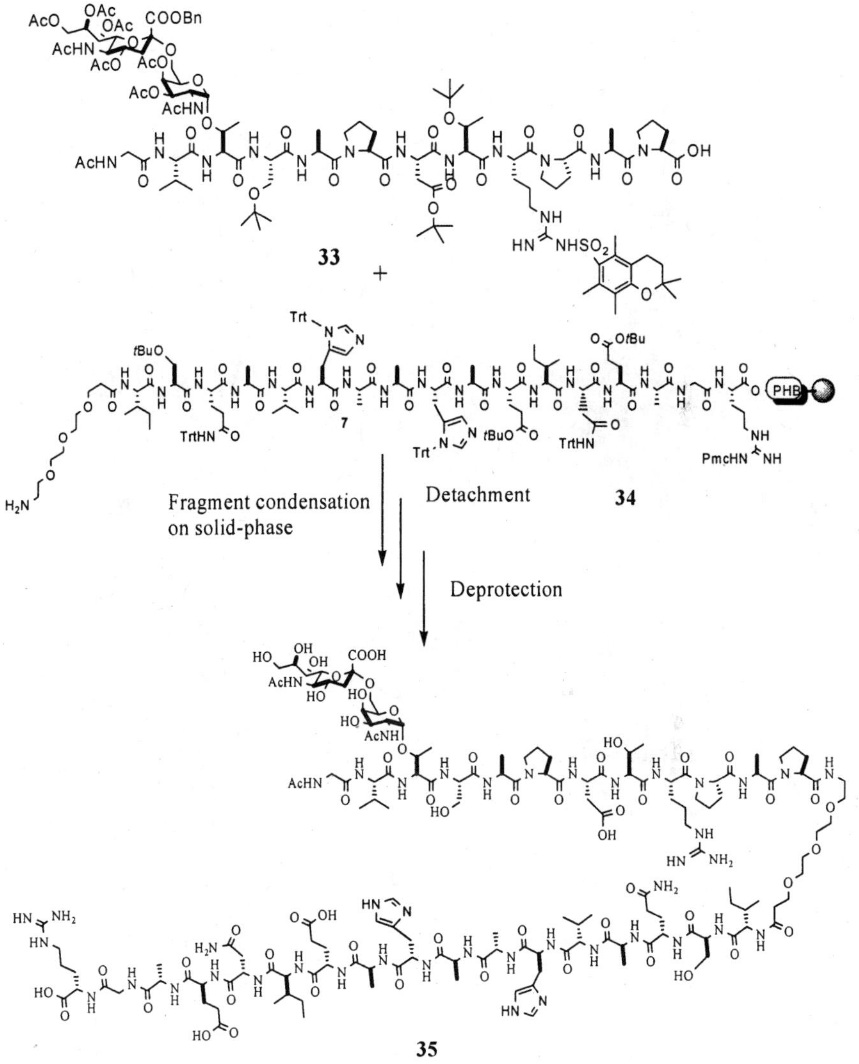

Figure 10. Synthetic vaccine combining MUC1 glycopeptide and T-cell epitope

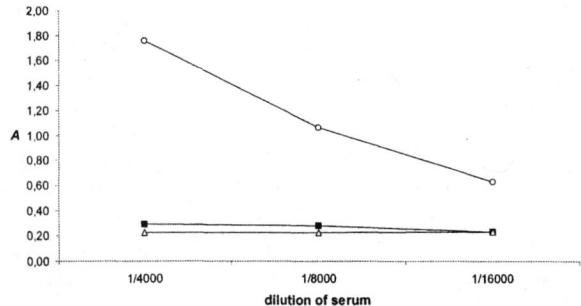

Figure 11. Neutralization of serum obtained from vaccination with 35: Without pre-incubation with the MUC1 glycopeptide; after pre-incubation with the MUC1 glycopeptide 36

of a very specific immune response elicited by the synthetic vaccine **35** containing the tumor-associated MUC1 tandem repeat sialyl T_N antigen glycopeptide. Such a specific immune reaction could not be induced by immunization with MUC1 glycoproteins isolated from tumor cell membranes, since these molecules always are microheterogenic in the saccharide side chains. The fact, however, that such a precise immune reaction is possible if a chemically pure vaccine such as **35** has been used for induction is considered particularly promising for the development of sufficiently tumor-selective anti-tumor vaccines.

References

1. Hynes, R.O. *Cell*, **1987**, *48*, 549.
2. Landsteiner, K. *Wien. Klein. Wochenschr.* **1901**, *14*, 1132.
3. Hakomori, S. *Ann. Rev. Immunol.* **1984**, *2*, 103.
4. Feizi, T.; Childs, R.A. *Trens Biochem. Sci.* **1985**, *10,* 24.

5. Springer, G.F. *Science*, **1984**, *224*, 1198.
6. Kunz, H.; Birnbach, S. *Angew. Chem. Int. Ed.* **1986**, *25*, 360.
7. a) Steinborn, A. *PhD. Thesis* **1990**, Universitaet Mainz; b) Dippold, W.; Steinborn, A.; Meyer zum Büschenfelde, K.-H. *Environ. Health Perspect.* **1990**, *88*, 255.
8. a) Fung, P.Y.S.; Madej, M.; Koganty, R.R.; Longenecker, B.M. *Cancer Res.* **1990**, *50*, 4308; b) Meinjohanns, E.; Meldal, M.; Schleyer, A.; Paulsen, H.; Bock, K. *J. Chem. Soc. Trans 1* **1996**, 985; c) Toyokuni, T.; Singhal, A.K. *Chem. Soc. Rev.* **1995**, *24*, 231; d) Iijima, H.; Ogawa, T. *Carbohydr. Res.* **1988**, *172*, 183; e) Koganty, R.R.; Reddish, M.A.; Longenecker, B.M. *Drug Discovery Today* **1996**, *1*, 190.
9. a) Kuduk, S.D., Schwarz, J.B.; Chen, X.T.; Glunz, P.W.; Sames, D.; Ragupathi, G.; Livingston, P.O.; Danishefsky, S.J. *J. Am. Chem. Soc.* **1998**, *120*, 12474; b) Danishefsky, S.J.; Allen J. R. *Angew. Chem. Int. Ed.* **2000**, *39*, 836.
10. a) Bielfeldt, T.; Peters, S.; Meldal, A.; Bock, K.; Paulsen, H. *Angew. Chem. Int. Ed.* **1992**, *31*, 857; b) St.Hilaire, P.M.; Meldal, A. *Angew. Chem. Int. Ed.* **2000**, *39*, 1162.
11. a) Gendler, S.; Taylor-Papadimitriou, J.; Duhig, T.; Rothbard, J.; Burchell, J. *J Biol. Chem.* **1988**, *263*, 12820; b) Taylor-Papadimitriou, J.; Burchell, J.; Miles, D.W.; Dalziel, M. *Biochim. Biophys. Acta* **1999**, 301; c) Taylor-Papadimitriou, J.; Burchell, J. M.; Plunkett, T.; Graham, R.; Correra, I.; Miles, D.; Smith, M. *J. Mammary Gland Biol. Neoplasia* **2002**, *7*, 209.
12. Briggs, S.; Price, M.R.; Tendler, S.J.B. *Eur. J. Cancer* **1993**, *29A*, 230.
13. a) von Mensdorff-Pouilly, S.; Suijdewind, F. G. M.; Verstraeten, A. A.; Verheijen, R.H.M.; Kenemans, P. *Int. J Biol. Markers*, **2000**, *15*, 343; b) Hilkens, J.; Lightenberg, M.J.-L.; Vos, H.L.; Litvinov, S.V. *Trends Biochem. Sci.* **1992**, *17*, 359.
14. a) Hanisch, F.G.; Peter-Katalinic, J.; Egge, H.; Dabrowski, U.; Uhlenbruck, G. *Glycoconj. J.* **1990**, *7*, 525; b) Hanisch, F.G.; Müller, S. *Glycobiology* **2000**, *10*, 439; c) Hanisch, F.G.; Stadie, T. R. E.; Deutzmann, F.; Peter-Katalinic, J. *Eur. J. Biochem.* **1996**, *236*, 318.
15. Swallow, D. M.; Gendler, S.J.; Griffith, B.; Corney, G.; Taylor-Papadimitriou, J. *Nature* **1987**, *328*, 82.
16. a) Wandall, H. H.; Hassan, H.; Mirgorodskaja, K.; Krostensen, A. K.; Roepstorff, P.; Bennett, E.P.; Nielsen, P. A.; Hollingworth, M. A.; Burchell, J.; Taylor-Papadimitriou, J. *J. Biol. Chem.* **1997**, *272*, 23503; b) Brockhausen, I. *Biochim. Biophys. Acta* **1999**, *1473*, 67; Brockhausen, I. *Biochem. Soc. Trans* **2003**, *31*, 318.
17. a) Brockhausen, I.; Yang, J.M.; Burchell, J.M.; Whitehouse, C.; Taylor-Papadimitriou, J. *Eur. J. Biochem.* **1995**, *233*, 607; b) Gisling, A.; Bartkova, J.; Burchell, J.; Gendler, S.; Taylor-Papadimitriou, J. *Int. J. Cancer* **1989**, *43*, 1072.

18. a) Dziadek, S.; Kunz, H. *The Chemical Record* **2004**, *3*, 308; b) Dziadek, S.; Espinola, C.G.; Kunz, H. *Aust. J. Chem.* **2003**, *56*, 519; c) Brocke, C.; Kunz, H. *Bioorg. Med. Chem.* **2002**, *2*, 3085.
19. Seitz, O.; Kunz, H. *Angew. Chem. Int. Ed.* **1995**, *34*, 803.
20. Leuck, M.; Kunz, H. *J. prakt. Chem.* **1997**, *339*, 322.
21. Braun, P.; Davies, G.M.; Price, M.R.; Williams, P.M.; Tendler; S.B.J.; Kunz, H. *Biorg. Med. Chem.* **1998**, *6*, 1531.
22. Live, D.H.; Williams, L.J.; Kuduk, S.D.; Schwarz, J.B.; Glunz, P.E.; Chen, X.-T.; Sames, D.; Kumar, R.A.; Danishefsky, S.J. *Proc. Natl. Acad. Sci. USA* **1999**, *96*, 3489.
23. Grinstead, J.S.; Koganty, R.R.; Krantz, M.J.; Longenecker, B.M.; Cambell, A.P. *Biochemistry* **2002**, *41*, 9946.
24. Elofson, M.; Salvador, L.A.; Kihlberg; J. *Tetrahedron* **1997**, *53*, 369.
25. a) Liebe, B.; Kunz, H. *Angew. Chem. Int. Ed.* **1997**, *36*, 618; b) Liebe, B.; Kunz, H. *Helv. Chim. Acta* **1997**, *80*, 1473.
26. Paulsen, H.; Adermann, K. *Liebigs Ann. Chem.* **1989**, 751.
27. Marra, A.; Sinaÿ, P. *Carbohydr. Res.* **1989**, *177*, 35.
28. Dasgupta, F.; Garegg, P.J. *Carbohydr. Res.* **1988**, *177*, c13.
29. Keil, S.; Claus, C.; Dippold, W.; Kunz, H. *Angew. Chem. Int. Ed.* **2001**, *40*, 366.
30. a) Brocke, C.; Kunz, H. *Synlett* **2003**, 2052; b) Dziadek, S.; Kunz, H. *Synlett* **2003**, 1623.
31. a) Dziadek, S.; Brocke, C.; Kunz, H. *Chem. Eur. J.* **2004**, *10*, 4150; b) Brocke, D.; Kunz, H. *Synthesis* **2004**, 525.
32. Dziadek, S., Kowalczyk, D.; Kunz, H. *Angew. Chem. Int. Ed.* **2005**, *44*, 7624.
33. Carpino, L.A.; Ionescu, D.; El-Faham, A. *J. Org. Chem.* **1996**, *61*, 2460.
34. Tietze, L.F.; Schröder, C.; Gabius, S.; Brinck, U.; Goerlach-Graw, A.; Gabius, H.-J. *Bioconjugate Chem.* **1991**, *2*, 148.
35. Murphy, K.M.; Heimberger, A.B.; Loh, Y.D. *Science*, **1990**, *250*, 1720.
36. Dziadek, S.; Hobel, A.; Schmitt, E.; Kunz, H. *Angew. Chem. Int. Ed.* **2005**, *44*, 7630.

Chapter 14

Glycopeptide-Based Cancer Vaccines: The Role of Synthesis and Structural Definition

R. Rao Koganty[1], Damayanthi Yalamati[1], and Zi-Hua Jiang[1,2]

[1]Biomira Inc., 2011 94 Street, Edmonton, Alberta T6N 1H1, Canada
[2]Current address: Department of Chemistry, Lakehead University, Thunderbay, Ontario P7B 5E1, Canada

Introduction

In 1798 Edward Jenner (*1*) ushered in the concept of vaccines to protect the population from the scourge of small pox. The traditional protective vaccines, which lacked the popularity of modern antibiotics, were ignored for almost two centuries. With small pox now largely eradicated, we may never know the specific molecular features of the small pox vaccine, which induced resistance to the virus. The current approaches to the vaccine development, particularly in the field of therapeutic vaccines, are vastly different from those of the earlier vaccines, which were devised from inactivated whole pathogens. Recent immunological advances have significantly increased our understanding of the mechanisms required to mount a successful immune response against infections. Majority of the data now available confirms that the immune system functions through a cooperation of its two major components, the innate and adaptive arms, which collaborate and initiate signalling process leading to immune defence. Immunological memory against invading pathogens is then established, which is the ultimate goal of vaccination. The two major components of the immune system recognize, through specific receptors, relatively small molecular structures of invading pathogens (pathogen associated molecular patterns or PAMPs). Therapeutic vaccines exploit this collaboration between the two arms of the immune system to fight against non-infectious disorders such as cancers. Consequently, a vaccine may be simplified to two distinct components; an adjuvant to alert innate immune complex and an antigen to trigger an adaptive immune response. All these immunological events must transpire through ligand-receptor interactions. Therapeutic vaccines to self antigens must not only overcome tolerance but also must not generate immune response that damages the healthy cells. In other words immune response must be extremely restrictive to diseased cells. Such highly specific vaccines may be formulated using well defined small molecular approach which can be better achieved through chemical synthesis. Several research groups have approached with similar concepts, which will re-define the vaccines. While describing our research and results in this chapter, we salute the research groups elsewhere who have made significant contributions for the advancement of vaccine design.

Receptors of the innate immunity

The mammalian immune system consists of innate (2,3) and adaptive arms of immunity, which operate through a complex array of receptors and signaling processes. While these two effector mechanisms of the immune system are distinct with different roles in protecting the host, they operate co-operatively (4). The innate immune system, as a front line host defense against invading pathogens, consists of receptors that recognize a wide range of chemical structures (PAMPs) such as bacterial lipo-polysaccharide (LPS) derived lipid-A, short sequences of bacterial DNA containing unmethylated CpG units and hydrophobic peptides known as flagellins from bacterial flagellae. PAMP receptors of innate immune system evolved and are conserved among many species including *Drosophila Melanogaster* (fruit fly), which could make its own medicines in the form of anti-bacterial peptides such as Drosocin (5-7), when attacked by pathogens. These receptors are known as a family of Toll Like Receptors or TLRs (8-12) which bind to bacterial and viral components such as unmethylated CpG containing bacterial DNA (TLR-9), bacterial lipid-A (TLR-4) and flagellin, a protein derived from flagellae of bacteria (TLR-5). In addition to about 10 well characterized TLRs there are other receptors of the innate immune system, the most notable of which is CD1d receptor (13) which binds to marine sponge derived α-Galactosyl Ceramide. Thus innate immune system has the ability to distinguish between host derived molecular patterns from pathogen derived. Once a pathogen derived specific molecule binds to a receptor and a danger signal is received, a cascade of events are initiated which include the expression of cytokines, co-stimulatory molecules that are essential to direct defensive (antigen specific) responses through adaptive immunity.

Induction of immune responses to cancer antigens, which are self antigens, is regulated by a number of processes that together make up a phenomenon known as immunological tolerance. This tolerance to self antigens is a major hurdle in the development of effective therapeutic vaccines for cancers. In order to enlist the participation of the innate immune system's help, molecular structures derived from pathogen cultures are artificially introduced along with a cancer associated antigen to which immune response must be elicited (Figure 1). As shown in figure 1, a bacterial lipid-A binds to toll like receptor 4 (TLR-4) present on the surface of an antigen presenting cell (APC). This initiates or up-regulates the expression of co-stimulatory molecules (*14,15*) CD 80/86.

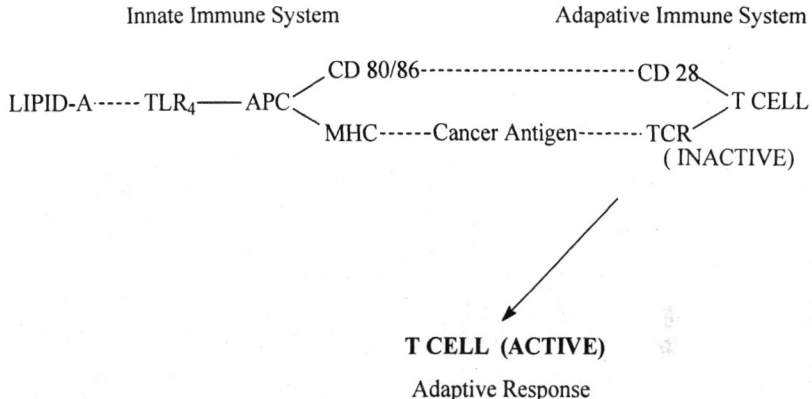

Figure 1. *A simplified mechanism for induction of immune responses against cancers by enlisting the participation of innate immune system. Dotted lines indicate receptor-ligand binding.*

In parallel, the cancer antigen is processed into specific short peptide sequences by the APC and presented on the surface in the context of major histocompatibility complex (MHC). All of this is verified by the adaptive system which consists of CD28 receptor of a T-cell that checks the presence of the CD80/86 and the T-cell receptor which recognizes the cancer antigen. These events result in the activation and maturation of antigen specific T-cells capable of recognizing the antigens on the surface of cancers.

We now have the ability to manipulate and direct our own immune system to respond against self antigens in the fight against diseases like cancers. At this juncture it is important to emphasize the consequences of generating immune responses to a self antigen, which can be beneficial as well as harmful by way of auto immune disorders. The differences of structural features of cancer and normal cells are subtle and often quantitative, particularly when the antigens are macromolecular such as MUC1, which is present on all epithelial cells, cancerous or normal. An ideal therapeutic cancer vaccine is composed of specific structural features (epitopes) that are exclusively associated with cancer. A cancer associated macromolecule or a cancer cell share many molecular structural features with normal cells. If a vaccine were to be devised using the whole cell or a macromolecule, there is also a risk of breaking immune tolerance to normal cells. This is where structural definition, which is accomplished by chemical synthesis, plays an important role. Once we know a dominant molecular feature that is more restricted to cancer than normal cells, that specific structure may be chemically synthesized eliminating the need for the whole macromolecule.

The Target antigen

Here, we describe examples of chemically synthesized peptide and glycopeptide vaccines formulated with synthetic carbohydrate based adjuvants and the role played by structural definition in engendering accuracy to the target, which is a cancer associated glycoprotein known as MUC1 mucin (*16-19*). MUC1 is a mucin like glycoprotein which is characterised by an extra-cellular domain consisting of a 20 amino acid sequence which repeats itself 60 to 80 times. The protein is heavily glycosylated in normal cells while on cancer cells glycosylation is less dense, often exposing the protein core. Vaccine development based on MUC1 mucin's up-regulated expression on cancers has so far proven feasible. The results from Phase II clinical trials of BLP25 liposome vaccine, now known as Stimuvax® (*20*), which is about to enter phase III clinical trials for the treatment of non small cell lung cancer (NSCLC), has validated MUC1 as a viable target. Liposomal Stimuvax® is formulated using an unglycosylated synthetic lipo-peptide **1** derived from the core sequence of MUC1 mucin. Stimuvax® has generated exclusively cell mediated immune response with no measurable antibodies to the peptide epitopes.

STAPPAHGVTSAPDTRPAPGSTAPPK(εN-Palmitoyl)G

1

Another MUC1 derived structure, BGLP40, consists of a glycosylated lipo-peptide **2**, designed to elicit both cellular and humoral responses to peptide epitopes and humoral responses to carbohydrate epitopes. To our knowledge, BGLP40 is the largest glycopeptide with molecular weight of over 5000, ever to be made entirely by chemical synthesis with all natural chemical linkages. BGLP40 represents two tandem repeats of MUC1 core protein backbone followed by two serines carrying lipid chains at the C-terminus, which terminates with leucine. The two tandem repeat linear sequence provides unglycosylated peptide epitopes and carbohydrates at pre-determined sites with Tn (*21-23*) and/or STn (*24,25*) determinants. L-BGLP40 vaccine is formulated in liposomes using the synthetic antigen **2** and an adjuvant, a synthetic mimic of lipid-A.

Chemical synthesis and structural definition

Our understanding of the immune system and its functioning has vastly improved over the last 15 years. The immune system's functioning is governed by a basic concept of interactions between molecules, more precisely, receptor-ligand type recognition, often followed by a specific patterns of signaling through a cascade of cytokines. This process ensures signaling by recognition rather than by random encounters. We now know that only small portions of a macromolecular antigen are recognized by receptors when presented by a complex of cell surface molecules known as human leukocyte antigens (HLA). In order to accomplish this, an antigen presenting cell must encounter and ingest the macromolecule (usually a protein or glycoprotein) and chop it to specific small sequences and present those on the surface HLA to the immune cells such as T-cells, which are involved in host defense.

Overcoming tolerance (*26*) to an entire macromolecule may not be an appropriate choice for the immunotherapy of cancers. Safety, stability and large scale production are important factors in the production of macromolecular vaccines. Normally, antigens originated from external pathogens that invade our body or those produced within (abnormal self antigens), are macromolecular structures. Large scale production of such antigens requires transfected cell cultures which require specially built facilities for extraction and purification. Purification of macromolecules requires an elaborate path to prevent contamination by other macromolecules which is one of the most serious problems encountered in this process. Immune response to a small but specific portion of the macromolecule, in the form of peptides and glyco-peptides (*27*), may be all that is required for immunotherapy.

MUC1 is over expressed on most epithelial cancers such as the cancers of the lung, breast, ovaries and colon to name a few. Though this mucin is present on normal epithelial cells, there are significant qualitative and quantitative differences in its expression on normal and cancers cells.

Synthetic MUC1 glyco-lipo-peptide

Chemically synthesized antigen engenders many advantages such as restriction of the size to an epitopic definition (*28,29*) and increased specificity, which avoids immune responses that do not offer benefits to diseased cells and may be detrimental to normal cells. Synthesis also offers an extraordinary benefit of placing the carbohydrate epitopes on a glycopeptide where needed and at a density (*30*) that is reflective of natural glycoprotein. Though biological macromolecules are too complex for chemical synthesis, proteins and glycoproteins may be an exception in the realm of vaccines since the whole macromolecule should not be required for therapeutic benefits, as long as the small molecule vaccine is processed through the immune system's mechanisms.

Unlike large organic molecules which must be synthesized through successive iterations, chemical synthesis of protein based molecules is fairly simple in the sense that smaller fragments may be separately and simultaneously synthesized,

$$\text{TSAPDTRPAPG}\overset{\text{Tn}}{\underset{|}{\text{S}}}\text{TAPPAHGV}\underset{\text{Tn}}{|}\quad \text{TSAPDTRPAPGSTAPPAHGVS}(l)\text{S}(l)\text{L}\underset{\text{OR}}{|}$$

$S(l)$ = Serine with a C_{14} lipid attached (see **5** in figure 2)

2a. R = H
2b. R = STn

purified and assembled into larger fragments and to the final product. An added advantage is that chemical synthesis gives the freedom to structurally pre-design a formulation specific antigen, for example, to be incorporated into the membrane of a liposome. Glyco-lipo-peptides **2a** and **2b** are examples of a structurally well defined molecular vaccine designed for liposome delivery.

The two tandem repeats accommodates two separate unglycosylated and glycosylated peptide and multiple carbohydrate epitopes, all of which are cancer associated. Solid phase synthesis is advantageous in facilitating rapid peptide synthesis, often through automation. This process, though efficient, offers no control through the entire synthesis, as deleted sequences are continuously formed during stepwise coupling reactions. Although the deleted peptide impurities are formed in very small quantities, their identities are unpredictable and present considerable challenge to the purification of final product.

The glyco-lipo-peptide **2** contains 40 amino acids from the antigen and two serine based lipids at the C-terminal, followed by a leucine. The carbohydrates **Tn** and STn (Sialyl-Tn) are synthesized as building blocks **3** and **4** (*31,32*) adaptable for glycopeptide synthesis in solution or solid phase using mild Fmoc chemistry (Figure 2). The lipid chains, pre-synthesized in the same format on serine **5**, are introduced as part of peptide synthesis.

These components together with block assembly approach (Schemes 1 and 2) may be considered as a significant advancement in vaccine design. Well

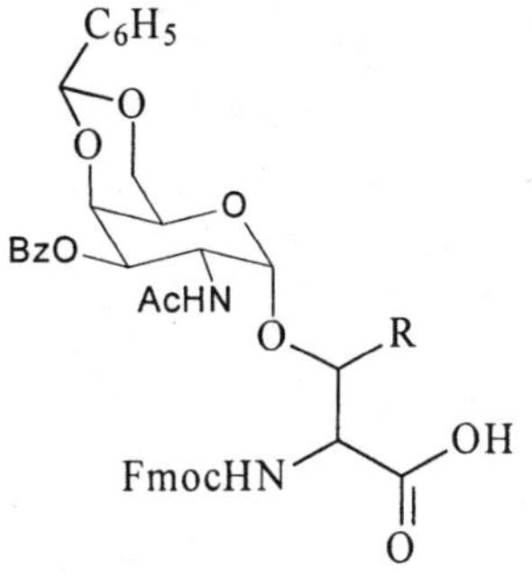

3a. R = H
3b. R = CH₃

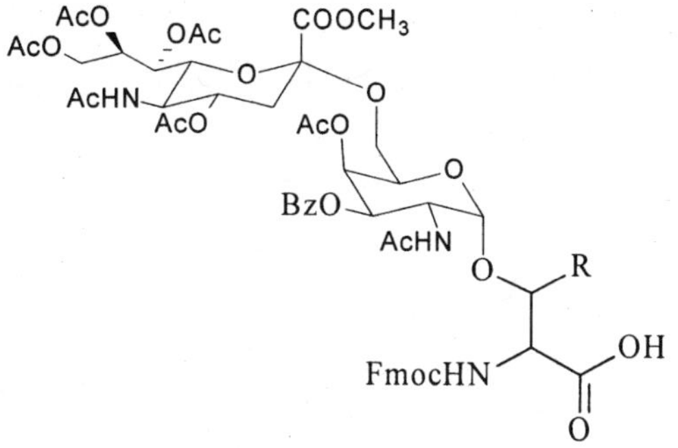

4a. R = H
4b. R = CH₃

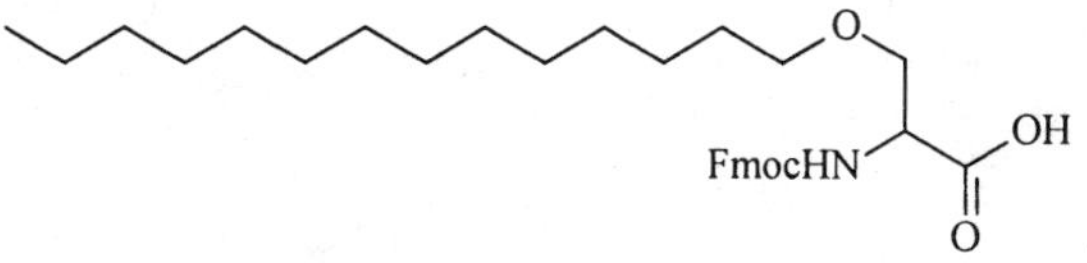

5

Figure 2. Protected serine and threonine 3-5 are modified for the synthesis of peptide blocks leading to the two configurations of antigens 2a and 2b.

controlled synthetic path, production, stability, and performance, are some of the most desirable characteristics for the vaccine development.

Instead of solid phase synthesis, both **2a** and **2b** are simultaneously produced in solution phase through progressive block building and coupling. Retro-synthetic analysis of glyco-lipo-peptide **2a** (Scheme 1) reveals that the molecule can be conveniently divided into smaller blocks which may be independently and simultaneously synthesized and assembled into larger blocks and final products.

The smaller blocks are assembled into intermediate larger blocks which are further assembled into the final protected product (Scheme 2). An added advantage of this approach is that several glycoforms of the antigen with same peptide sequence can be synthesized using the primary blocks **6-12** combinatorially. The number of primary blocks can be further expanded to increase the number of final configurations. Such freedom to produce several products is useful in determining an effective vaccine structure. Scheme 2 shows the synthesis of two glycoforms simultaneously. Purified blocks **6-12, 13-16** and **17-19** can be produced at commercial manufacture scale according to specifications, quality control, and documentation. The stability of the blocks allows them to be stored for a long term for future use.

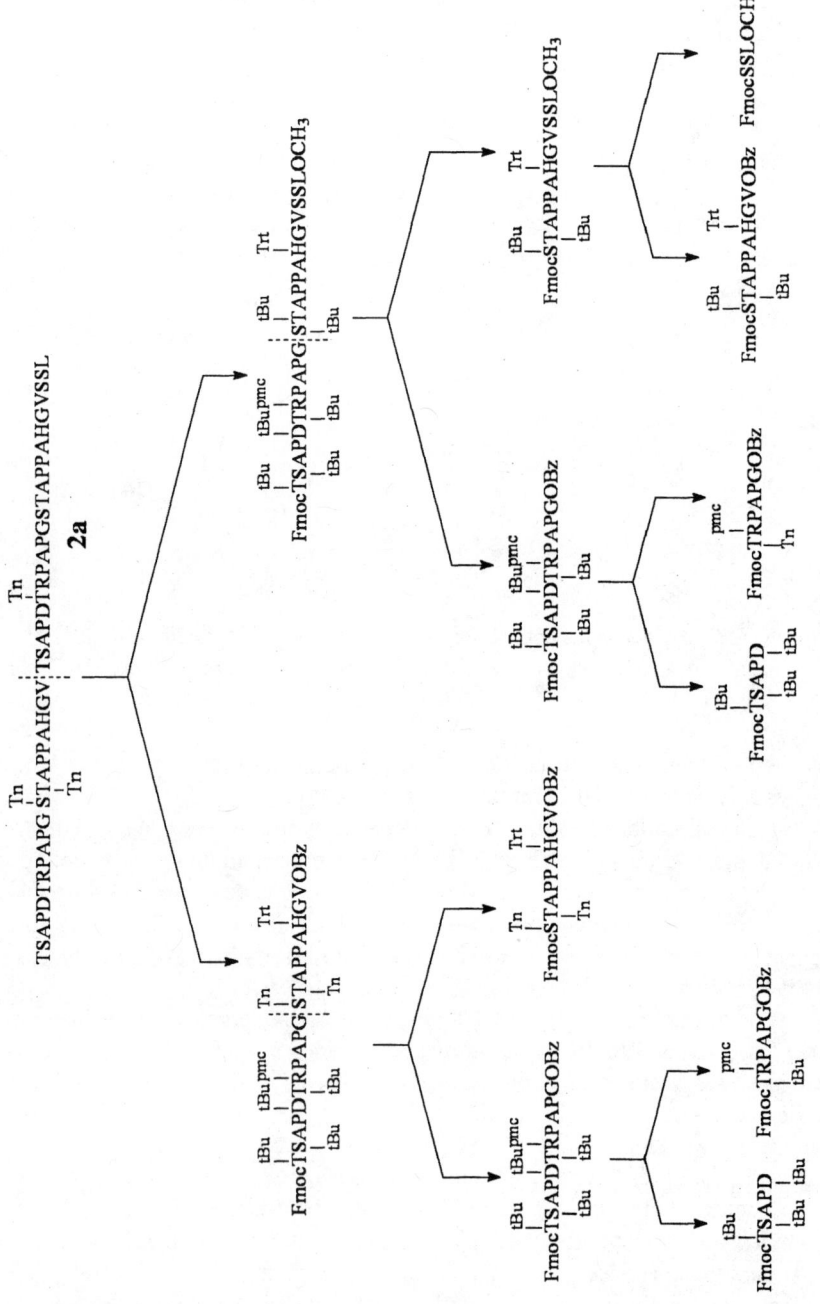

Scheme 1. *Convenient linking sites with least hindrance to coupling are identified for assembling.*

Purification is accomplished at every step of every block synthesis and their purity and structure can be established using routine analytical methods such as HPLC, MS and NMR. Excellent yields are obtained even when the larger blocks **17-19** are coupled. Considerable molecular weight differences between coupled and uncoupled blocks facilitate better chromatographic separation and purification. Final de-blocking with a normal cocktail of cleaving agents to remove acid sensitive protecting groups on peptide, followed by NaOMe treatment in methanol to remove acetate and benzoate protecting groups on carbohydrates and finally methyl ester is hydrolyzed by reacting with 0.1N NaOH in aqueous DMF for 1.5 hrs at 0° C to avoid β-elimination at serine and threonine as well as racemization. Final purification is carried out on preparative reverse phase chromatography. The product is obtained in almost 100% pure form with no detectable impurities. While this process appears to be slow, it offers a robust chemistry for scale-up to much larger quantities than can be possible with solid phase synthesis. Besides, this synthesis gives an excellent quality control at every step and provides a well defined Chemistry, Manufacture and Control (CMC) section for regulatory submissions.

Synthetic immune stimulants

Most immune stimulants function by binding to a specific receptor of the innate immune system. Many receptors have been identified and well characterized, which recognize pathogen associated molecules such as lipid-A derived structures.

To date most widely used vaccine adjuvants are natural products such as bacterial lipid-A derivatives, as an undefined mixture of chemically modified and purified from bacterial cultures. Mono phosphoryl lipid-A or MPL® (Corixa Corporation) is currently being used as an adjuvant in many experimental vaccines (*33,34*). As a natural product derived from macromolecular bacterial cell wall, MPL® is a detoxified admixture of several major and some minor components consisting of structurally similar disaccharides with lipids varying in number and size. Manufacture of safe and consistently effective adjuvants is much dependent on achieving of a good balance of all major components. Lipid-A derivatives can be chemically synthesized with excellent structural integrity as well as purity (*35,36*). Being a single molecule the complexity of composition associated with natural preparations is easily avoided. Synthetic structures fashioned after lipid-A or their partial mimics have proven to be comparable in immune stimulating activity to their natural counter parts (*37,38*). Manufacture of synthetic compounds is a fairly routine operation with far superior control of synthetic chemistry and quality.

In order for a vaccine to be effective, the participation by the innate immune system with its receptors is essential. Most of the receptors recognize pathogen

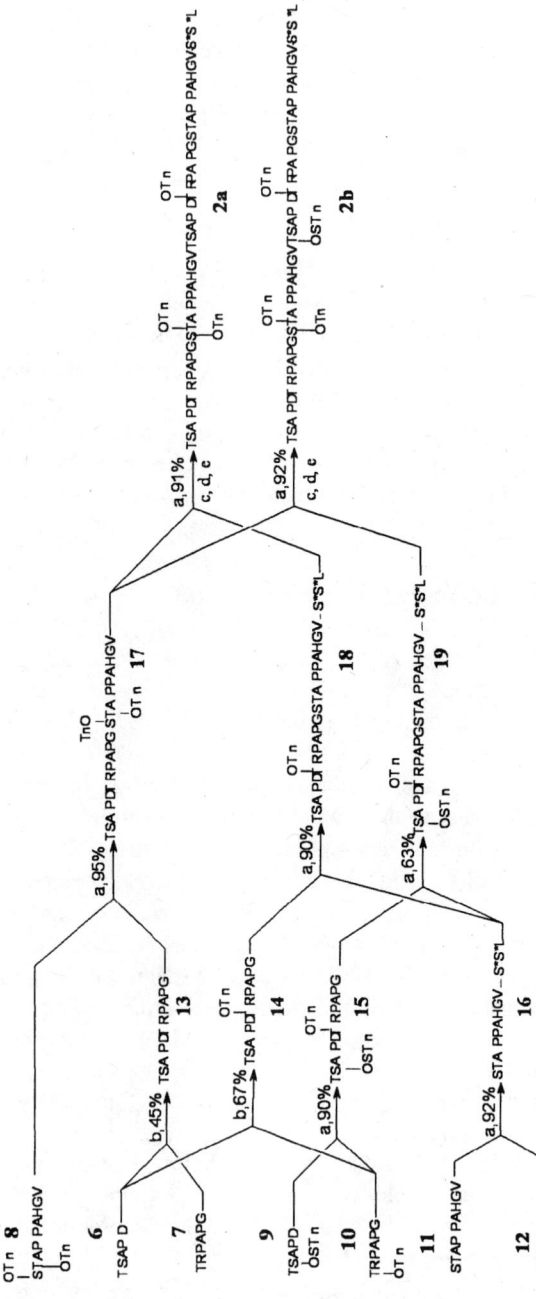

Scheme 2. *Blocks can be used combinatorially to produce several synthetic glycoforms of the antigen to enable the determination of the most effective glycosylation pattern, in terms of variety, extent and location.*

a. DCC/HOBt/CH$_2$Cl$_2$/RT/2-4 hrs.; b. DCC/NHS/THF/RT/2-4 hrs.; c. TFA/iPr$_3$SiH/PhSH.; d. NaOCH$_3$/MeOH/3 hrs.; e. NaOH(0.1N)/H$_2$O/DMF/0°C /2.5 hr.

associated molecules. Structurally simple and single molecules such as those shown in figure 3 are sufficient to initiate signaling and activation of innate immunity through a receptor interaction.

Structural mimics of lipid-A analogues are potent and stable, two of the properties that enable TLR related biological investigations. Partial mimics such as **21** containing a pentaerythritol amine (*37*) can accommodate an additional phosphate group as shown in **23** and additional lipids on the free hydroxyl group, which mimics the 3-OH of glucosamine. Functionally, if not structurally, pentaerythritol amine is the closest mimic of glucosamine (*37*). The stereo chemistry of the lipids such as that of the 3-hydroxy myristic acid, which is commonly present in the R form in most bacterial lipid-A structures, can be reproduced with a less expensive *l*-aspartic acid analogue as shown at the 2' position of **22**. The amide linkages of such lipids not only add stability to the molecule but also increase the hydrophobic properties of the molecule, which seem to play profound role in modulating immune stimulatory properties and possibly the affinity to receptor binding.

Structurally, the differences between agonistic and antagonistic activities of either synthetic structures or naturally derived molecules may be subtle and likely to be related to the hydrophobic properties of the molecule. Many therapeutic applications based on agonistic and antagonistic activities towards the family of TLRs are being explored (*39-41*). The differences in TLR agonistic activities displayed by all three structures **20, 21** and **22** are largely related to their solubility. Both **20** and **21** behaved similarly when used alone or formulated into liposome. **22** did not show stimulatory activity when used alone, but when formulated in liposome it has proven to be as potent as the other two demonstrating that delivery of TLR ligand to the receptor is an important factor in determining the agonistic and antagonistic activities.

Chemically synthesized structures of biological origin or their mimics provide impeccable structural definition and purity, which are essential for consistent performance and safety. Quality Control (QC) is perhaps the most critical aspect in drug development which defines manufacturing by both chemical synthesis as well as by biological process. By nature, chemically synthesized products must follow a step wise process, which allows verification and quality control at every step of the entire process. Accurate and sensitive qualitative and quantitative methods are available, which not only determine the integrity of every intermediate but also the detection, identity and the extent of even minor impurities. Synthetic compounds described here have been repeatedly produced with excellent process consistency, stability, purity and performance.

Properly protected fragments **24-32** (Figure 4) such as the donors, acceptors and the lipid complexes may be combinatorially used to rapidly produce several analogues for evaluation. Additional fragments maybe created by varying the number and the type of lipid substituents to enable the synthesis of a large library of compounds individually in pure form.

21

20

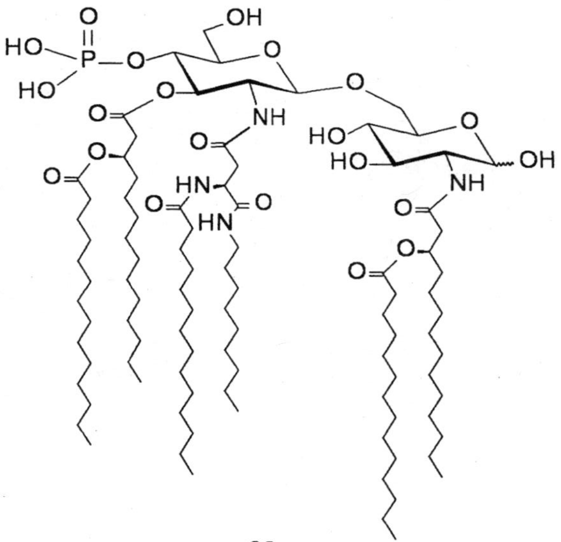

Figure 3. Synthetic structures that are either similar to natural lipid-A (20), a partial glucosamine mimic of lipid-A (21) and carrying an unnatural lipid complex based on aspartic acid (22) to exploit the stereochemistry.

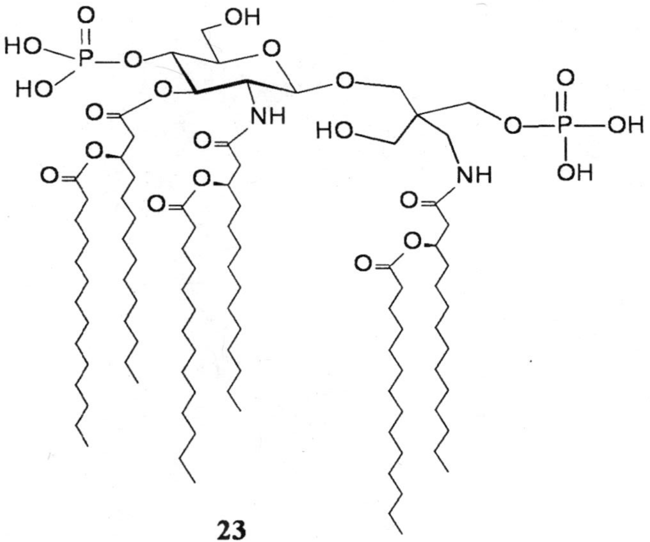

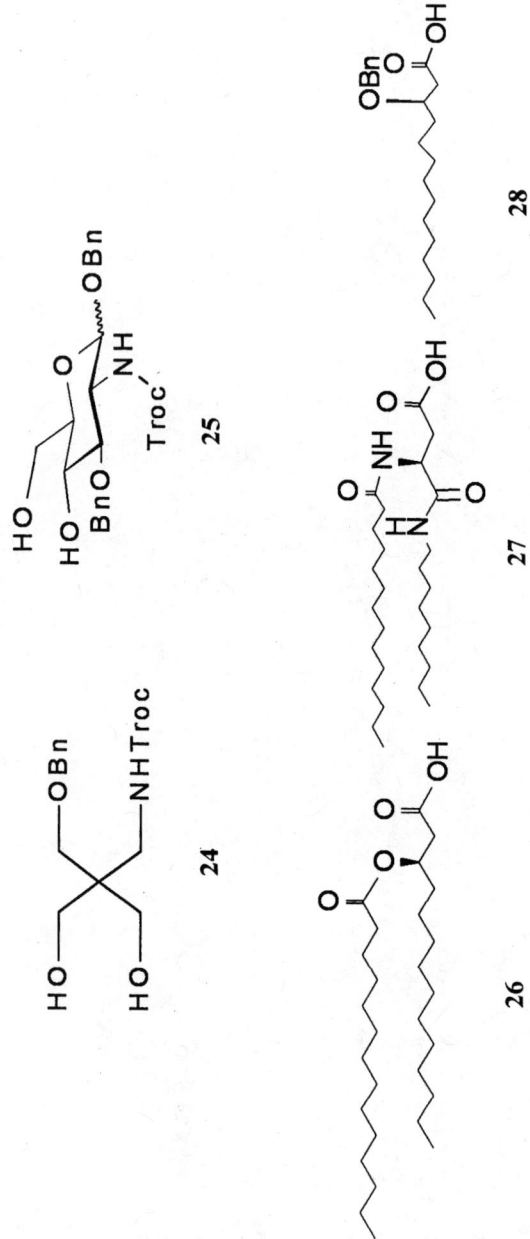

325

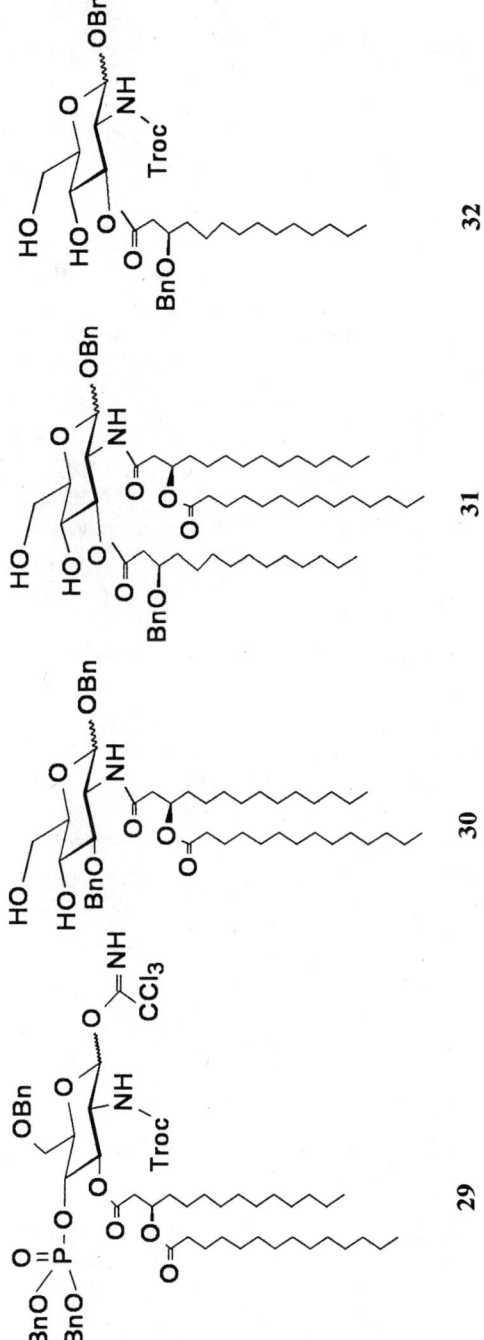

Figure 4. Components are strategically protected to allow synthesis of large number of lipid-A analogues and mimics for screening.

These intermediate fragments are fairly stable under normal storage conditions and can be produced in large scale for final assembly under current good manufacture practices (cGMP), required for clinical testing.

Vaccine delivery

Vaccine formulation and delivery (42) have become critical factors in ensuing safety and effectiveness of vaccination, particularly against self antigens such as MUC1. Protective vaccines, as poorly defined as they are, contain highly effective antigens of pathogen origin and have acceptable history of safety. These vaccines protect the host against infections and are given only once with boosters in certain cases.

Small molecules such as peptides are not very immunogenic since they are easily cleared by enzymatic degradation. Strongly hydrophobic molecules such as 22 are poorly soluble in aqueous media and must be delivered to TLR through a suitable formulation. In a simplistic mechanism where TLRs control adaptive responses (43) to an antigen, a vaccine must deliver both antigen and TLR ligand to the antigen presenting cell as effectively as possible. In order to be processed by antigen presenting cells, the antigen must either be in a macromolecular particulate form or must be linked to a macromolecule, which are normally taken up and processed by macrophages and dendritic cells.

Strongly hydrophobic lipids like small molecules have self assembling properties and form lipid bi-layers that are similar to macromolecular particles (44). These self assembled lipid layers can accommodate molecules with similar hydrophobic properties in their membrane assembly and may be converted to multi-lamellar liposomes. Synthetic antigens such as BLP25 and BGLP40 and adjuvants such as synthetic lipid-A structures contain hydrophobic chains and hydrophilic surfaces and can easily participate in the assembly of lipid bi-layers. These liposomes closely resemble in size a pathogen (45) or a mammalian cell and are strongly immunogenic as they are ingested and processed by macrophages and other antigen presenting cells. These macromolecule like particles, under appropriate conditions, exhibit physical and chemical properties of the original small molecules and can be analyzed for composition. Thus the self-assembled liposomes (figure 5) display the immunological properties of a macromolecule as well physico-chemical properties of small molecules.

Figure 5 is a pictorial representation of a liposome vaccine displaying various components that constitute the membrane organization. In the cut out the lipids chains of the bi-layer can be seen as yellow strings and attached to the lipids are the hydrophilic components represented by colored spheres on the outer and inner surface. Such liposome formulated vaccines, being entirely synthetic and of epitopic precision, generate targeted immune responses, which

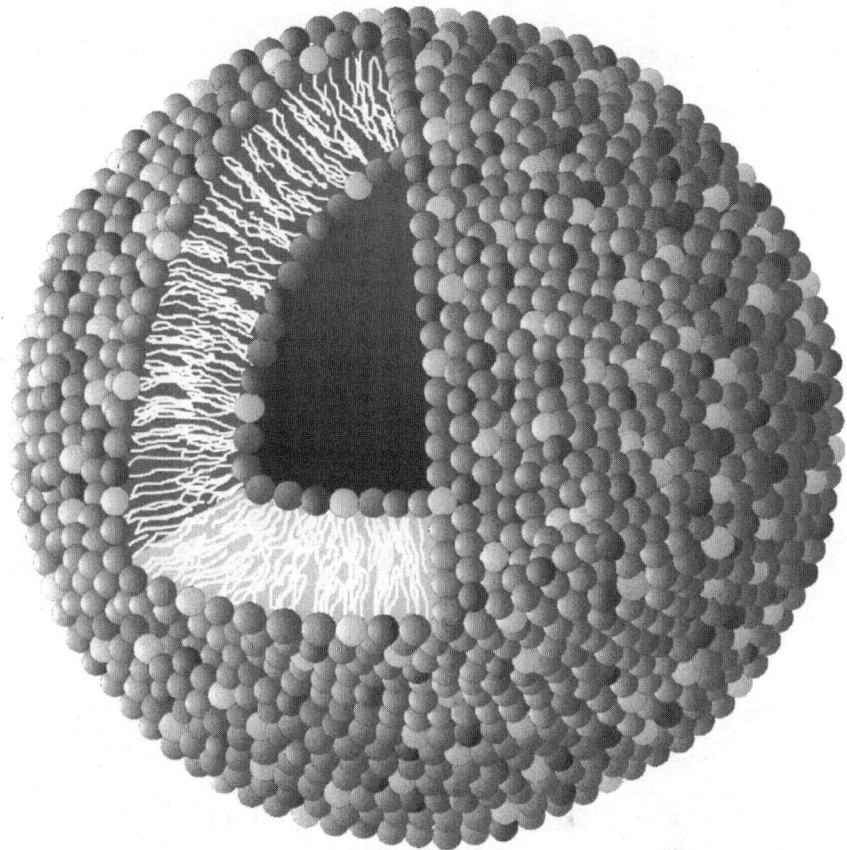

Figure 5. A pictorial of liposome with various lipidated components such as the antigens and adjuvants in red and yellow embedded in membrane organization, with the cut-out showing lipid bi-layer. (See page 1 of color inserts.)

are safe and effective against their targets. The empty inner core may contain water in the hydrated or aqueous formulation while in the freeze dried form the liposomes are empty.

Pre-clinical studies

Self antigens such as MUC1 must overcome immune tolerance in order to be effective as a cancer vaccine. In a phase IIB randomized clinical trial, NSCLC patients, the antigen BLP25 (Figure 1) in Stimuvax® has generated cellular responses and has proven to be effective in enhancing patient survival as well as quality of life. Structural definition of the epitope restricted the immune response. The rationale behind the development of **2a** and **2b** is to design and synthesize a small molecule vaccine, which is capable of generating both cellular and humoral responses to several cancer associated epitopes. A glycopeptide is an ideal mimic of cancer associated MUC1 mucin. Structure **2a** contains two tandem repeats of MUC1 core protein and three α-N-acetyl galactosamines attached to serines and threonines, which are considered carcinoma antigens (*21-23*). Antibodies to glycopeptides are proven to have better affinity in binding to cancer cells expressing the mucin (*28*). Consequently glycopeptides that closely mimic the native mucin can be much more effective for the immunotherapy of cancers, than either peptide or carbohydrate vaccines alone.

Immune tolerance can be a serious problem for targeting self antigens such as MUC1 mucin in spite of differences in glycosylation patterns of normal and cancer expressed mucins. Often gene transfected mouse models are used to test the immune responses. Primates such as macaques display sequence homology with human proteins. Macaque mucin has about 75% homology with human MUC1. Immune responses to **2a** in macaques are a practical model to study immune tolerance as well as the nature of responses to various epitopes contained in **2a**.

TSAPDTSAAPGSTGPPARVVTSAPDTSAAPGSTGPPAHVVS(*l*)S(*l*)L

33

S(*l*) = A C_{14} lipid is attached to Serine shown as in **5** (figure 2)

A synthetic macaque sequence of similar length **33** is used as a solid phase to test the affinity of **2a** antibodies towards macaque mucin reactivity.

Two groups of four healthy animals each were chosen to test two separate formulation of **2a**, one formulated with natural lipid-A preparation and another with a synthetic lipid-A, **20**. Both antigen and the adjuvant are formulated in liposome in which both components are incorporated into liposome membrane. Macaques were given 8 immunizations two weeks apart. Responses were seen after two immunizations and the responses were sustained through the immunization schedule and beyond.

This study is expected to address the comparison of performance of natural and synthetic immune stimulants in all aspects of immune response, both qualitative and quantitative. In early pre-clinical animal studies as well as immune responses in Macaque primates, BGLP40 with three Tn epitopes confirmed high titers of IgM and IgG antibodies to peptide and moderate titers to carbohydrate epitopes. Additionally, moderate levels of peptide specific T-cell

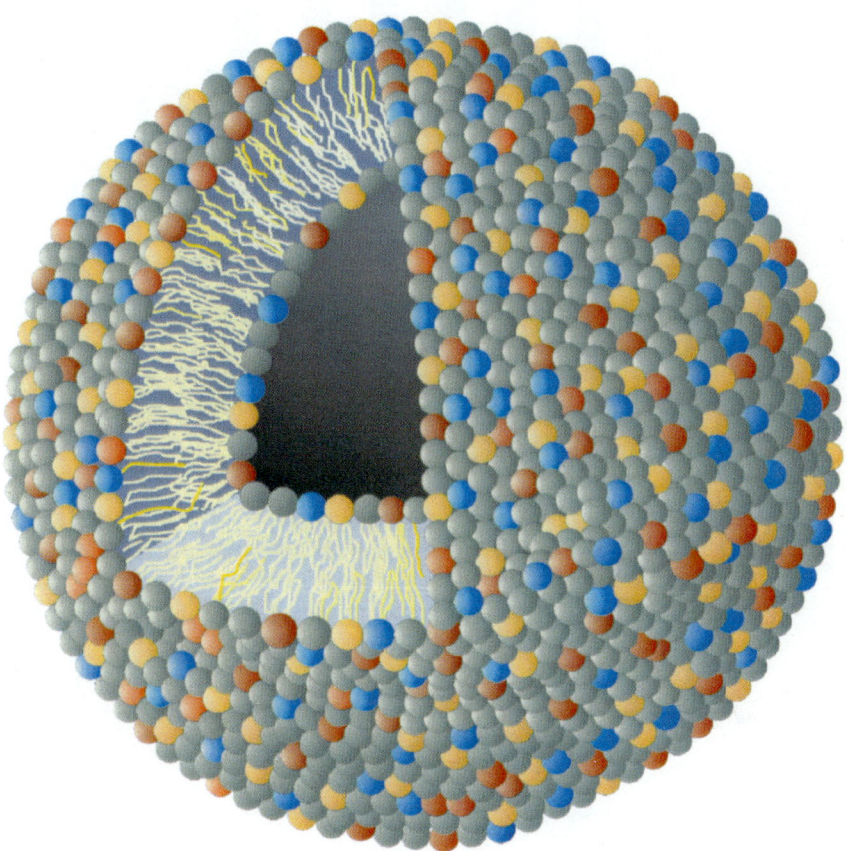

Figure 14.5. A pictorial of liposome with various lipidated components such as the antigens and adjuvants in red and yellow embedded in membrane organization, with the cut-out showing lipid bi-layer.

2 - *Color inserts*

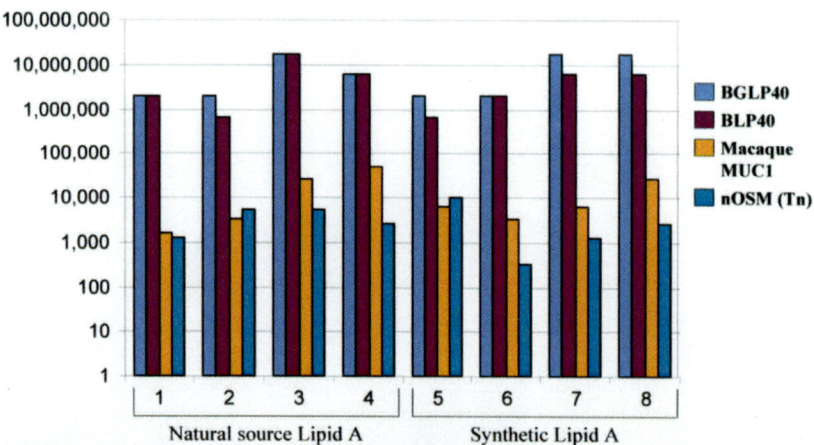

*Figure 14.6. IgG titers response in healthy Macaque monkeys. Liposome formulations of **2a** (BGLP40) and either with natural lipid-a preparation (left panel) or synthetic lipid-A analogue **20** (right panel) were used as vaccine. The colored bars represent various solid phases used to determine specificity of titers.*

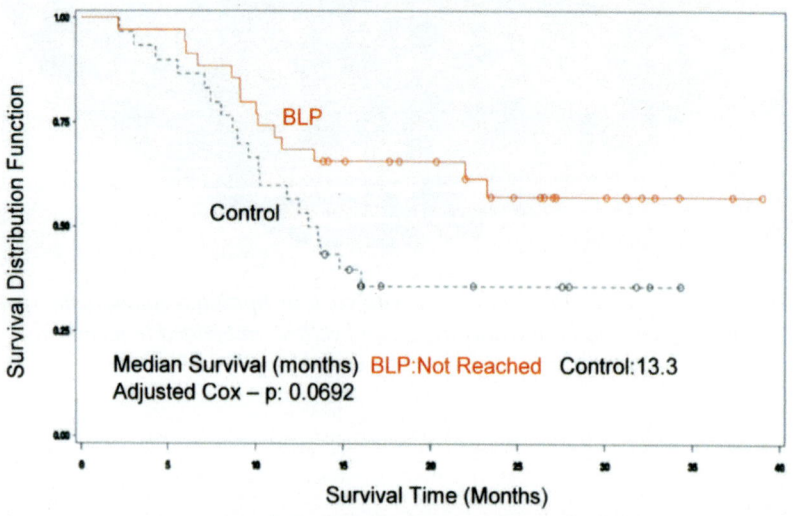

Figure 14.7 Survival graphs of patients on vaccine + best standard care arm and best standard care alone as control. Median survival is in fact reached at 30.6 months.

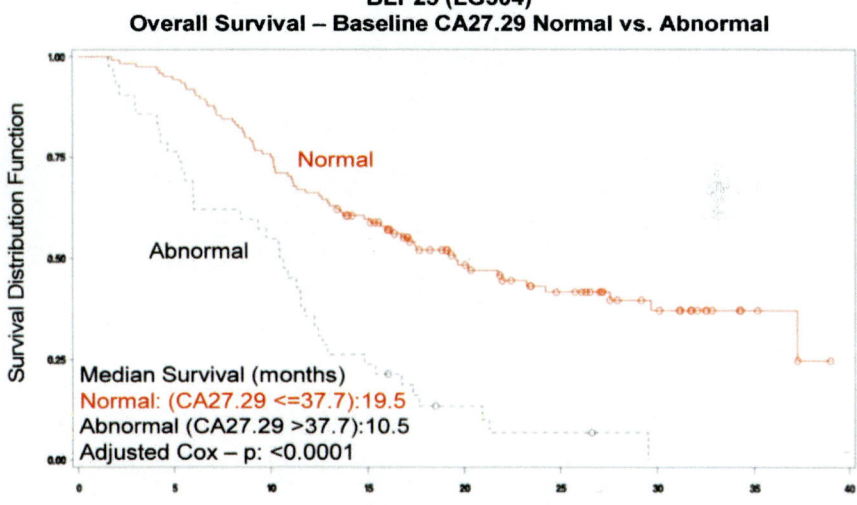

Figure 14.8. Survival of all patients by normal and abnormal expression of MUC1 mucin as estimated by anti MUC1 antibody, CA 27.29.

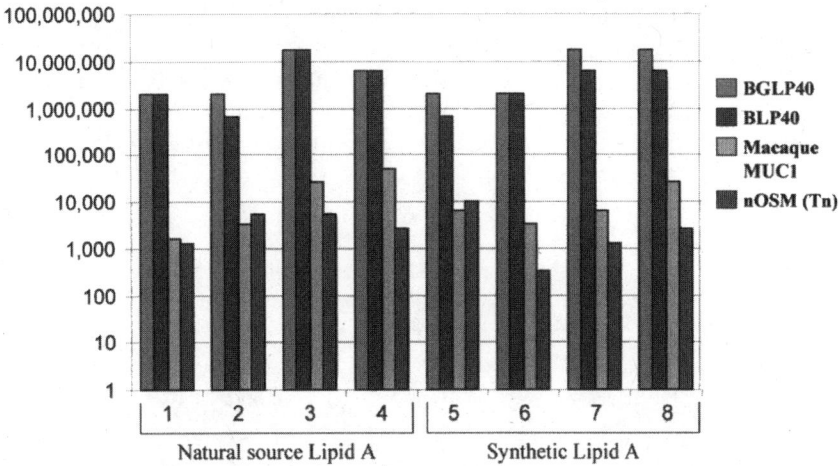

*Figure 6. IgG titers response in healthy Macaque monkeys. Liposome formulations of **2a** (BGLP40) and either with natural lipid-a preparation (left panel) or synthetic lipid-A analogue **20** (right panel) were used as vaccine. The colored bars represent various solid phases used to determine specificity of titers.*
(See Page 2 of color inserts.)

responses were also observed. Figure 6 shows the IgG titers which were characterized against various solid phases. Cross reactivity to macaque antigen was an indirect but not an ideal confirmation of overcoming tolerance to self antigen. When neuraminidase treated ovine sub-maxillary mucin (nOSM) was used as solid phase, moderate reactivity was seen indicating antibodies to N-acetylgalactosamine. Significantly, antibody titers of similar magnitude were seen to synthetic macaque antigen **33** as solid phase.

Clinical Testing

BLP25 (Stimuvax®) is one of the earliest synthetic small molecule cancer vaccines to enter clinical trials. In a phase IIB randomized trial of patients with stage IIIB and IV non small cell lung cancer (NSCLC), half the patients (n = 88) were treated with Stimuvax® in addition to the best supportive care (BSC) and patients in the control group (n = 83) received the best support care alone (Figure 7). A 4.4 month advantage in the median survival was observed in patients receiving Stimuvax® in addition to BSC compared with the patients receiving BSC alone. In a subset of patients (n = 65) with stage IIIB loco-regional disease, the vaccine group (n = 35) showed a strong trend towards more than two year survival. The median survival reached by the vaccine group was found to be 30.6 months, a remarkable figure compared to 13.3 months median survival for the control group (n = 30).

The extent of MUC1 expression on epithelial cancers is considered a prognostic indicator. Increased MUC1 expression is thought to influence growth and metastatic potential of tumor cells. In the Stimuvax® (BLP25) clinical trial,

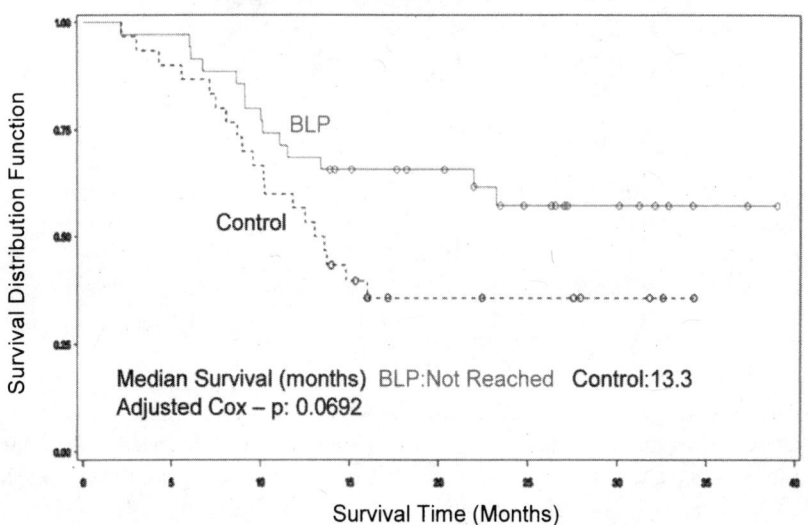

Figure 7. Survival graphs of patients on vaccine + best standard care arm and best standard care alone as control. Median survival is in fact reached at 30.6 months. (See page 2 of color inserts.)

a further analysis of survival vs. MUC1 expression of all patients in both groups has shown significant survival differences (Figure 8) between normal and abnormal levels of mucin expression (as a function of their serum levels measured with an anti MUC1 antibody CA27.29). Patients with normal expression of MUC1 lived longer with a median survival of 19.5 months against median survival of 10.5 months for patients with abnormal expression, with an adjusted cox-p: <0.0001. This supports the view that MUC1 over expression is associated with poor prognosis in cancer patients.

From the safety point of view, this vaccine has been consistent in maintaining quality of life for the patients. While the antigen BLP25 in

Stimuvax® is a synthetic molecule, it is formulated with lipid-A preparation from bacterial cultures. Vaccines such as BGLP40 formulated with a synthetic adjuvant may be termed as totally synthetic vaccine since none of the com-

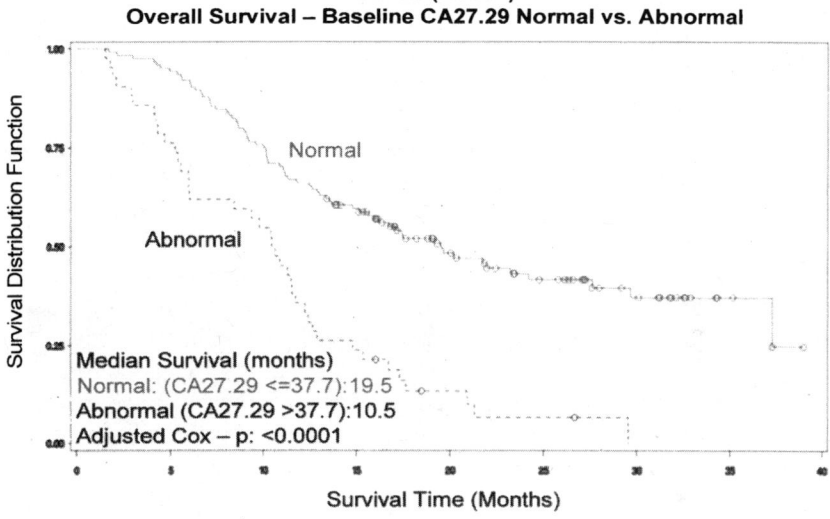

Figure 8. Survival of all patients by normal and abnormal expression of MUC1 mucin as estimated by anti MUC1 antibody, CA 27.29. (See page 3 of color inserts.)

ponents that interact with the immune system are from natural or recombinant sources. This vaccine is expected to build on the safety profile of Stimuvax® and enhance the quality of life in addition to the survival benefits to cancer patients.

The results from phase IIB clinical trial demonstrate significant advancements in vaccine design and clinical benefits, particularly to the lung cancer patients. Glycopeptide based cancer vaccines are expected to enhance the benefits seen with peptide based vaccine such as Stimuvax®. Tumor specific carbohydrates that are added to the peptide core, complete with their natural linkages, can simulate the natural environment of the cancer associated glycoproteins such as the MUC1 mucin. We hope that immunizing cancer patients with a multitude of precise cancer associated structures (epitopes) and the resultant array of cellular and humoral responses may attack cancers more effectively.

Summary

Vaccines are an important aspect of public health, perhaps comparable to law and order in a civilized society, which we take for granted. The threat of bioterrorism increased the need for preparedness with quick remedies in the form of protective or rapid response therapeutic vaccines to deal with the devastating after effects. Precision is essence of safety (*46*) in the realm of immunotherapeutics. Synthetic adjuvants and antigens, coupled with proper design for vaccine formulation, not only eliminate undesired immune responses and the need for carrier proteins but also play a significant role in consistency, safety and stability. The clinical benefits of well defined molecular vaccines may be harder to assess in the absence of comparison to an appropriate equivalent. However, judging from the clinical results of Stimuvax® vaccine and the early animal experiments with synthetic antigen **2a** (BGLP40), we can conclude that these well defined vaccines are expected to enhance safety profile and restrict immune responses to specific targets. Obviously, more work is necessary before BGLP40 can be validated as a vaccine for human testing. Full characterization, of the various antibody responses through inhibition ELISAs, antibody dependent cell killing (ADCC) and serum binding to tumor cell lines expressing MUC1 mucin, as well as comparison of immune responses in transgenic and wild type mouse models, is under progress.

Routine validation experiments apart, of greater importance are target specificity and restriction of immune responses to disease associated structures. Chemical synthesis alone contributes to the structural definition, not to speak of superior control of production and performance of such vaccines.

Acknowledgements

The authors wish to thank Drs. Peter Emtage, Guy Ely, Marita Hobman and Ms. Marilyn Olsen for their valuable suggestions and help with immunology and clinical sections of this manuscript and Heather Jenkins and Sarah Lehman for their help with figures 5-8.

References

1. Mercer A. J. 'Smallpox and epidemiological-demographic change in Europe: The role of vaccination', *Population Studies* **1985**, *39*, 287-307. (A note: *Inhalation of dry powdered cow pox, as a protection against small pox, had been practiced for centuries earlier in ancient India and China.*)

2. Janeway, C.A. *Annu. Rev. Immunol.* **2002**, *20*, 1-28.
3. Carroll, M.C.; Janeway, C.A. *Curr. Opin. Immunol.* **1999**, *11*, 11-12.
4. Medzhitov, R.; Janeway, C.A. Jr. *Cold Springs Harb. Quant. Biol.* **1999**, *64*, 429-435.
5. Immler, J.L.; Bulet, P. *Chem. Immunol. Allergy (Review)* **2005**, *86*, 1-21.
6. Otvos, L. Jr. *J. Pept. Sci.* **2000**, *10*, 497-511.
7. Otvos, L. Jr. *Cell Mol. Life Sci.* **2002**, *59*, 1138-1150.
8. Krutzik, S.R.; Sieling, P.A.; Modlin, R.A. *Cur. Opinion in Immunol.* **2001**, *13*, 104-108.
9. Akira, S.; Takeda, K.; Kaisho, T. *Nature Immunology* **2001**, *2*, 675-680.
10. Koganty, R.R. *Expert Review of Vaccines (Editorial)* **2002**, *1*, 123-124.
11. Wagner, H. *Immunity* **2001**, *14*, 499-502.
12. Jiang, Z-H.; Koganty, R.R. *Curr. Med. Chem.* **2003**, *10*, 1423-1439.
13. Kawano, T.; Cui, J.; Koezuka, Y.; Toura, I.; Kaneko, Y.; Motoki, K.; Ueno, H.; Nakagawa, R.; Sato, H.; Kondo, E.; Koseki, H.; Taniguchi, M. *Science* **1997**, *278*, 1626-1629.
14. Janeway, C.A. Jr.; Goodnow, H.G.; Medzhitov, R. *Current Biology* **1996**, *6*, 519-522.
15. Banchereau, P.; Briere, F.; Caux, C.; Davoust, J.; Lebecque, S.; Liu, Y.J.; Pulendran, B.; Palucka, K. *Annu. Rev. Immunol.* **2000**, *18*, 767-811.
16. Taylor-Papadimitriou, J.; Burchell, J.; Miles, D.W.; Dalziel, M. *Biochim. Biophys. Acta.* **1999**, *1455*, 301-313.
17. Miles, D.W.; Taylor-Papadimitriou, J. *Pharamacol. Ther.* **1999**, *82*, 97-106.
18. Apostolopoulos, V.; Sandrin, M.S.; McKenzie, I.F. *J. Mol. Med.* **1999**, *77*, 427-436.
19. Apostolopoulos, V.; Pietersz, G.A.; McKenzie, I.F. *Curr. Opin. Mol. Ther.* **1999**, *1*, 98-103.
20. Butts, C. et.al *J. Clin. Oncol.* **2005**, *23*, 6674-6681.
21. Springer, G.F. *Science* **1984**, *224*, 1198-1206.
22. Springer, G.F. *Crit. Rev. Oncog.* **1995**, *6*, 57-85.
23. Springer, G.F. *J. Mol. Med.* **1997**, *75*, 594-602.
24. Itzkowitz, S.H.; Bloom, E.J.; Kokal, W.A.; Modin, G.; Hakomori, S.; Kim, Y.S. *Cancer* **1990**, *66*, 1960-1966.
25. Kobayashi, H.; Toshihiko, T.; Kawashima, T. *Clin. Oncol.* **1992**, *10*, 95-101.
26. Pardoll, D.M. *Nature Reviews. Immunology* **2002**, *2*, 227-238.
27. Koganty, R.R.; Reddish, M.A.; Longenecker, B.M. '*Glycopeptides in the immunotherapy of cancers*' in *Glycopeptides and related compounds*, (Large, D.G., Warren, C.D., Eds.) Marcel Dekker, Inc. New York, **1997**, 707-743.
28. Liu, X.; Sejbal, J.; Kotovych, G.; Koganty, R.R.; Reddish, M.A.; Jackson, L.; Gandhi, S.S.; Mendonca, A.J.; Longenecker, B.M. *Glycoconjugate J.* **1995**, *12*, 607-617.

29. Grinstead, J.S.; Koganty, R.R.; Krantz, M.J.; Longenecker, B.M.; Campbell, A.P. *Biochemistry* **2002**, *41*, 9946-9961.
30. Raghupathi, G.; Howard, L.; Capello, S.; Koganty, R.R.; Qiu, D.; Reddish, M.A.; Longenecker, B.M.; Lloyd, K.O.; Livingston, P.O. *Cancer Immunol. Immunother.* **1999**, *48*, 1-8.
31. Yule, J.E.; Wong, T.C.; Gandhi, S.H.; Qiu, D.; Riopel, M.A.; Koganty, R.R. *Tetrahedron Lett.* **1995**, *36*, 6839-6842.
32. Qiu, D.; Gandhi, S.S.; Koganty, R.R. *Tetrahedron Lett.* **1996**, *37*, 595-598.
33. Ulrich, J.T.; Myers, K.R. *Pharm. Biotechnol.* **1995**, *6*, 495-524.
34. Persing, D.H.; Coler, R.N.; Lacy, M.J.; Johnson, D.A.; Baldridge, J.R.; Hershberg, R. M.; Reed, S.G. *Trends in Microbiol.* **2002**, *10*, 32-37.
35. Johnson, D.A.; Keegan, D.S.; Sowell, C.G.; Livesay, M.T.; Johnson, C.L.; Taubner, L.M.; Harris, A.; Myers, K.R.; Thompson, J.D.; Gustofson, G.L.; Rhodes, M.J.; Ulrich, J.T.; Ward, J.R.; Yorgensen, Y.M.; Cantrell, J.L.; Brookshire, V.G. *J. Med. Chem.* **1999**, *42*, 4640-4649.
36. Johnson, D.A.; Sowell, C.G.; Johnson, C.L.; Livesay, M.T.; Keegan, D.S.; Rhodes, M.J.; Ulrich, J.T.; Ward, J.R.; Cantrell, J.L.; Brookshire, V.G. *Bioorg. Med. Chem. Lett.* **1999**, *9*, 2273-2278.
37. Jiang, Z-H.; Budzynski, W.A.; Skeels, L.N.; Krantz, M.J.; Koganty, R.R. *Tetrahedron* **2002**, *58*, 8833-8842.
38. Jiang, Z-H.; Bach, M.V.; Budzynski, W.A.; Krantz, M.J.; Koganty, R.R.; Longenecker, B.M. *Bioorg. Med. Chem. Lett.* 2002, *12*, 2193-2196.
39. Zuany-Amorim, C.; Hastewell, J.; Walker, C. *Nature Reviews. Drug Discovery* **2002**, *1*, 797-807.
40. Corey, D.B. *Nature Reviews. Drug Discovery* **2002**, *1*, 55-64.
41. Peng, G.; Guo, Z.; Kiniwa, Y.; Voo, K.; Peng, W.; Fu, T.; Wang, D.Y.; Li, Y.; Wang, H.Y.; Wang, R-F. *Science* **2005**, *309*, 1380-1384.
42. Guan, H.; Budzynski, W.; Koganty, R.R.; Krantz, M.J.; Reddish, M.A.; Rogers, J.A.; Longenecker, B.M.; Samuel, J. *Bioconjugate Chem.* **1998**, *9*, 451-458.
43. Schnare, M.; Barton, G.M.; Holt, A.C.; Takeda, K.; Akira, S.; Medzhitov, R. *Nature Immunology* **2001**, *10*, 947-950.
44. Koganty, R.R.; Qiu, D.; Budzynski, W.; Gandhi, S.S.; Yule, J.E.; Reddish, M.A.; Krantz, M.J.; Longenecker, B.M. *Self-assembling molecules as cancer vaccines. In: Biomedical functions and biotechnology of natural and artificial polymers (Yalpani M., Ed)*. Science Publishers, Frontiers in Biomedicine and Biotechnology, ATL Inc. **1996**, *3*, 105-114.
45. Berzofsky, J.A.; Ahlers, J.D.; Belyakov, I.M. *Nature Review. Immunology* **2001**, *1*, 209-219.
46. a. Wilson, C.B.; Marcuse, E.M. *Nature Reviews. Immunology* **2001**, *1*, 160-165.; b. Koganty, R.R. *Expert Review of Vaccines(Editorial)* **2003**, *2*, 725-727.

Chapter 15

Peptide Mimics of Bacterial Polysaccharides: Potential for Discriminating Vaccines

Silvia Borrelli, Margaret A. Johnson, Rehana B. Hossany, and B. Mario Pinto[*]

Department of Chemistry, Simon Fraser University, Burnaby, British Columbia V5A 1S6, Canada

An alternative approach to vaccine design is the use of molecules that mimic the immunogenic element of interest. Carbohydrate-mimetic peptides, as surrogate ligands for traditional carbohydrate vaccines, have the potential to generate more discriminating immune responses. In this context, peptide-mimics of Group A and B *Streptococcus,* and *Shigella flexneri Y* polysaccharides have been identified, their interaction with anti-carbohydrate antibodies studied by NMR techniques, molecular modeling, or X-ray crystallography, and compared to the corresponding interactions with polysaccharide or component oligosaccharides. The immunological responses of the carrier-protein conjugates of peptide mimics corresponding to Group A and B *Streptococcus* polysaccharides have been investigated in mice, and yielded cross reactive responses against the polysaccharides. A working hypothesis that peptide turn-conformations present important epitopes is advanced.

© 2008 American Chemical Society

Introduction

Bacterial infections are still a major global health problem *(1, 2)*. Antimicrobial therapeutics, although very successful, are not the ideal solution to handle many of the bacterial infections today. Drug resistance is rather common, and the development of new, effective drugs is still expensive *(2, 3)*. In addition, the impact on society from the cost of drug administration and the "down-time" due to sickness is a significant issue. Thus, the search for new, effective vaccines to prevent these types of infections is still an enviable goal, especially in developing countries *(1- 3)* .

The host responds to a bacterial infection by triggering a specific immune response, both B-cell or humoral, with the production of specific antibodies, and T cell or cellular, with a clear role of T-cells in immunological memory functions. The identification of antigens capable of inducing protective and long lasting immunity is the major goal for effective vaccine development *(4)*.

The surface of many bacterial species is covered by several structurally variable and sometimes cryptic polysaccharides that can be in the form of capsules, glycoproteins or glycolipids, among other molecules, that can present antigens to the host immune system. Polysaccharides are classified as thymus independent (TI) antigens because they do not required mature T-cells to elicit a humoral response in vivo, although they do require a late developing subset of B-cells. TI antigens generally fail to elicit a memory response, (characterized by high affinity antibodies), and usually do not show affinity maturation. The response to polysaccharide TI antigens is different from that to most protein antigens *(5)*. The conversion of a TI antigen to a thymus dependent (TD) form by chemically coupling the antigen to a protein carrier alters the response to antigen because of the TD help conferred by the carrier protein. This strategy has been very successful with bacterial capsular polysaccharides, and glycoconjugate vaccines against *Neisseria meningitidis* groups A, C, W-135, and Y, *Haemophilus influenzae* type b, *Streptococcus pneumoniae* (23 serotypes), and *Salmonella typhi* are commercially available *(2, 3, 6)*.

Proteins and peptides are TD antigens that require immune stimulation from helper T-cells to elicit an immune response. Such antigens are presented to T-cells in the context of Major Histocompatibility Complex (MHC) molecules on the macrophages, B-cells, or dendritic cells (antigen presenting cells) following bacterial infection *(5, 7)*. The following activation of T-cells induces cytokine production and a display of immunologic effects. TD antigens are effective at inducing a lasting immune response, forming memory B and T-cells and producing high affinity antibodies of multiple isotypes.

An alternative approach to vaccine design is the use of molecules that mimic the immunogenic element of interest *(4, 8)*. Carbohydrate-mimetic peptides have potential as surrogate ligands for traditional carbohydrates vaccines providing more discriminating immune responses *(4, 8)*. Peptides, which are intrinsically

TD antigens, can be processed by antigen-presenting cells and presented to T-cells by MHC molecules. Peptides are simpler molecules than carbohydrates, easier and cheaper to produce, and they also have the potential to focus the immune response on protective epitopes *(9)*. However, peptides have poor chemical stability under physiological conditions, and subsequent lower antigenicity. Small peptides are weak immunogens requiring conjugation to a carrier protein and should be co-administered with an adjuvant. The use of anti-idiotypic antibodies and the screening of phage-displayed peptide libraries with carbohydrate-specific antibodies have identified carbohydrate-mimetic peptides with demonstrated antigenic and immunogenic potential for inducing cross-reactive immune responses in certain cases (reviewed in *(8)*).

We have focused for several years on two bacterial carbohydrate peptide-mimics: DRPVPY (compound 2, Figure 1), identified as a mimic of the *Streptococcus* Group A cell-wall polysaccharide (compound 1, Figure 1) and MDWNMHAA (compound 4, Figure 2), which mimics the *Shigella flexneri* Y *O*-antigen (compound 3, Figure 2). We have also investigated the interaction of a third peptide mimic, FDTGAFDPDWPA (compound 6, Figure 3), of the Group B streptococcal type III (GBS) capsular polysaccharide (compound 5, Figure 3) with a protective anti-GBS monoclonal antibody. Our progress and insights on the development of experimental vaccines based on peptide mimics are reviewed herein.

Group A *Streptococcus*

Streptococcus pyogenes (Group A *Streptococcus*, GAS) is a Gram-positive extracellular bacterium, which colonizes the throat or skin and is responsible for a number of suppurative infections and their non-suppurative, immunologically-mediated sequelae including acute rheumatic fever, acute glomerolunephritis and reactive arthritis *(10)*. GAS has been recognized as a common cause of bacterial pharyngitis (strep throat), skin abscesses, impetigo, and scarlet fever. More recently, GAS has been recognized as the causative agent of toxic shock-like syndrome, and necrotizing fasciitis which invades skin and soft tissues and in severe cases leaves infected tissues destroyed *(10-12)*. Acute rheumatic fever and rheumatic heart disease are the most serious autoimmune sequelae of GAS infection affecting children worldwide. After the advent of antibiotics, diseases caused by *S. pyogenes* declined in incidence and severity in developed countries *(13)*. In the last 20 years, these diseases have re-emerged in industrial countries and a dramatic increase of their frequency is observed in developing countries *(14, 15)*. Mass vaccination would therefore be of interest for efficient prevention of these infections *(10, 16)*.

The Lancefield classification scheme of serologic typing distinguished the beta-hemolytic streptococci based on their Group A cell-wall polysaccharide

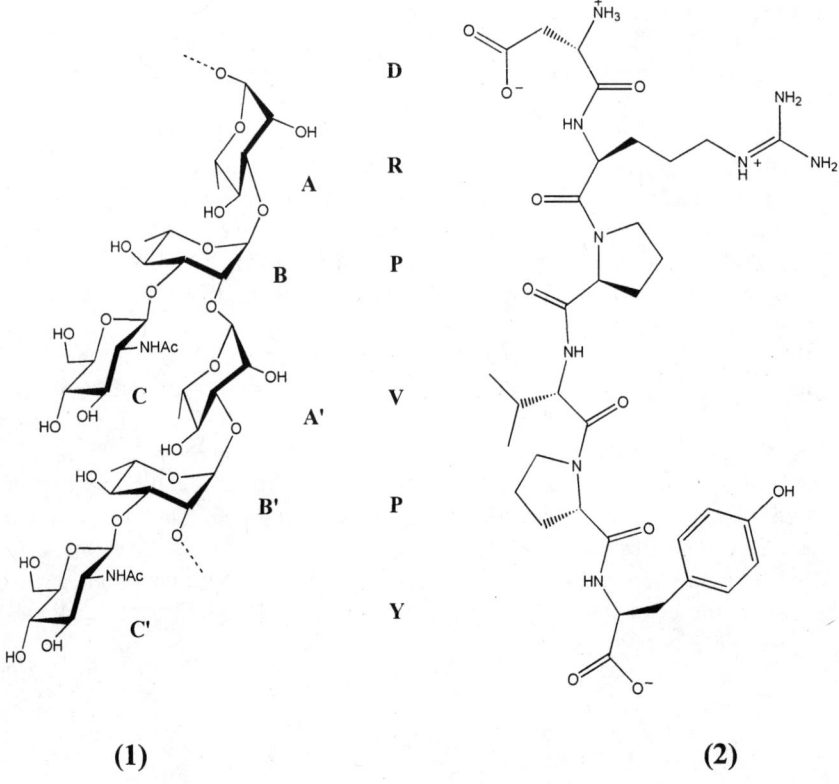

Figure 1. Structures of the cell-wall polysaccharide (CWPS) of Group A Streptococcus (compound 1, with monosaccharide residues of the branched trisaccharide repeating unit labeled A, B, and C); and of a peptide mimic of CWPS, the hexapeptide DRPVPY (compound 2, single letter amino acid code).

(CWPS), composed of *N*-acetylglucosamine linked to a rhamnose polymer backbone (compound 1, Fig. 1). Vaccines based on the CWPS and its well-defined oligosaccharide-protein conjugates are under investigation *(17)*, among other antigens *(10)*. Several years ago, we proposed an alternative approach, based on carbohydrate-mimetic peptides *(18)*.

Identification of a GAS-CWPS-mimetic peptide

In the present context, a GAS carbohydrate-mimetic peptide, DRPVPY (compound 2, Figure 1), cross-reactive with an anti-CWPS monoclonal antibody, SA-3, was identified by screening a phage-displayed peptide library

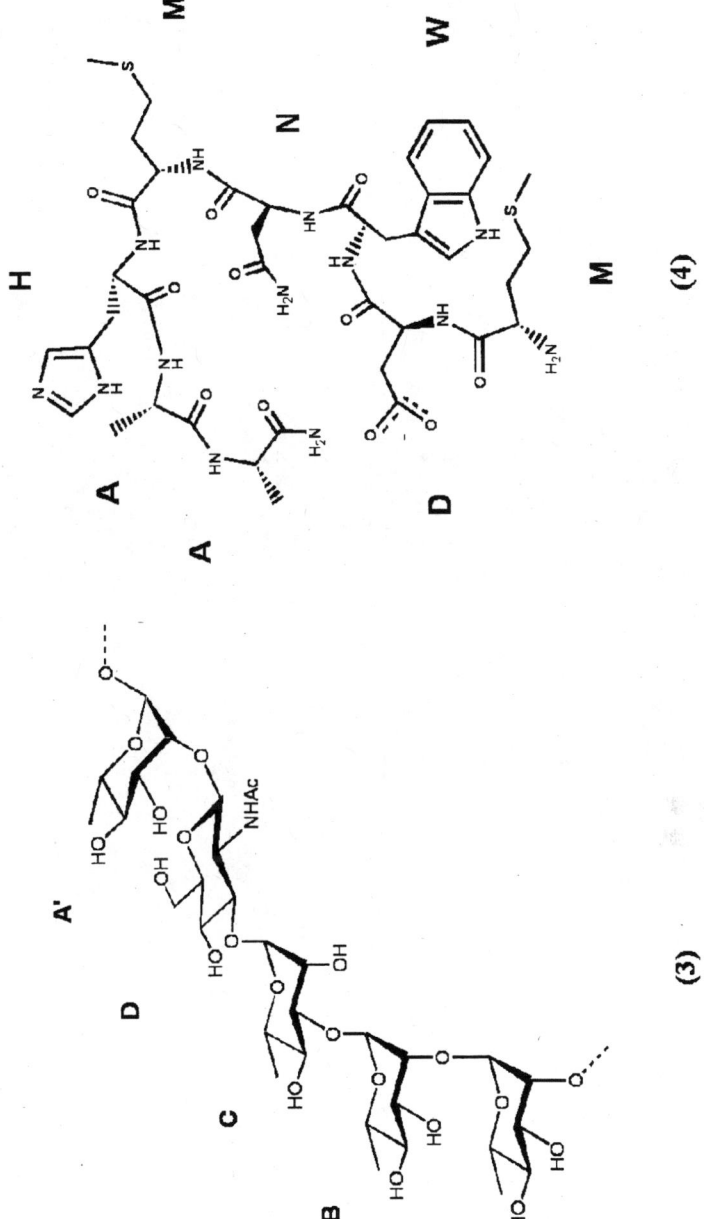

Figure 2. Structures of the Opysaccharide of Shigella flexneri Y (compound 3, with monosaccharide residues of the repeating unit labeled A, B, C, and D); and of a peptide mimic, the octapeptide MDWNMHAA (compound 4, (single letter amino acid code).

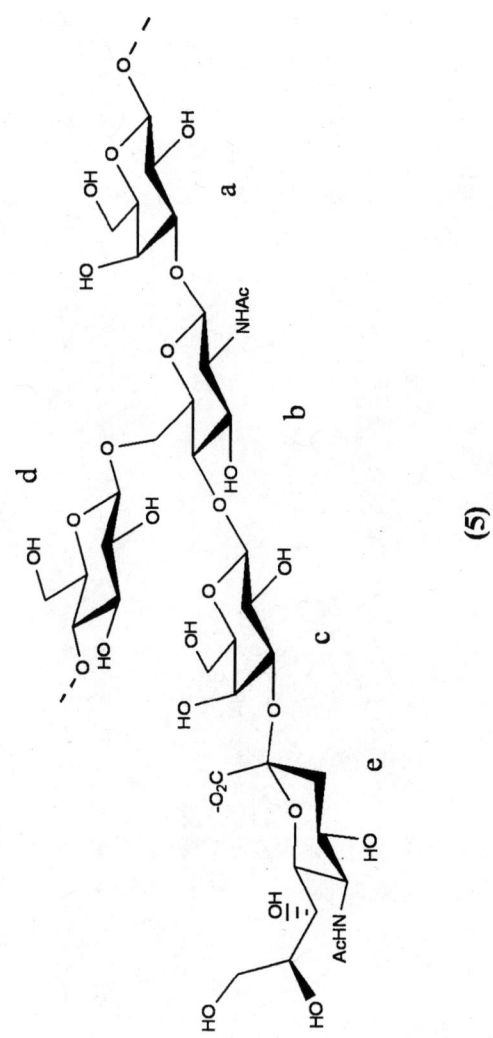

(5)

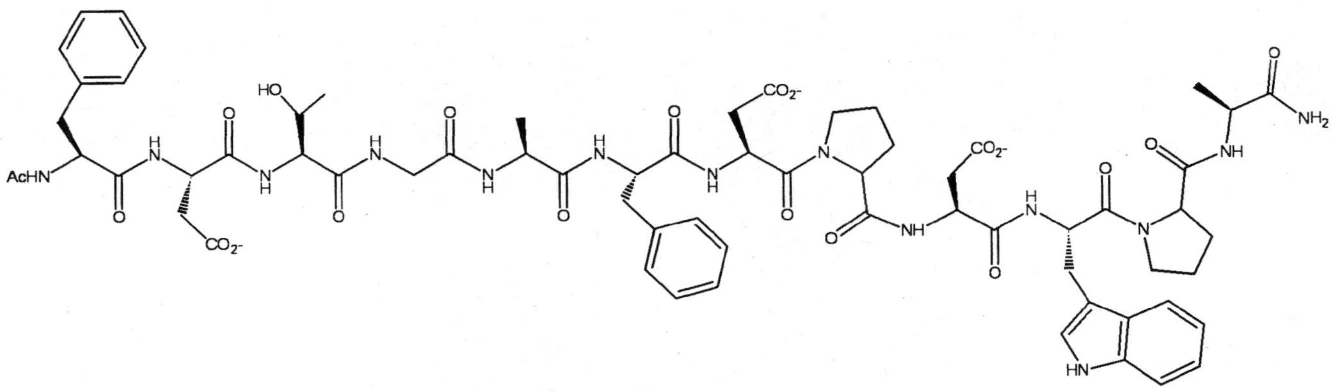

(6)

Figure 3. Structures of the Group B Streptococcus type III capsular polysaccharide (GBS) repeating unit (compound 5, with monosaccharide residues of the branched trisaccharide repeating unit labeled a, b, c, d, and e); and of a peptide mimic of GBS, the dodecapeptide FDTGAFDPDWPA (compound 6, single letter amino acid code).

(18). SA-3 recognizes a branched trisaccharide repeating unit L-Rha-α-(1→2)-[D-GlcNAc-β-(1→3)]-α-L-Rha (compound 1, Figure 1) as an epitope *(19)*. Several closely related antibodies that recognized a similar branched trisaccharide epitope of GAS selected very different peptides from phage-displayed libraries. Thus, the peptide that bound to each antibody did not bind to the others. This lack of cross-reactivity was observed with both monoclonal and polyclonal antibodies. Detailed immunochemical studies of the recognition of DRPVPY by SA-3 showed that the mechanism of peptide binding differs from that of the carbohydrate *(18)*. These observations imply that the observed molecular mimicry is not based on *structural* mimicry of the carbohydrate by the peptide; what is important is the complementary fit of a ligand to the antibody surface, irrespective of whether the two ligands engage the antibody in the same way *(18)*. This feature also suggests that peptides will act as more discriminating ligands for an antibody receptor, and will lead, in turn, to more discriminating immune responses when incorporated in vaccines.

Bound conformation and epitope mapping of a peptide mimic of the GAS-CWPS

NMR spectroscopic techniques in combination with molecular modeling disclosed details of the conformation adopted by the peptide when bound to a murine IgM monoclonal antibody SA-3. Transferred nuclear Overhauser effect (trNOE) experiments were performed at 800 MHz to investigate the bound conformation of DRPVPY. This hexapeptide adopts a tight turn conformation in the bound state, with close contacts between the side chains of valine and tyrosine, suggesting a likely peptide epitope *(20, 21)* (Figure 4). Interestingly, the pentapeptide, DRPVP did not bind to the SA-3 mAb. *(18)* More information about the mode of peptide binding was provided by saturation transfer difference NMR (STD-NMR) experiments. Enhancements of nearly all residues of the peptide were observed with the exception of Asp-1, suggesting that this would be an excellent position at which to conjugate the peptide to proteins for the preparation of conjugate vaccines *(21)*.

Synthesis and immunochemical characterization of protein conjugates of a GAS-CWPS mimetic peptide as experimental vaccines.

The synthesis of DRPVPY-based conjugates, to bovine serum albumin, BSA (compound 7, Figure 5) and to tetanus toxoid, TT, (compound 8, Figure 5) together with their immunochemical evaluation with SA-3 was reported recently *(22)*. The CWPS-peptide mimetic DRPVPY was synthesized using the Fmoc solid-phase strategy and linked via the amino terminus to a bifunctional linker,

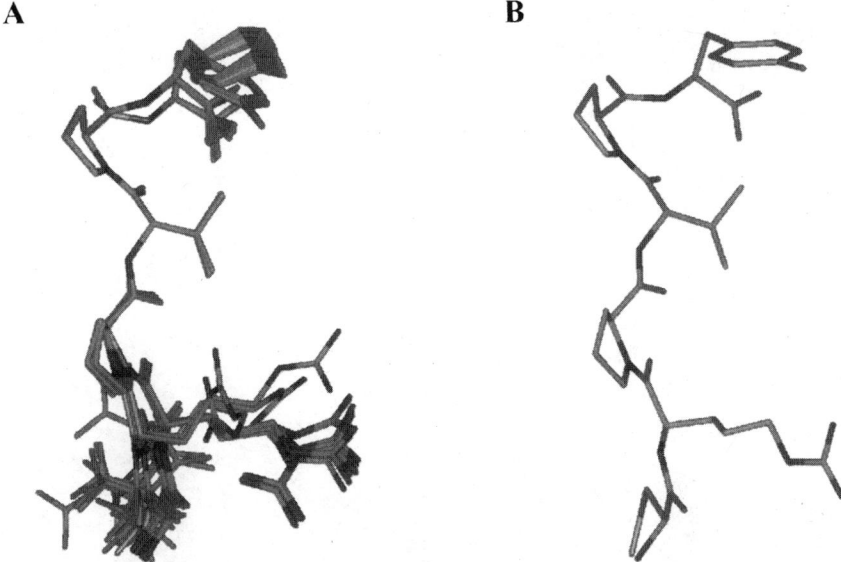

Figure 4. Bound conformation of the hexapeptide DRPVPY (compound 2) in complex with the monoclonal anti-carbohydrate SA-3 Ab. (A) A family of 27 calculated structures, with backbone atoms of residues 4-6 superimposed. (B) Average of 27 structures shown in panel (A). Residues 4-6 adopt a well defined tight turn. Reproduced from reference 20. Copyright 2002.

diethylsquarate, and then conjugated to TT or BSA as immunogenic carriers, as described (22) (compounds 7 and 8, Figure 5). Confirmation of strong binding to SA-3 was also demonstrated.

Immunogenicity of the GAS-CWPS mimetic peptide conjugate

The immunogenicity of DRPVPY-TT conjugate was recently examined by immunizing BALB/c mice, with alum as adjuvant (23). Antibody titers were determined by ELISA with GAS bacteria and DRPVPY-BSA as solid phase antigens. All mice primed with DRPVPY-TT responded with high IgG anti-peptide and anti-GAS titers. The specificity of the binding of polyclonal anti-peptide antibodies to GAS was assessed by competitive-inhibition ELISA, using CWPS, peptides, and synthetic oligosaccharide corresponding to CWPS as inhibitors. These results were further confirmed upon titration of anti-peptide sera with synthetic oligosaccharide-BSA conjugates. In addition, all mice primed with CWPS-TT and mice primed and boosted with GAS developed high IgG

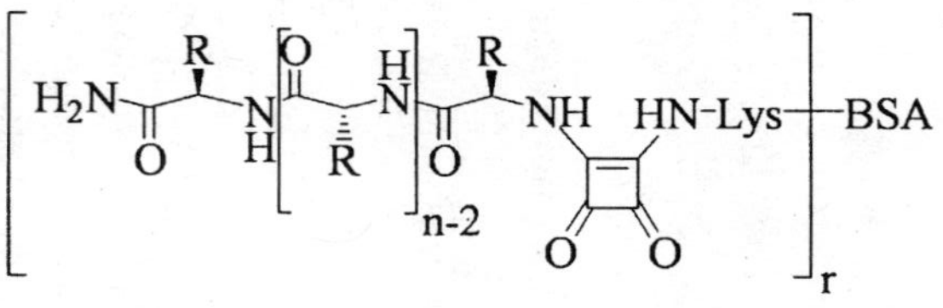

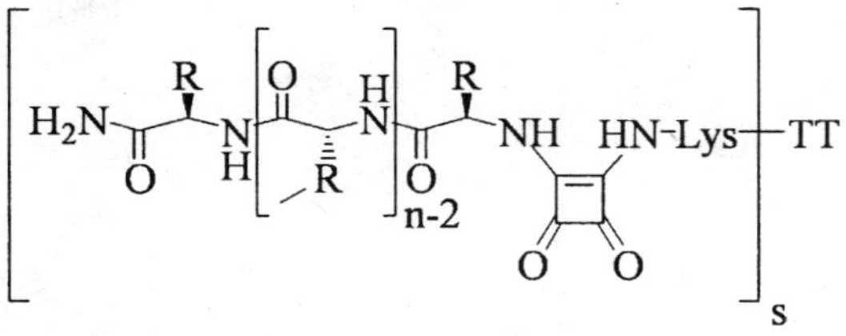

n	peptide hapten	r	
6	DRPVPY	28	**(7)**
8	MDWNMHAA	28	**(9)**

n	peptide hapten	s	
6	DRPVPY	39	**(8)**
8	MDWNMHAA	45	**(10)**

Figure 5. Synthetic peptide conjugates: DRPVPY-BSA (compound 7), DRPVPY-TT (compound 8), MDWNMHAA-BSA (compound 9), and MDWNMHAA-TT (compound 10).

anti-peptide titers. These data demonstrated conclusively the cross-reactivity of the immune responses and supported the hypothesis of antigenic mimicry of the GAS-CWPS by the hexapeptide DRPVPY.

However, mice boosted with DRPVPY-TT, showed a decrease in IgG anti-GAS titers, but an increase in IgG anti-peptide titers, suggesting carrier-induced suppression of the response to polysaccharide. Thus, modification of the carrier in the next-generation DRPVPY-conjugate may be necessary to develop an effective vaccine against Group A *Streptococcus (23)*.

Group B *Streptococcus*

Group B *Streptococcus* (GBS) is a Gram-positive bacterium that causes life-threatening infections in newborn infants, such as sepsis and meningitis, and is a common cause of pneumonia. More often, pregnant women who carry the bacteria can unknowingly transmit GBS to their newborns at birth *(24)*.

GBS can also cause serious diseases in pregnant women, the elderly, and adults with other illnesses, weakened immune systems and other health problems, such as diabetes or cancer. GBS infection is fatal in about 20% of infected men and non-pregnant women, and in about 5% to 15% of infected newborns. Babies who survive can be left with speech, hearing, and vision problems as well as mental retardation *(24)*.

Infections in both newborns and adults are usually treated with antibiotics given intravenously *(24)*. Efforts to prevent this disease include immunization to induce protective antibodies *(25)*. Initial vaccines containing only GBS capsular polysaccharides were not effective *(26)*. The GBS type III capsular polysaccharide contains terminal sialic acid residues, which are not immunogenic because of their presence in human tissues. Capsular polysaccharide conjugation to protein carriers, produced a much more effective vaccine *(6, 24)*.

An alternative approach, based on a carbohydrate-mimetic peptide has been demonstrated *(27)*. Thus, a protective monoclonal antibody (S9), directed against GBSP type III (compound 5, Figure 3), was developed *(27)*. Peptides acting as specific ligands for antibody S9 were isolated by screening a phage-displayed peptide library. A high-affinity peptide ligand FDTGAFDPDWPAC (compound 6, Figure 3), competitively inhibited the binding of GBS not only to monoclonal antibody S9 but also to polyclonal anti-GBS antibodies. Immunization of mice with peptide-protein conjugates produced an anti-polysaccharide immune response *(27)*.

Bound conformations and epitope mapping of a peptide mimic of the capsular polysaccharide of type III Group B *Streptococcus* (GBS)

Transferred nuclear Overhauser effect (trNOE) NMR spectroscopy was used to investigate the binding of the dodecapeptide, FDTGAFDPDWPA (compound 6, Figure 3) to the anti-carbohydrate antibody S9 *(28)*. The peptide was found to adopt a type I β-turn conformation spanning residues Asp-7-Trp-10; in addition, STD-NMR measurements revealed the involvement of Trp-10 and several other residues in this region in contact with the antibody (Figure 6A). Importantly, this type I β-turn was also significantly populated in the absence of antibody, indicating a strong intrinsic propensity to adopt the bioactive conformation (Figure 6B). This strong preference likely contributes to the peptide's immunological effectiveness *(27, 28)*.

Shigella flexneri Y

Diarrhoeal diseases are an important global problem, especially in the developing world. One prominent diarrhoeal disease, and the most infectious, is shigellosis caused by the *Shigella* species. Shigellosis is a public health concern in developing countries, particularly for young children who make up 69% of all cases *(29)*. Bacteria are transmitted via the fecal oral route and require as little as 100 organisms to cause disease *(30)*. *Shigella flexneri*, the species of *Shigella* responsible for the highest mortality rate, is endemic in most developing countries. The 13 serotypes of *S. flexneri*, with the exception of serotype 6, result from structural modifications of the O-antigen polysaccharide, part of the lipopolysaccharide (LPS). The basic O-antigen polysaccharide repeat is referred to as serotype Y. A murine IgG$_3$ monoclonal antibody, SYA/J6, specific for the polysaccharide O-antigen of the *Shigella flexneri* Y LPS was developed *(31)*. The O-antigen polysaccharide Y is a linear heteropolymer [→2)-α-L-Rha-(1→2)-α-L-Rha-(1→3)-α-L-Rha-(1→3)-β-D-GlcNAc-(1→2)-α-L-Rha-(1→] (compound 3, Figure 2) and is recognized by the antibody SYA/J6 *(32)*.

Identification of a carbohydrate-mimetic peptide of the *Shigella flexneri* Y O-antigen.

A carbohydrate-mimetic peptide of the *Shigella flexneri* Y O-antigen polysaccharide, MDWNMHAA, (compounds 3 and 4, Figure 2), cross-reactive with the anti-*S. flexneri* Y O-polysaccharide monoclonal antibody, SYA/J6, was identified by phage library screening *(18)*. Comparison of the thermodynamics of binding of the antibody to the octapeptide to that of its binding to the native pentasaccharide, revealed that the enthalpy of binding to the octapeptide was favorable, but it was offset by an unfavorable entropy of binding *(33)*.

Structural basis of peptide-carbohydrate mimicry in the antibody-combining site.

The structures of complexes of the antibody SYA/J6 Fab fragment with synthetic deoxytrisaccharide and pentasaccharide ligands, related to the *S. flexneri* Y O-antigen, and with a carbohydrate-mimetic peptide have been determined by X-ray crystallography *(33-35)*. This is the only example of a carbohydrate and its peptide mimic bound in the same antibody combining site and provides a unique opportunity for comment on the nature of mimicry. The structure of the Fab complex with MDWNMHAA (compound 4, Figure 2) revealed interesting differences, and few similarities, with respect to the oligosaccharide complexes *(33)*. Octapeptide binding complemented the shape of the combining site groove much better than pentasaccharide binding (Figure 7A and 7B). Moreover, the peptide made a much greater number of contacts (126), which were mostly van der Waals interactions, with the Fab than the saccharide (74). Twelve water molecules mediated hydrogen bonds between residues within the peptide or the peptide and Fab. Of note, one of the rhamnose residues (C) in the pentasaccharide (ABCDA; A, B, C = Rha; D = GlcNAc) penetrated into a deep hole at the bottom of the combining site; this hole was not occupied by the peptide; rather three immobilized water molecules provided the complementarity to the Fab through a hydrogen-bonding network. These results *(33)*, provide the first concrete evidence that the modes of binding of the pentasaccharide and octapeptide differ considerably, and that few aspects of structural mimicry exist. This insight suggests that the design of peptide mimics with improved affinity for use as vaccines will derive from a detailed knowledge of the complementarity of the ligand within the combining site, and not necessarily from mimicking the ligand itself. In addition, we suggest this strategy as a general paradigm for discovering new ligands for protein receptors. We have described our initial attempts using molecular modeling to design surrogate peptide ligands that bind to SYA/J6 *(36)*.

Epitope mapping of the carbohydrate-mimetic peptide of the *Shigella flexneri* Y O-polysaccharide.

The comparison of epitope mapping data obtained by STD-NMR spectroscopy *(37)* and those obtained by X-ray crystallography *(33)* is especially important both in validating the STD technique and also identifying differences that may exist between complexes in the solution phase and in the crystal. STD-

A)

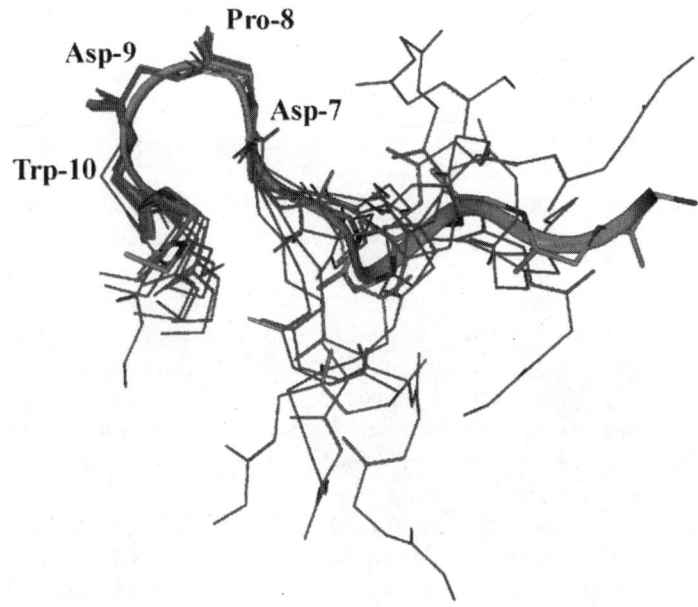

Figure 6. Views of calculated bound structure of the peptide FDTGAFDPDWPA (compound 6). (A) Backbone (N, Cα, C, O) atoms of the 10 structures with lowest NOE restraint energy produced by simulated annealing refinement, superimposed using backbone atoms of residues 7-10. The lowest-energy structure is colored by atom (C, green; N, blue; O, red) and a ribbon representation is shown in green, while the other structures are shown in light blue. Residues 7-10 comprise a type I β-turn and are labeled. (B) Close-up views of residues 6-12 of the lowest-energy calculated structure (colored by atom). Backbone atoms only are shown for residues 6 and 12. In the left panel, the turn is shown in a similar orientation to the structures in (A); in the right panel, it is shown in an orthogonal orientation, so that the exposed position of the Trp-10 side chain, on the face of the turn, is visible. (C) Close-up views of residues 6-12, as in (B), but in a typical type I β-turn conformation is very similar to the calculated bound conformation (B). Reproduced with permission from reference 28. Copyright © 2003 American Society for Biochemistry and Molecular Biology, Inc.

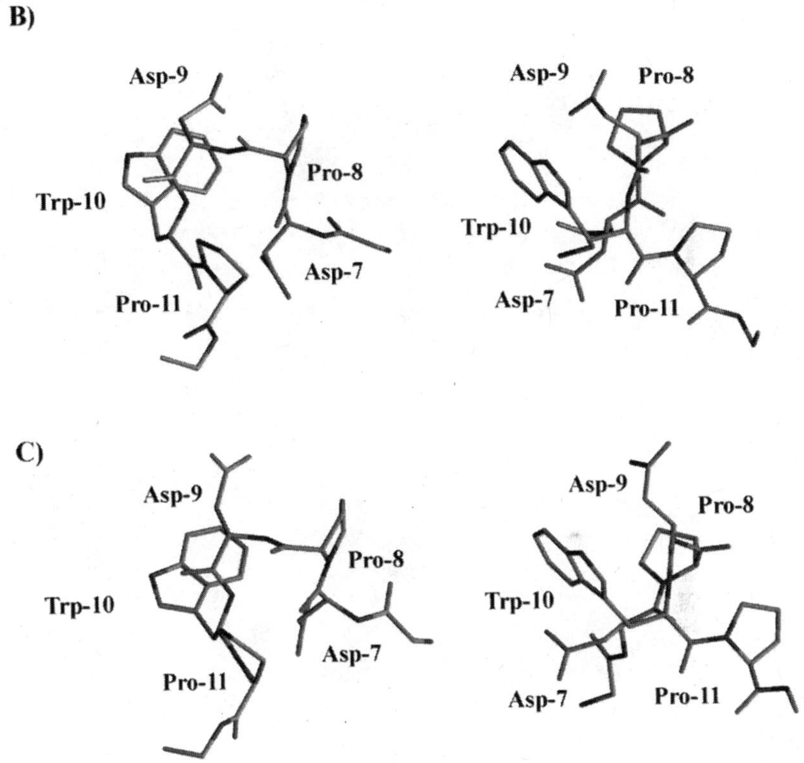

Figure 6. Continued.

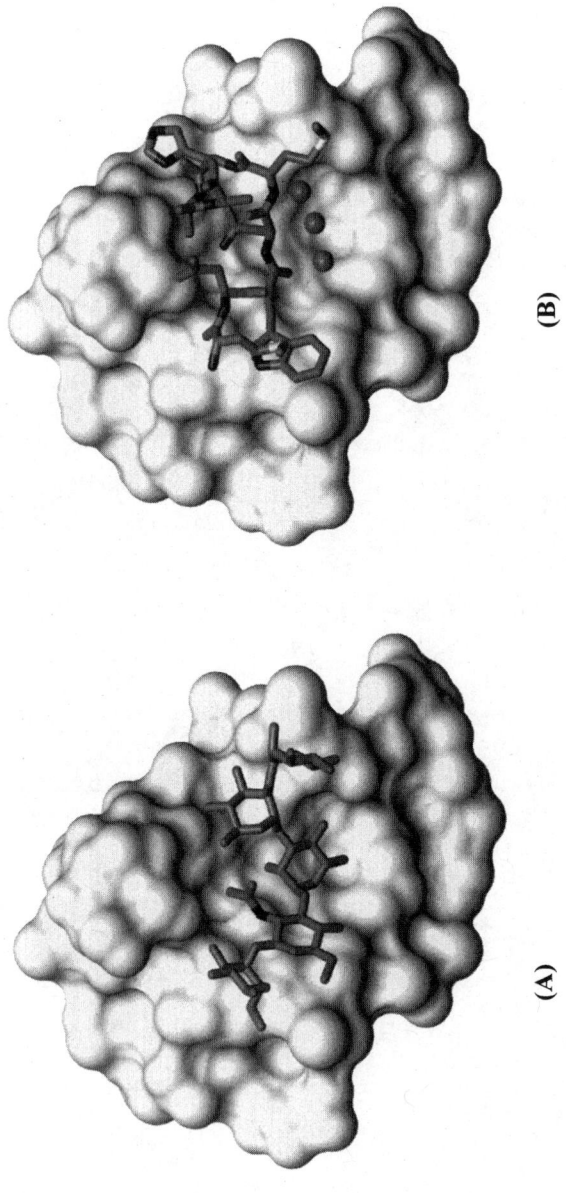

Figure 7. Structure of the Fab fgment of SYA/J6 antibody with bound pentasaccharide (compound 3) (A) and bound octapeptide MDWNMHAA (compound 4) (B). The red spheres represent three immobilized water molecules.

NMR epitope mapping of the carbohydrate-mimetic peptide MDWNMHAA in complex with the monoclonal antibody SYA/J6, revealed certain features consistent with the high-resolution crystal structure, and certain features that were inconsistent *(37)*. The presence of water in the crystal structure forming the hydrogen-bonded bridges involving amide protons of the peptide, suggested that water may be responsible for the lack of STD effects, or alternatively, that the peptide conformation in solution differed somewhat from that observed in the crystal structure *(21)*. Furthermore, the STD-NMR data indicated that the N-terminus of Met-1, since it does not contact the antibody surface, could likely be used to link the MDWNMHAA peptide to protein carriers to generate vaccines *(22)*.

Synthesis and immunochemical characterization of protein conjugates of the *S. flexneri* Y O-antigen-mimetic peptide as experimental vaccines.

The synthesis of MDWNMHAA-based conjugates, to bovine serum albumin, BSA (compound 9, Figure 5) and to tetanus toxoid, TT, (compound 10, Figure 5), together with their immunochemical evaluation with SYA/J6 was reported recently *(22)*. The *S. flexneri* Y O-antigen peptide mimic, MDWNMHAA, was synthesized using the Fmoc solid-phase strategy and linked via the amino terminus to a bifunctional linker, diethylsquarate, and then conjugated to TT or BSA as immunogenic carriers, as described *(22)*. Confirmation of strong binding to SYA/J6 was also demonstrated. Investigation of the immunogenicity of these conjugates is presently in progress.

Design and synthesis of a glycopeptide chimera of *Shigella flexneri* Y O-polysaccharide and its peptide mimic MDWNMHAA.

In order to maximize the recognition elements and to eliminate the unfavorable entropy of binding due to immobilization of water molecules, we have designed a chimeric molecule that combines certain elements of the oligosaccharide and the peptide mimic. Specifically, we have chosen the rhamnose trisaccharide A-B-C (compound 3 Figure 2) to ensure that ring C penetrates into the deep hole at the bottom of the binding site, thus obviating the engagement of the water molecules that provide complementarity with the antibody combining site in binding of the peptide MDWNMHAA *(33)*. In addition, we have chosen to keep the MDW motif to maintain the favorable hydrophobic interaction of the W moiety with the antibody combining site. Elimination of the NMHAA unit should also reduce any conformational entropy differences that result from imposition of the α-turn at the C-terminus of the

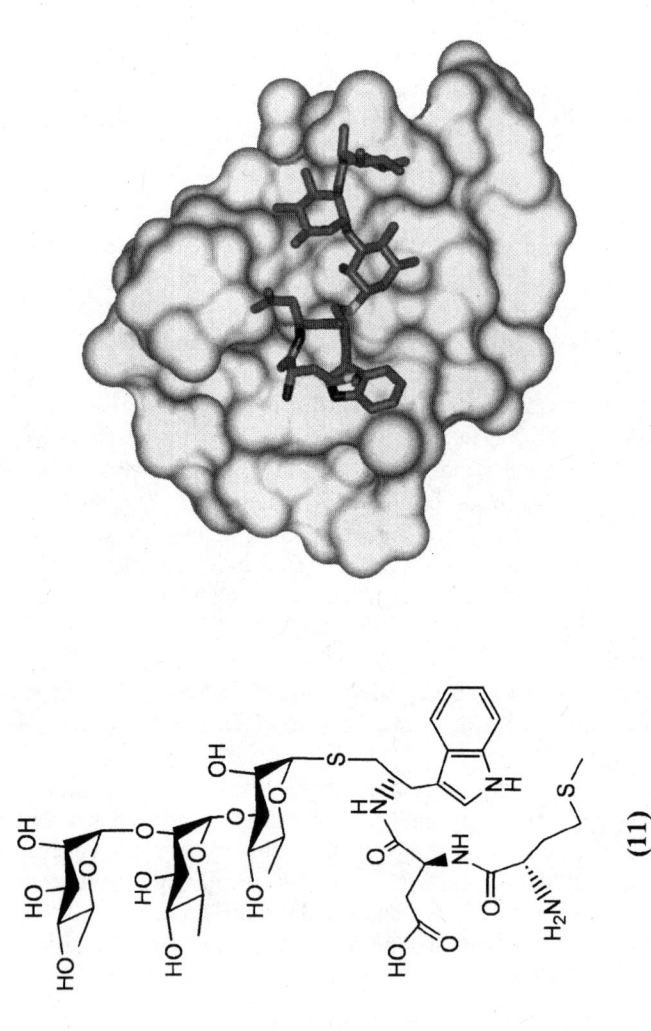

Figure 8. Modeled structure of the Fab fragment of SYA/J6 antibody with bound glycopeptide (compound 11)

parent octapeptide upon binding *(8, 33)*. The glycopeptide was designed with a thioglycosyl linkage (Figure 8) and was modeled into the FAb fragment *(33)* using Sybyl (Figure 8). The synthesis and immunochemical evaluation of this candidate are in progress, and will be reported in due course *(38)*.

Conclusions

Promising results with implications for the design of vaccines based on peptide-mimetics of the GAS and GBS polysaccharides have been obtained. The primary response to the GAS-CWPS peptide-mimetic immunogen had high titers of mature antibody isotype, IgG, showing participation of both cellular and humoral immune responses. The antibodies generated were cross-reactive with carbohydrate epitopes displayed on GAS bacteria, and this interaction was inhibited by CWPS and its oligosaccharide fragments. However, a long term, stable response against GAS could not be maintained after boosting injections due to a carrier-suppression effect. Further investigation with different carriers will be required for the design of effective anti-GAS vaccines based on the mimetic-peptide DRPVPY. For the GBS capsular polysaccharide mimetic-peptide, a strong, cross-reactive primary response was achieved, but no booster injections were given. Both peptides, displayed tight turn conformations, which we propose to be critical epitopes. In this context, the *S. flexneri* Y polysaccharide mimetic-peptide also displays an α-turn upon binding. Since this structural constraint appears not be present in the free peptide, it will be of interest to test whether this peptide conjugate elicits a cross-reactive response.

References

1. Grandi, G. *Trends Biotechnol.* **2001**, *19*, 181-188.
2. Roy, R. *Drug Discovery Today: Technologies* **2004**, *1*, 327-336.
3. Lindberg, A. A. *Vaccine* **1999**, *17*, S28-S36.
4. Monzavi-Karbassi, B.; Cunto-Amesty, G.; Luo, P.; Kieber-Emmons, T. *Trends Biotechnol.* **2002**, *20*, 207-214.
5. Mond, J. J.; Lees, A.; Snapper, C. M. *Ann. Rev. Immunol.* **1995**, *13*, 655-92.
6. Jennings, H. J. *Curr. Top. Microbiol. Immunol.* **1990**, *150*, 97-127.
7. Lesinski, G. B.; Westerink, M. A. J. *J. Microbiol. Meth.* **2001**, *47*, 135-149.
8. Johnson, M. A.; Pinto, B. M. *Aust. J. Chem.* **2002**, *55*, 13-25.
9. Irving, M. B.; Pan, O.; Scott, J. K. *Curr. Opin. Chem. Biolog.* **2001**, *5*, 314-324.
10. Cunningham, M. D. *Clin. Microbiol. Rev.* **2000**, *13*, 470-511.

11. Bisno, A. L., In *Principles and Practice of Infectious Diseases*; Mandell, G.L.; Douglas, R.G.; Bennet, J.E., Eds.; Wiley: New York, N. Y., 1985; p. 1133-1142.
12. Stevens, D. L. *Infect. Med.* **2003**, *20*, 483-+.
13. Quinn, R. W. *Rev. Infect. Dis.* **1989**, *II*, 928-952.
14. Stevens, D. L.; Tanner, M. H.; Winship, J.; Swarts, R.; Ries, K. M.; Schlievert, P. M.; Kaplan, E. *A. N. Engl. J. Med.* **1989**, *321*, 1-7.
15. Stollerman, G. H. *Arch. Intern. Med.* **1988**, *148*, 1268-1270.
16. Kehoe, M. A. *Vaccine* **1991**, *9*, 797-806.
17. Michon, F.; Moore, S. L.; Kim, J.; Blake, M. S.; Auzanneau, F.-I.; Johnston, B. D.; Johnson, M. A.; Pinto, B. M. *Infect. Immun.* **2005**, *73*, 6383-6389.
18. Harris, S. L.; Craig, L.; Mehroke, J. S.; Rashed, M.; Zwick, M. B.; Kenar, K.; Toone, E. J.; Greenspan, N.; Auzanneau, F. I.; MarinoAlbernas, J. R.; Pinto, B. M.; Scott, J. K. *Proc. Natl. Acad. Sci. U.S.A.* **1997**, *94*, 2454-2459.
19. Reimer, K. B.; Gidney, M. A. J.; Bundle, D. R.; Pinto, B. M. *Carbohydr. Res.* **1992**, *232*, 131-142.
20. Johnson, M. A.; Rotondo, A.; Pinto, B. M. *Biochemistry* **2002**, *41*, 2149-2157.
21. Johnson, M. A.; Pinto, B. M. *Carbohydr. Res.* **2004**, *339*, 907-928.
22. Hossany, R. B.; Johnson, M. A.; Eniade, A. A.; Pinto, B. M. *Bioorg. Med. Chem.* **2004**, *12*, 3743-3754.
23. Borrelli, S.; Hossany, R. B.; Findlay, S.; Pinto, B. M. *Infect. Immun.* **2005** submitted.
24. Baker, C. J.; Kasper, D. L., In *Infectious diseases of the fetus and newborn infant.*; Remington, J.S.; Klein, J.O., Eds.; Saunders: Philadelphia, P. A., 1983; p. 820-881.
25. Maione, D.; Margarit, I.; Rinaudo, C. D.; Masignani, V.; Mora, M.; Scarselli, M.; Tettelin, H.; Brettoni, C.; Iacobini, E. T.; Rosini, R.; D'Agostino, N.; Miorin, L.; Buccato, S.; Mariani, M.; Galli, G.; Nogarotto, R.; Dei, V. N.; Vegni, F.; Fraser, C.; Mancuso, G.; Teti, G.; Madoff, L. C.; Paoletti, L. C.; Rappuoli, R.; Kasper, D. L.; Telford, J. L.; Grandi, G. *Science* **2005**, *309*, 148-150.
26. Baker, C. J.; Rench, M. A.; Edwards, R. J.; Carpenter, B. M.; Hays, B. M.; Kasper, D. L. *N. Engl. J. Med.* **1988**, *319*, 1180-1187.
27. Pincus, S. H.; Smith, M. J.; Jennings, H. J.; Burritt, J. B.; Glee, P. M. *J. Immunol.* **1998**, *160*, 293-298.
28. Johnson, M. A.; Jaseja, M.; Zou, W.; Jennings, H. J.; Copie, V.; Pinto, B. M.; Pincus, S. H. *J. Biol. Chem.* **2003**, *278*, 24740-24752.
29. Kotloff, K. L.; Winickoff, J. P.; Invanoff, B.; Clemens, J. D.; Swerdlow, D. L.; Sansonetti, P. J. *Bull. World Health Organ.* **1999**, *77*, 651-665.
30. DuPont, H. L.; Levine, M. M.; Hornick, R. B.; Formal, S. B. *J. Infect. Dis.* **1989**, *159*, 1126-1128.

31. Carlin, N. I.; Gidney, M. A.; Lindberg, A. A.; Bundle, D. R. *J. Immunol.* **1986**, *137*, 2361-2366.
32. Carlin, N. I.; Lindberg, A. A.; Bock, K.; Bundle, D. R. *Eur. J. Biochem.* **1984**, *139*, 189-194.
33. Vyas, N. K.; Vyas, M. N.; Chervenak, M. C.; Bundle, D. R.; Pinto, B. M.; Quiocho, F. A. *Proc. Natl. Acad. Sci. U.S.A.* **2003**, *100*, 15023-15028.
34. Vyas, M. N.; Vyas, N. K.; Meikle, P. J.; Sinnott, B.; Pinto, B. M.; Bundle, D. R.; Quiocho, F. A. *J. Mol. Biol.* **1993**, *231*, 133-136.
35. Vyas, N. K.; Vyas, M. N.; Chervenak, M. C.; Johnson, M. A.; Pinto, B. M.; Bundle, D. R.; Quiocho, F. A. *Biochemistry* **2002**, *41*, 13575-13586.
36. Johnson, M. A.; Eniade, A. A.; Pinto, B. M. *Bioorg. Med. Chem.* **2003**, *11*, 781-788.
37. Johnson, M. A.; Pinto, B. M. *Bioorg. Med. Chem.* **2004**, *12*, 295-300.
38. Hossany, R. B.; Johnston, B. D.; Wen, X.; Borrelli, S.; Pinto, B. M. *Manuscript in preparation* **2005**.

Indexes

Author Index

Becker, Torsten, 293
Bencomo, Vicente Verez, 1, 71
Borrelli, Silvia, 335
Brade, H., 239
Bundle, David R., 163
Carmenate, Tania, 1
Chaudhuri, Siddhartha Ray, 199
Danishefsky, Samuel J., 258
Dziadek, Sebastian, 293
Evans, S. V., 239
Fernandez-Santana, Violeta, 71
Fraser-Reid, Bert, 199
Garcia, Ernesto, 71
Heynngnezz, Lazaro, 71
Hölemann, Alexandra, 137
Hossany, Rehana B., 335
Icart, Luis Pena, 1
Jansen, Wouter T. M., 85
Jayaprakash, K. N., 199
Jiang, Zi-Hua, 311
Johnson, Margaret A., 335
Jones, Christopher, 21
Kamerling, Johannis P., 85
Kim, Hyo-Sun, 184
Koganty, R. Rao, 311
Kosma, P., 239
Kubler-Kielb, J., 36
Kunz, Horst, 293
Lowary, Todd L., 184

Lu, Jun, 199
Medina, Ernesto, 71
Mulard, Laurence A., 105
Ng, Ella S. M., 184
Nitz, M., 163
Ouerfelli, Ouathek, 258
Phalipon, Armelle, 105
Pinto, B. Mario, 335
Pozsgay, V., 36
Rodriguez, Maria C., 71
Roy, René, 1, 71
Sadowska, Joanna M., 163
Schriemer, David C., 184
Seeberger, Peter H., 137
Sirois, Suzanne, 1
Snippe, Harm, 85
Sosa, Ivan, 71
Stocker, Bridget L., 137
Valdes, Yury, 71
Veloso, Roberto C., 1
Villar, Annete, 71
Warren, J. David, 258
Whittal, Randy M., 184
Wilson, Rebecca M., 258
Wittrock, Sven, 293
Wu, Xiangyang, 163
Yalamati, Damayanthi, 311
Zheng, Ruixiang Blake, 184

Subject Index

A

Adipic acid-derived linkers, coupling carbohydrate and protein, 53, 54, 55
Adipic acid dihydrazide (ADH), coupling carbohydrate and protein, 53, 54, 55
Antibody-ligand complexes
 crystal structures, 248, 250, 252
 See also Chlamydia
Anticancer vaccines. *See* Antitumor vaccines; Glycopeptides for anticancer vaccines
Antigen presenting cells (APCs), carbohydrate antigens, 2, 4*f*
Antigen studies
 Streptococcus pneumoniae type 3 mono- to heptasaccharides, 86–87
 See also O-antigen (O-Ag)
Antimony medications, leishmaniasis, 139
Anti-tuberculosis vaccines
 antigen of mycobacterial cell wall, 185–186
 capping motifs in mycobacterial lipoarabinomannan (LAM), 187*f*
 experimental, 194–196
 fragments of LAM for neoglycoconjugate generation, 188
 LAM, 185–186
 low-temperature glycosylation approach for hexasaccharide, 190, 191*f*
 mannopyranose-capped fragments, 188–189
 potential of LAM for development of, 187
 proposing protein binding to hexasaccharide, 192
 screening oligosaccharide fragments of LAM for CS-35 antibody, 191–192, 193*f*
 structural domains in mycobacterial LAM, 186*f*
 synthesis of branched hexasaccharide motif, 190
 synthesis of oligosaccharide fragments of LAM, 187–188, 191, 192*f*
 synthetic fragments of manLAM, 189*f*
 terminating structures in arabinan domains of mycobacterial LAM, 186*f*
Antitumor vaccines
 anticancer, 259, 260*f*
 approaches to multivalent vaccines, 274, 275*f*
 biological evaluations of first-generation pentavalent constructs, 283
 cassette approach to glycosylamino acid synthesis, 279, 282*f*, 284*f*, 285*f*
 conjugate vaccines in use, 272*f*, 273*f*
 design of synthetic carbohydrate-based anticancer vaccine, 260*f*
 five antigens in triplicate on polypeptide backbone, 283, 290*f*
 future research, 283, 289*f*, 290*f*
 Globo-H monomer synthesis and conjugation, 270*f*, 271*f*
 hexavalent vaccine with GM2 and Le^y antigens, 283, 290*f*

methods for preparation of glycosylamino acids, 276, 277*f*
modified pentavalent construct, 283, 289*f*
synthesis of Globo-H glycal: ABC trisaccharide, 262*f*, 263*f*
synthesis of Globo-H glycal: coupling ABC and DEF domains, 268*f*, 269*f*
synthesis of Globo-H glycal: DEF trisaccharide, 266*f*, 267*f*
synthesis of monomeric Globo-H-KLH construct, 259–265
synthesis of pentavalent vaccine, 279, 286*f*, 287*f*, 288*f*
tumor-associated carbohydrate-based antigens, 261*f*
tumor cells, 259
unimolecular multivalent vaccine constructs, 265, 274–279, 283
unimolecular pentavalent vaccine construct, 276, 279, 280*f*, 281*f*
unimolecular trivalent vaccine construct, 276, 278*f*
See also Glycopeptides for anticancer vaccines
Arabinofuranose domain lipoarabinomannan (LAM), 201, 203, 220–221
See also *Mycobacterium tuberculosis*
Automated approach lipophosphoglycan (LPG) tetrasaccharide cap, 153, 156, 157
oligosaccharide synthesis, 138
oligosaccharide synthesis for malaria vaccine, 141–142
outlook, 156
synthesis of malaria hexasaccharide, 145
synthesis of malarial toxin, 146
synthesis of malarial vaccine candidates, 147
synthesis of tetra-mannan cap, 144

Azide-alkyne [3+2] cycloaddition, conjugation for synthetic vaccines, 58, 60, 61–62

B

Bacille Calmette–Guerin (BCG) vaccine, tuberculosis, 185
Bacterial infections, global health, 336
Bacterial polysaccharides immune response, 2
See also Carbohydrate-mimetic peptides
Batch release, glycoconjugate vaccines, 27–28
Bioconjugation, oxime linkages for synthetic vaccines, 46, 52, 53
Bioterrorism threat, vaccines, 331–332
N,N'-Bis-hydroxysuccinimide ester of adipic acid, coupling carbohydrate and protein, 55, 56–57
Block assembly, vaccine design, 317, 318–320
BLP25 liposome vaccine
clinical testing, 329–331
Stimuvax®, 314
See also Glycopeptide based cancer vaccines
Bone marrow-derived B cells, immunology, 86
Bordetella pertussis, periodate oxidation for protein-carbohydrate conjugation, 41
Bovine serum albumin (BSA)
adipic acid derivative for polysaccharide-protein conjugates, 55, 57
bioconjugation with oxime linkages for synthetic vaccines, 46, 52, 53
conjugates of Group A *Streptococcus* mimetic peptide, 342–343, 344*f*

conjugation through thioether linkages, 41, 46, 47, 48, 49, 50, 51
conjugation with reductive amination, 38, 40
ELISA titration of rabbit serum on trisaccharide-BSA coated plates, 175, 181*f*
tumor-associated T antigen structures on BSA, 295*f*
vaccines by conjugation of mucin MUC1 glycopeptides to, 304, 306*f*

C

Cancer
  induction of immune responses to, antigens, 312–314
  mechanism for induction of immune responses against, 313*f*
  See also Antitumor vaccines; Glycopeptide based cancer vaccines; Glycopeptides for anticancer vaccines
Candida albicans
  alternate synthetic approach for conjugate vaccines, 171, 176–177
  β1,2-mannan antigen for protective immunity, 165, 166*f*
  binding specificity of monoclonal antibodies to oligosaccharides, 165–166
  disaccharide synthesis, 168, 169
  efficient large scale synthesis of β1,2-linked mannans, 171, 176–177
  ELISA inhibition by synthetic oligosaccharides for mouse IgG monoclonal antibody C3.1, 167*f*
  experimental, 180–181
  glycosylation product, 168, 170
  preferred strategy for conjugation of oligosaccharide to protein carrier, 175, 178–179
  rabbit immunization with tetanus toxoid conjugates and titration of immune sera, 175, 180*f*, 181*f*
  synthesis of oligosaccharide epitopes, 168, 171, 175, 180
  synthesis of pentasaccharide and hexasaccharide (1→2)-β-D-mannopyranan oligomers, 171, 173
  synthesis of thioglyoside mimetic of (1→2)-β-D-mannopyranotetraose, 171, 174
  tetanus toxoid conjugates of deprotected tetrasaccharide hapten, 175
  trisaccharide intermediates, 168, 172
Candidiasis, *Candida albicans*, 165
Capsular polysaccharides (CPSs)
  oligosaccharide fragments of, of *Streptococcus pneumoniae* species, 92*t*, 93*t*
  poly(ribosyl-ribitol-phosphate) (PRP), of *Haemophilus influenzae* type b (Hib), 73
  subdividing pneumococci, 87, 89
  survey of CPSs structures of *S. pneumoniae* species, 88*t*
  synthesized amino-spacered fragments of, of *S. pneumoniae* species, 94*t*, 95*t*, 96*t*
  See also Haemophilus influenzae type b (Hib); *Streptococcus pneumoniae* serotypes
Capsular polysaccharides (CPSs) of pneumococci, immunogenic in adults, 36
Carbohydrate antigens
  classical mechanisms, 2–3
  cytosolic pathway, 2–3, 10*f*
  endocytic pathway, 3, 4*f*

immunochemistry and
immunology, 164
major histocompatibility complex
(MHC)-I pathway, 9–17
MHC-II pathway, 3–8
role in recognition of foreign and
self-antigens, 2–3
tumor-related, 259, 261f
vaccine development, 72
Carbohydrate assembly, solid-phase
automated synthesis, 138
Carbohydrate-based vaccines
anti-malarial, 140
design of synthetic anticancer, 259,
260f
See also Antitumor vaccines;
Leishmaniasis; Malaria
Carbohydrate epitopes
chemical structure of *Chlamydia*-
specific Kdo trisaccharide, and
C. psittaci specific
oligosaccharides, 244f
chlamydial lipid A, 241, 243f
Kdo residues, 239, 241–245
lipid A species from *C. trachomatis*
serotypes L_2, E and F, 243f
structure and synthesis of
Chlamydia-specific, 241–245
synthesis of *C. psittaci* specific
Kdo tetrasaccharide, 246f, 247f
See also Chlamydia
Carbohydrate-mimetic peptides
bound conformation and epitope
mapping of peptide mimic of
Group A *Streptococcus*–cell-
wall polysaccharide (GAS-
CWPS), 342, 343f
bound conformations and epitope
mapping of peptide mimic of
Group B *Streptococcus* (GBS),
346, 348f, 349f
design and synthesis of
glycopeptide chimera of
Shigella flexneri Y O-
polysaccharide and peptide
mimic, 351, 352f, 353
epitope mapping of, of *S. flexneri* Y
O-polysaccharide, 347, 351
Group A *Streptococcus* (GAS),
337–345
Group B *Streptococcus* (GBS),
345–346
identification of, of *S. flexneri* Y O-
antigen, 346–347
identification of GAS-CWPS
mimetic peptide, 338, 342
immunogenicity of GAS-CWPS
mimetic peptide conjugate, 343,
345
potential as ligands for
carbohydrates vaccines, 336–
337
S. flexneri Y, 346–347, 351–353
structural basis of mimicry in
antibody-combining site, 347,
350f
structures of CWPS of GAS and
peptide mimic (DRPVPY), 337,
338f
structures of GBS type III capsular
polysaccharide and peptide
mimic (FDTGAFDPDWPA),
337, 340f, 341f
structures of O-polysaccharide of *S.
flexneri* Y and peptide mimic
(MDWNMHAA), 337, 339f
synthesis and immunology of
conjugates of GAS-CWPS
mimetic peptide as experimental
vaccines, 342–343, 344f
synthesis and immunology of
conjugates of *S. flexneri* Y O-
antigen-mimetic peptide as
experimental vaccines, 351
CD1 antigen
crystal structure of human CD1d
bound to α-galactosylceramide,
16f

glycolipids presentation, 13–17
pathway of T cell stimulation, 3
processing and presentation
  pathway, 15f
Cell wall phosphomannan, *Candida albicans*, 165, 166f
Center for Biological Evaluation and Research (CBER)
  guidance, 23
  vaccine batch testing, 28
Chemical degradation, preparation of polysaccharide fragments, 90
Chemoselectivity
  convergent assembly strategy, 231–232
  n-pentenyl furanosyl derivatives, 221, 228, 229
  synthesis of 26-mer-arabinofuranan, 233–234
Chlamydia
  antigenic structures of chlamydial epitopes, 245, 248
  binding environments of oligosaccharides to antibodies, 250f, 254f, 255f
  crystal structures of antibody-ligand complexes, 248, 250, 252
  developmental cycle, 242f
  diseases, 240–241
  major *Chlamydia*-specific and cross-reactive epitopes, 249f
  specificity of monoclonal antibodies for Kdo epitopes, 245t
  structure and synthesis of *Chlamydia*-specific carbohydrate epitopes, 241–245
  structures of oligosaccharides to antibodies, 251f, 253f
  treatment of infections, 240
  See also Carbohydrate epitopes
Code of Federal Regulations Title 21 (21 CFR), pharmaceuticals licensing, 23

Committee on Human Medicinal Products (CHMP), process and advice, 29
Conjugate vaccines
  efficacy of glycoconjugate vaccines, 30–31
  introduction, 72
  polysaccharide-protein, 89–90
  See also Candida albicans; *Haemophilus influenzae* type b (Hib); *Streptococcus pneumoniae* serotypes
Conjugation methods
  adipic acid-derived linkers, 53, 54, 55, 56–57
  attaching polysaccharides to proteins, 36–38
  azide-alkyne [3+2] cycloaddition, 58, 60, 61–62
  bioconjugation through oxime linkages, 46, 52, 53
  cycloaddition methods, 58–61
  Diels–Alder reaction, 58, 59
  free oligosaccharides with carrier proteins, 91
  glutardialdehyde, 55
  homobifunctional linkers, 53–55
  olefin metathesis, 64, 66
  reductive amination, 38, 39, 40–41, 42–43, 44, 45, 46
  squarate method, 53, 54
  Staudinger ligation, 60, 63
  thioether linkages, 41, 46, 47, 48, 49, 50, 51
  transglycosylation, 60, 64, 65
Convergent assembly, dodecaarabinofuranan, 221, 229, 231–232
Cow mosaic virus, coupling azido-modified, to polymer, 58, 60, 61–62
Cycloaddition
  azide-alkyne [3+2], 58, 60, 61–62
  Diels–Alder, 58, 59

Cytosolic proteins, preferred peptides for MHC class I molecules, 9–10

D

2-Deoxy-oxulosonic acid (Kdo)-containing polysaccharide fragments, conjugation, 40, 45
2-Deoxy-oxulosonic acid (Kdo) epitopes
  binding environment of bisphosphorylated pentasaccharide to monoclonal antibody, 252, 255f
  binding environment of Kdo trisaccharide to fragment of monoclonal antibody, 248, 250f, 254f
  crystal structures of antibody-ligand complexes, 248, 250, 252
  specificity of monoclonal antibodies for, 245t
  structures of Kdo disaccharides to monoclonal antibody, 250, 251f, 253f
  See also Chlamydia
Diels–Alder reaction, conjugation through cycloaddition, 58, 59
Diptheria toxin, carrier protein, 164
Dodecasaccharide lipomannan, *Mycobacterium tuberculosis*, 220, 224–225
Donor/acceptor MATCH, Paulsen's, 199, 202, 204, 205

E

Endoplasmic reticumum (ER), MHC-II and Li assemblying into, 3–4
Epitope mapping, protective carbohydrate epitopes, 116, 118
Epitopes. *See* Carbohydrate epitopes
Escherichia coli

conjugation with reductive amination, 40, 45
periodate oxidation for protein-carbohydrate conjugation, 40–41
European Medicines Evaluation Agency (EMEA)
  guidance, 23
  licensing procedures, 29
European Pharmacopoeia (EP), glycoconjugate vaccine standards, 23
European Union (EU), licensing procedures, 28–29

F

Food and Drug Administration (FDA), non-inferiority concept, 29

G

Glutardialdehyde, coupling carbohydrate and protein, 55
Glycan chains, glycoconjugate vaccine with size-reduced, 25f
Glycan identity, efficacy of glycoconjugate vaccines, 30–31
Glycoconjugate vaccines
  byproducts from conjugation reactions, 31
  carbohydrate haptens, 119–120, 121f, 122–123
  carriers and their glycoconjugates, 124–126
  characterization methodology, 24, 26
  conjugation chemistry, 124
  controls during manufacture, 27–28
  crosslinked matrix structure, 25f
  design and conception, 118–126
  development, 36
  efficacy factors, 30–31
  framework for regulation, 22, 32

free, unconjugated saccharide, 31
guidelines, 22–23
individual glycan and carrier
  protein components, 30–31
interrupting transmission of
  disease, 22
license variations, 30
licensing, 28–29
mechanism of immune response,
  37
molecular size of final conjugate,
  31
Phase 1 and Phase 2 studies, 26
Phase 3 clinical trial, 27
Phase 4 and post licensing
  surveillance, 29–30
polysaccharide-protein ratio, 31
preclinical development, 24
protective immunity of anti-
  Shigella flexneri 2a
  liposaccharide antibody, 129,
  131*f*
release tests, 27–28
saccharide content of vial, 31
saccharide identity and protein
  components in conjugate, 31
structural types, 25*f*
vaccine and manufacturing process
  design, 24
vaccine with size-reduced glycan
  chains, 25*f*
See also Conjugation methods;
  Haemophilus influenzae type b
  (Hib); *Shigella flexneri* 2a
  infection
Glycolipid
  cell surface, of lipoarabinomannan
    (LAM), 200, 201, 202
  chemists rendition, 203
  *See also Mycobacterium
    tuberculosis*
Glycopeptide based cancer vaccines
  block assembly approach, 317,
    318, 320

BLP25 liposome vaccine
  Stimuvax®, 314, 329–330, 331*f*,
  332
chemical synthesis and structural
  definition, 315
clinical testing, 329–331
convenient linking sites with least
  hindrance to coupling, 317, 318
IgG antibody titers response in
  healthy macaque monkeys, 328,
  329*f*
innate immune system, 312–314
liposome pictorial showing lipid bi-
  layer, 327*f*
mechanism of immune responses
  against cancers, 313*f*
mono phosphoryl lipid-A (MPL®)
  as adjuvant in experimental
  vaccines, 319
patient survival by normal and
  abnormal expression of MUC1
  mucin, 331*f*
pre–clinical studies, 327–329
protected serine and threonine
  modified for synthesis of
  peptide blocks, 317*f*
purification at every step, 319
quality control, 321
small block assembly into
  intermediate blocks for final
  glycoform, 317, 319, 320
strategy for synthesizing lipid-A
  analogues and mimics for
  screening, 324*f*, 325*f*
synthetic immune stimulants, 319,
  321, 326
synthetic mucin MUC1 glyco-
  lipopeptide, 315–319
synthetic structures exploiting
  stereochemistry, 322*f*, 323*f*
target antigen, 314–315
vaccine delivery, 326–327
Glycopeptides, MHC-II binding, 6, 7*f*
Glycopeptides for anticancer vaccines

biomimetic synthesis of silalyl T and silalyl T$_N$ antigen amino acid building blocks, 302, 303*f*
biosynthesis of mucin O-glycans in malignant epithelial cells, 296*f*
combination of MUC1 glycopeptide and T-cell epitope for vaccine, 304, 306, 307*f*, 308
conformations of glycopeptides in water, 297, 298*f*
conjugation of MUC1 tandem repeat glycopeptides to bovine serum albumin (BSA) for vaccine, 304, 306*f*
early experiments of tumor-associated saccharide antigens, 294–295
immunological evaluation, 300–301
serum neutralization by synthetic vaccine with and without pre-incubation with MUC1 glycopeptide, 308*f*
silalyl T$_N$ glycopeptide sequence of MUC1, 300*f*
solid-phase synthesis of sialyl T and sialyl T$_N$ antigen glycopeptides, 305*f*
synthesis of silalyl T$_N$ threonine building block, 299*f*
synthesis of tumor-associated sialyl T threonine conjugates, 303*f*
synthetic glycopeptide antiges with silalyl T$_N$ and silalyl T structure, 297–301
synthetic glycopeptide partial sequences of MUC1, 298*f*
synthetic vaccine of MUC1 glycopeptide and T-cell epitope, 301*f*
tumor-associated mucin MUC1, 295–297
tumor-associated T antigen structures, 295*f*

See also Antitumor vaccines

Glycoproteins
biological recognition, 294
cytosolic proteins as source for major histocompatibility complex (MHC) class I, 9–10
MHC class I bound to glycopeptide antigen, 11*f*
ubiquitin-tagged, 9

Glycosylphosphatidylinositols (GPIs), malarial toxin, 141, 142*f*

Good Manufacturing Practices (GMP), vaccine manufacture, 27–28

Group A *Streptococcus* (GAS)
conformations and epitope mapping of peptide mimic, 342, 343*f*
identification of GAS–cell-wall polysaccharide (CWPS)–mimetic peptide, 338, 342
immunogenicity of GAS–CWPS mimetic peptide conjugate, 343, 345
infections, 337–338
structures of CWPS of, and peptide mimic, 338*f*
synthesis of protein conjugates, 342–343, 344*f*

See also Carbohydrate-mimetic peptides

Group B *Streptococcus* (GBS)
conformations and epitope mapping of peptide mimic of capsular polysaccharide of type III, 346, 348*f*, 349*f*
infections in newborns and adults, 345
structures of GBS type III capsular polysaccharide and peptide mimic, 340*f*, 341*f*

See also Carbohydrate-mimetic peptides

H

Haemophilus influenzae type b (Hib)
  adipic acid dihydrazide for polysaccharide-protein conjugates, 53, 54, 55
  anti-Hib conjugate vaccines, 73
  capsular polysaccharide, poly(ribosyl-ribitol-phosphate) (PRP), 73*f*
  chain elongation of oligosaccharide, 76–77
  classical synthesis of PRP-fragments, 73, 74*f*
  conjugate vaccines, 72, 81–83, 164
  disaccharide intermediates synthesis, 77*f*
  mice immunogenicity of pentamer-OMP conjugate vs. commercial vaccine, 78*f*
  new strategy for synthesis of PRP fragments, 78–81
  ribitol unit, 74
  ribosyl unit and disaccharide, 75–76
  synthetic oligosaccharide vaccine in Cuba, 98
Hapten-carrier complexes, immunology, 86
Heterobifunctional spacers, reductive amination, 40
Huisgen-cycloaddition, conjugation for synthetic vaccines, 58, 60, 61–62
Human serum albumin (HSA), conjugation through thioether linkages, 46, 50

I

Immune system
  mammalian, 312–314
  synthetic immune stimulants, 319, 321, 326
Immunogenicity
  capsular polysaccharides (CPSs), 36
  Group A *Streptococcus* cell-wall polysaccharide (GAS-CWPS) mimetic peptide conjugate, 343, 345
  malarial glycosylphosphatidylinositol (GPI) glycan, 148
  Shigella flexneri 2a oligosaccharide conjugates in mice, 126–129
Immunology
  hapten-carrier complexes, 86
  T-cell/B-cell cooperation, 86
Innate immunity. *See* Immune system
International Conference on Harmonization of Technical Requirements for Registration of Pharmaceuticals for Human Use (ICH)
  acceptance specifications, 28
  licensing, 28
  markets in US, Europe, and Japan, 22–23

K

Kala azar, visceral leishmaniasis, 139
Kdo (2-deoxy-oxulosonic acid)-containing polysaccharide fragments, conjugation, 40, 45
Kdo (2-deoxy-oxulosonic acid) epitopes
  binding environment of bisphosphorylated pentasaccharide to monoclonal antibody, 252, 255*f*
  binding environment of Kdo trisaccharide to fragment of monoclonal antibody, 248, 250*f*, 254*f*
  crystal structures of antibody-ligand complexes, 248, 250, 252

specificity of monoclonal
antibodies for, 245*t*
structures of Kdo disaccharides to
monoclonal antibody, 250, 251*f*,
253*f*
See also Chlamydia
Keyhole limpet hemocyanin (KLH)
appending carbohydrate antigens
on, 259
design of carbohydrate-based
anticancer vaccine, 260*f*
immunodominant portion, 37
leishmaniasis vaccine constructs,
153, 155
neoglycoconjugate preparation, 40,
44
synthesis of monomeric Globo-H-
KLH construct, 259–265
See also Antitumor vaccines

L

Large-scale production,
macromolecular vaccines, 315
Leishmaniasis
antimony medications, 139
automated synthesis of leishmania
cap tetrasaccharide, 157
carbohydrate-based vaccine
constructs against, 153, 156
epitope lipophosphoglycan (LPG)
on leishmania parasites, 149,
152*f*
formation of synthetic and
semisynthetic vaccine
constructs, 155
forms in humans, 139
outlook, 156
solution phase synthesis of fully
deprotected leishmania cap, 154
treatment, 139
vaccine design considerations, 149,
153
See also Malaria

Licensing
glycoconjugate vaccines, 28–29
post, surveillance, 29–30
variations for vaccines, 30
Lipid-A analogues
pictorial of liposome showing lipid
bilayer, 327*f*
strategic protection for
synthesizing, 324*f*, 325*f*
synthetic structure and
stereochemistry, 321, 322*f*, 323*f*
Lipoarabinomannan (LAM)
antigen of mycobacterial cell wall,
185–186
anti-tubuculosis vaccines using
fragments of LAM, 191–192
capping motifs of mycobacterial
LAM, 187*f*
cartoons of, from *Mycobacterium
tuberculosis*, 201
cell surface glyolipid, 200, 202
mannopyranose-capped fragments,
188–189
oligosaccharide fragments, 187–
188
potential for anti-tuberculosis
vaccines, 187
screening LAM fragments as
ligands for CS-35 antibody,
191–192, 193*f*
structural domains of
mycobacterial LAM, 186*f*
See also Anti-tuberculosis
vaccines; *Mycobacterium
tuberculosis*
Lipophosphoglycans (LPGs)
carbohydrate-based vaccine
constructs against leishmaniasis,
153, 156
leishmania vaccine design, 149,
153
structure, 152*f*
Lipopolysaccharide antigens
antigenic structures of chlamydial
epitopes, 245, 248

crystal structures of antibody-
ligand complexes, 248, 250, 252
specificity of monoclonal
antibodies for 2-deoxy-
oxulosonic acid (Kdo) epitopes,
245*t*
structure and synthesis of
Chlamydia-specific
carbohydrate epitopes, 241–245
structure of *C.*-specific Kdo
trisaccharide epitope and *C.
psittaci* specific
oligosaccharides, 244*f*
structure of lipid A species from *C.
trachomatis* serotypes L_2, E and
F, 243*f*
synthesis of *C. psittaci* specific
Kdo tetrasaccharide, 246*f*,
247*f*
See also Chlamydia
Lithium, role in MHC class II
molecules, 3–4
Lot release, glycoconjugate vaccines,
27–28

M

Macromolecular vaccines, factors in
large-scale production, 315
Major histocompatibility complex
(MHC)-I pathway
activation of T-cell receptor (TCR),
10–12
CD1-mediated glycolipids
presentation, 13–17
crystal structure of human CD1d
bound to α-galactosylceramide,
16*f*
mechanisms of
presentation/binding, 2–3
molecule bound to glycopeptide
antigen, 11*f*
processing and presentation of
antigens by CD1 pathway, 15*f*

processing and representation, 9–10
T-cell activation and consequences,
10–13
toll-like receptors (TLRs)
recognizing microbial
components, 13, 15*f*
top view of portion of MHC class I
H-2KB, 12*f*
Major histocompatibility complex
(MHC)-II pathway
fragments of polysaccharides, 6
glycopeptides, 6, 7*f*
model of complex between MHC-II
and glycopeptide, 7*f*
processing and presentation, 3–8
T-cell activation and consequences,
8
T-cell receptor (TCR) and MHC
class II complex with antigen
peptide, 5*f*
top view of truncated MHC-II
antigen binding site, 7*f*
zwitterionic polysaccharides (ZPS),
6, 8
Malaria
anti-malarial carbohydrate-based
vaccines, 140
automated oligosaccharide
synthesis for vaccine against,
141–142
automated synthesis of tetra-
mannan cap, 144
experimental, 158–159
glycosylphosphatidylinositols
(GPIs), 141, 142*f*
identification of malarial toxin,
140–141, 142*f*
immunogenic testing, 148
limited treatment, 139
mechanisms of, pathogenesis, 139–
140
outlook, 156
Plasmodium species, 138
retrosynthesis of GPI malaria toxin,
143

synthesis of malaria hexasaccharide, 145
synthesis of malarial toxin 1, 146
synthesis of malarial vaccine candidates, 147
synthesis of second generation vaccine candidates, 149, 150f, 151f
target, GPI glycans of differing lengths, 151f
See also Leishmaniasis
Mammalian immune system, carbohydrate antigens, 164
Mammals, immune system, 312–314
Manufacturing design, vaccine, 24
Marketing Authorization Applications (MAA), licensing, 28–29
Masamune's concept, MATCH, 202, 204, 205
MATCH approach
concepts, 204
discovery, 208, 214
evidence for reciprocal donor acceptor selectivity (RDAS), 214–215
example of RDAS, 212–213
family of n-pentenyl donors, 208, 210–211
Masamune's concept, 204
Paulsen's concept, 204
selectivities and concepts, 202, 205
See also Mycobacterium tuberculosis
Mechanisms
immune response of glycoconjugate vaccines, 37
induction of immune responses against cancer, 313f
malaria pathogenesis, 139–140
processing carbohydrate antigens, 2–3
Meningococcal liposaccharide (LPS), conjugation with reductive amination, 38, 40, 46

Methyl glycosides
biological repeat unit (RU) of *Shigella flexneri* 2a, 110
block synthesis of fragments of *S. flexneri* 2a O-antigen (O-Ag), 117
convergent strategies to pentasaccharides and larger fragments, 115–116
representing fragments of O-Ag, 110–116
See also Shigella flexneri 2a infection
Mimics, peptide. *See* Carbohydrate-mimetic peptides
Mucin MUC1 glycopeptide. *See* Glycopeptides for anticancer vaccines
Mycobacterial lipoarabinomannan. *See* Anti-tuberculosis vaccines
Mycobacterium tuberculosis
arabinofuranose domain, 220–221
cartoons of lipoarabinomannan (LAM) from, 201f
challenges to synthesis of LAM, 202, 203
complex lipoarabinomannan cell surface glycolipid, 200, 202
discovery of MATCH, 208, 214
dodecasaccharide lipomannan component, 220, 224–225
donor/acceptor concept of MATCH, 199, 204
evidence for reciprocal donor acceptor selectivity (RDAS), 214–215
example of RDAS, 212–213
lanthanide triflates and n-pentenyl donors, 218–219
n-pentenyl family of donors, 208, 210–211
n-pentenyl orthoesters in synthesis of LAM sub-domains of, 217, 220–221
oligosaccharide synthesis, 205, 206

phosphatidyl inositol mannoside (PIM) moiety, 217, 222–223
preparation of n-pentenyl furanosyl donors, 226–227
primary vs. secondary selectivity, 215, 216
probing chemoselectivity in n-pentenyl furanosyl derivatives, 221, 228, 229
random glycosidation, 208, 209
reagent combination for unparalleled regioselectivity, 215, 217
regioselectivity problems in protection and glysosidation, 205, 207, 208
selectivities and concepts of MATCH, 202, 205
selectivities in glycoside coupling, 205
strategy for convergent assembly, 231–232
synthesis of 26-mer-arabinofuranan, 233–234
synthesis of dimannosylated phosphoinositide, 222–223
synthesis of dodecaarabinofuranan, 229, 230

N

National control laboratories (NCL), vaccine batch testing, 28
National Institute for Biological Standards and Control (NIBSC), assay standardization, 23
Neisseria meningitis
 conjugate vaccine, 164
 coupling pentamer to, for conjugate vaccine, 77, 78*f*
Neoglycoconjugate vaccine
 characterization, 26
 glycoconjugate vaccine type, 25*f*
 oligosaccharide fragments of lipoarabinomannan (LAM), 188
 See also Glycoconjugate vaccines
Neoglycopeptides, major histocompatibility complex (MHC-II) binding, 6
4-Nitrophenyl diester of adipic acid, coupling carbohydrate and protein, 55, 57
Non-inferiority, concept of FDA, 29

O

O-antigen (O-Ag)
 repeating unit (RU), 106, 110
 synthetic methyl glycosides representing fragments of, 110–116
 target of protective immune response, 106
Official Medicines Control Laboratories (OMCLs), vaccine batch testing, 28
Olefin metathesis, conjugation for synthetic vaccines, 64, 66
Oligosaccharides
 approach to multi-gram synthesis of oligo-mannan epitopes, 171, 176–177
 automated approach to synthesis, 138
 automated synthesis for malaria vaccine, 141–142, 143, 144, 145, 146, 147
 binding native O-linked antigens, 165
 conjugated oligosaccharide-proteins inducing protective (anti-PS) responses in mice and rabbits, 97*t*
 conjugation of functionalized, to proteins, 175, 178–179
 immunogenicity of *Shigella flexneri* 2a oligosaccharide-

based conjugates in mice, 126–129
inhibiting binding to immobilized native mannan, 165–166, 167f
optimum, length for polysaccharide with conformational epitopes, 98
recognition by T cells, 6
synthesis of, epitopes, 168, 171, 175, 180
synthesis of amino-spacered, fragments from monosaccharides, 90–91
synthesized amino-spacered fragments of capsular polysaccharides of *Streptococcus pneumoniae* species, 94t, 95t, 96t
synthetic, in pneumococcal vaccines, 91, 96
See also *Candida albicans*
O-specific polysaccharide (O-SP) target of protective immune response, 106
See also O-antigen (O-Ag)
Oxime linkages, bioconjugation for synthetic vaccines, 46, 52, 53

P

PADRE. See Pan human leukocyte-associated antigens DR-binding epitope (PADRE)
Pan human leukocyte-associated antigens DR-binding epitope (PADRE)
carrier and glycoconjugates, 125–126
glycoconjugates with, as carrier, 126
Parasitic infections
health concerns, 138–139, 156
See also Leishmaniasis; Malaria
Pattern recognition receptors (PRR), immune system, 2
Paulsen's concept, MATCH, 202, 204, 205
Peptides. See Carbohydrate-mimetic peptides
Pharmacopeia of United States of America (USP), glycoconjugate vaccine standards, 23
Phase 1 studies, glycoconjugate vaccines, 26
Phase 2 studies, glycoconjugate vaccines, 26
Phase 3 clinical trial, glycoconjugate vaccines, 27
Phase 4 trials, carriage for glycoconjugate vaccines, 29–30
Phosphatidyl inositol mannoside (PIM), moiety, 217, 222–223
Phosphomannan, cell wall, of *Candida albicans*, 165, 166f
Plasmodium species
malarial infection, 138
See also Malaria
Pneumococci, immunogenicity of capsular polysaccharides (CPSs), 36
Poly(ribosyl-ribitol-phosphate) (PRP)
anti-Hib conjugate vaccines, 73
chain elongation, 76–77
classical synthesis of PRP-fragments, 73, 74f
disaccharide intermediates synthesis, 77f
new strategy for synthesis of PRP fragments, 78–81
pentamer coupling to *Neisseria meningitidis*, 77
repeat unit, 73f
ribitol unit, 74
ribose synthesis from D-glucose, 75f
ribosyl unit and disaccharide, 75–76

strategy for synthesis of ribitol
derivatives, 74f
Polysaccharide (PS) based vaccines
development, 36
PS-protein conjugate vaccines, 89–90
23-valent PS vaccine, 89
See also Conjugation methods;
Streptococcus pneumoniae
serotypes
Polysaccharide-protein ratio, efficacy of glycoconjugate vaccines, 31
Polysaccharides
fragment preparation by chemical degradation, 90
major histocompatibility complex (MHC-II) binding to fragments of, 6
thymus-independent (TI) antigens, 86
See also Carbohydrate-mimetic peptides
Preclinical development, prototype vaccine, 24
Process design, vaccine and manufacturing, 24
Production, macromolecular vaccines, 315
Protective carbohydrate epitopes
mapping, 116, 118
serotype-specific monoclonal antibodies (mAbs), 108, 109f
synthetic methyl glycosides representing fragments of O-antigen (O-Ag), 110–116
See also Shigella flexneri 2a infection
Public health, vaccines, 331–332

Q

Quality control, macromolecular vaccines, 321
Quinine, malaria, 139

R

Reciprocal donor acceptor selectivity (RDAS)
evidence, 214–215
example, 212–213
See also MATCH approach
Reductive amination
conjugation for synthetic vaccines, 38–41, 42–43, 44, 45, 46
free oligosaccharide conjugation with carrier proteins, 91
Regioselectivity
family of n-pentenyl donors, 208, 210–211
problems in protection and glycosidation, 205, 207, 208
random glycosidation, 209
reagent combination for unparalleled, 215, 217
Release tests, glycoconjugate vaccines, 27–28
Respiratory bacterial infections, *Streptococcus pneumoniae*, 87

S

Saccharides, efficacy of glycoconjugate vaccines, 30–31
Safety, macromolecular vaccines, 315
Salmonella typhimurium, periodate oxidation for protein-carbohydrate conjugation, 40–41
Selectivities
glycoside coupling, 205
MATCH concept, 202, 205
oligosaccharide synthesis, 205, 206
primary vs. secondary, 215, 216
regioselectivity problems in protection and glycosidation, 205, 208
Shigella dysenteriae type 1
conjugate vaccine against, 37

conjugation with reductive amination, 40, 42
Shigella flexneri
  carbohydrate-mimetic peptide of, Y O-antigen, 346–347
  design and synthesis of glycopeptide chimera of, Y O-polysaccharide and peptide mimic, 351, 352f, 353
  diarrhoeal diseases, 346
  epitope mapping of carbohydrate-mimetic peptide, 347, 351
  protein conjugates of carbohydrate-mimetic peptide as experimental vaccines, 351
  structural basis of peptide-carbohydrate mimicry in antibody-combining site, 347, 350f
  structures of O-polysaccharide of, and peptide mimic, 339f
  See also Carbohydrate-mimetic peptides
Shigella flexneri 2a infection
  anti-liposaccharide antibodies (Abs), 106
  carbohydrate haptens, 119–120, 121f, 122–123
  carriers and their glycoconjugates, 124–126
  carriers tetanus toxoid (TT) and diphtheria toxoid (DT), 124
  conjugation chemistry for vaccines, 124
  convergent strategies to pentasaccharides and larger, 115–116, 117
  design and conception of glycoconjugates, 118–119
  diarrheal disease, 106
  disaccharide EC, 110–111, 112
  epitope mapping, 116, 118
  glycoconjugates with pan human leukocyte-associated antigens (HLA) DR-binding epitope (PADRE) as carrier, 126
  homologous protection by subclasses of monoclonal IgG (mIgG) specific for, 109f
  IC_{50} of binding of protective mIgGs to *S. flexneri* 2a LPS by synthetic methyl glycosides, 118t
  identification of protective carbohydrate epitopes, 108
  immunogenicity of, oligosaccharide-based conjugates in mice, 126–129, 130f
  monosaccharide D, 114, 115, 116f
  monosaccharides A and B, 111, 113, 115
  neoglycopeptides, 125–126
  neoglycoproteins, 124–125
  O-antigen (O-Ag) repeating unit (RU), 106, 110
  oligosaccharide-peptide conjugates, 125–126
  oligosaccharide-protein conjugates, 124–125
  prevalence and virulence, 106–107
  protective capacity of anti-*S. flexneri* LPS Ab induced by oligosaccharide-TT glycoconjugates, 129, 131f
  protective serotype-specific monoclonal antibodies (mAbs), 108, 109f
  semi-synthetic glycoconjugates with TT as carrier, 127–129
  strategy for potential vaccines, 107, 129, 131
  synthetic methyl glycosides representing fragments of O-Ag, 110–116
Sialylated forms
  biomimetic synthesis, 302, 303f
  silalyl T_N glycopeptide sequence of MUC1, 300f

solid-phase synthesis of, antigen glycopeptides, 305f
synthesis of silalyl T$_N$ threonine building block, 299f
tumor-associated antigen glycopeptides, 297–301
See also Glycopeptides for anticancer vaccines
Small pox, vaccines, 311
Solid-phase automated synthesis, carbohydrate synthesis, 138
Squarate method, coupling carbohydrate and protein, 53, 54
Stability, macromolecular vaccines, 315
Staudinger ligation, conjugation for synthetic vaccines, 60, 63
Stereochemistry, synthetic structures of lipid-A analogues, 321, 322f, 323f
Stimuvax® vaccine
　BLP25 liposome vaccine, 314, 332
　clinical testing, 329–331
　See also Glycopeptide based cancer vaccines
Streptococcus pneumoniae serotypes
　capsular polysaccharides (PS), 87, 89
　conjugate vaccine, 164
　conjugations with carrier proteins, 91
　free oligosaccharide fragments of capsular PSs of, 92t, 93t
　induction of antibodies against, by oligosaccharide fragments, 93f
　minimal oligosaccharide fragments conjugated to carrier proteins inducing anti-PS responses in mice and rabbits, 97t
　optimum chain length in oligosaccharide for PS, 98
　preparation of PS fragments via chemical degradation, 90
　PS-protein conjugate vaccines, 89–90
　survey of capsular PS structures of, 88t
　synthesis of oligosaccharide fragments with amino-spacers from monosaccharides, 90–91
　synthesized amino-spacered fragments of capsular PSs of, 94t, 95t, 96t
　synthetic oligosaccharides in pneumococcal vaccines, 91, 96
　23-valent PS vaccine, 89
　See also Group A *Streptococcus* (GAS); Group B *Streptococcus* (GBS)
Streptococcus pneumoniae type 14, adipic acid derivative for polysaccharide-protein conjugates, 55, 56–57
Synthetic vaccines. *See* Conjugation methods; Glycopeptides for anticancer vaccines

T

T-cell activation
　major histocompatibility complex (MHC I)-peptide complexes, 10–13
　MHC-II pathway, 8
T-cell immunity
　bacterial polysaccharides, 2
　major histocompatibility complex (MHC-I) pathway, 2–3, 9–17
　mechanisms for processing carbohydrate antigens, 2–3
　MHC-II pathway, 3–8
T-cell receptors (TCRs)
　complex between MHC-loaded peptides and, 8
　models explaining activation, 10–12
　recognition, 2, 4, 5f
Tetanus toxoid (TT)
　advantages as carrier candidate, 81

carrier and glycoconjugates, 124–125
carrier protein, 164
conjugates of Group A *Streptococcus* mimetic peptide, 342–343, 344*f*
oligosaccharide-TT glycoconjugates against *Shigella flexneri* 2a, 129, 131*f*
rabbit immunization with TT conjugates, 175, 180*f*, 181*f*
semi–synthetic glycoconjugates with TT as carrier, 127–129, 130*f*
thiolated, conjugated to polysaccharide for vaccine candidate, 81–82
vaccine testing, 82–83
Thioether linkages, conjugation for synthetic vaccines, 41, 46, 47, 48, 49, 50, 51
Thymus-dependent (TD) antigens
oligosaccharides, 86–87
polysaccharides, 336
proteins and peptides, 336
Thymus-derived T cells, immunology, 86
Thymus-independent (TI) antigens, polysaccharides, 86, 336
Toll-like receptors (TLRs), recognition of microbial components, 13
Transglycosylation, conjugation for synthetic vaccines, 60, 64, 65
Tuberculosis
antibody-mediated immunity, 185
Bacille Calmette–Guerin (BCG) vaccine, 185
morbidity and mortality, 200
treatment, 200
See also Anti-tuberculosis vaccines; *Mycobacterium tuberculosis*

Tumor-associated mucin MUC1
biosynthesis of mucin O-glycans in malignant epithelial cells, 296*f*
early experiments, 294–295
expression on epithelial tissues, 295–297
See also Glycopeptides for anticancer vaccines
Tumor cells
malignantly transformed, 259
See also Antitumor vaccines
Tumor necrosis factor (TNF)
role in severe malaria, 140–141
T-helper (Th) lymphocytes, 8

U

Unimolecular multivalent vaccine constructs. See Antitumor vaccines

V

Vaccines
bioterrorism threat, 331–332
chemical synthesis and structural definition, 315
development approaches, 311–312
formulation and delivery, 326–327
license variations, 30
regulation for development of, 22
small pox, 311
See also Antitumor vaccines; Conjugation methods; Glycoconjugate vaccines; *Haemophilus influenzae* type b (Hib); *Shigella flexneri* 2a infection
Visceral leishmaniasis, kala azar, 139

W

World Health Organization (WHO), glycoconjugate vaccine recommendations, 23

Z

Zwitterionic polysaccharides (ZPS), major histocompatibility complex (MHC-II) binding, 6, 8